T0225872

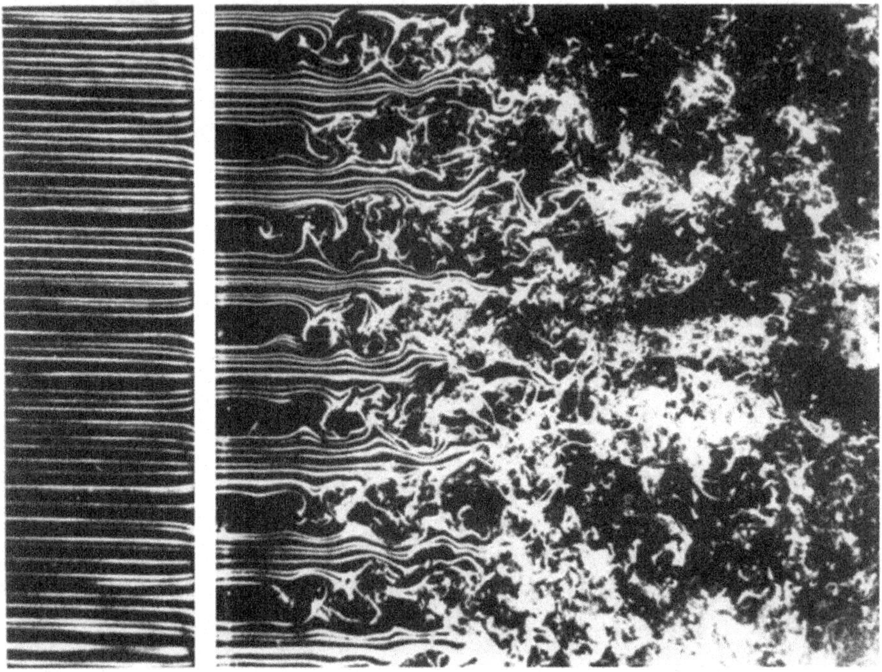

Generation of turbulence by a grid. A uniform laminar stream passing through a grid. Instability of the shear layers leads to turbulent flow downstream. (Photograph by Thomas Corke and Hassan Nagib.)

Murray Rosenblatt

Stationary Sequences and Random Fields

Birkhäuser
Boston · Basel · Stuttgart
1985

Author:
Murray Rosenblatt
Department of Mathematics
University of California, San Diego
La Jolla, CA 92093 (USA)

AMS Subject Classification (1980)
Primary 62M10, 62M15, 60F05
Secondary 60G10, 60G15, 60G25, 60G35, 60G60,
62E20, 62F12, 76F05, 86A15, 47B35

Library of Congress Cataloging in Publication Data

Rosenblatt, Murray.
Stationary sequences and random fields.
Bibliography: p. ....
Includes index.
1. Time-series analysis. 2. Stationary sequences
(Mathematics). 3. Random fields. I. Title.
QA280.R67 1985 519.5'5 84-20321.
ISBN 0-8176-3264-6

CIP-Kurztitelaufnahme der Deutschen Bibliothek

Rosenblatt, Murray:
Stationary sequences and random fields /
Murray Rosenblatt. –
Boston ; Basel ; Stuttgart : Birkhäuser, 1985.
ISBN 3-7643-3264-6
ISBN 0-8176-3264-6

All rights reserved.
No part of this publication may be reproduced, stored in a retrieval system, or transmitted, in any form or by any means, electronic, mechanical, photocopying, recording or otherwise, without prior permission of the copyright owner.

© 1985 Birkhäuser Boston, Inc.
Printed in Switzerland

ISBN 0-8176-3264-6
ISBN 3-7643-3264-6

To David

Preface

This book has a dual purpose. One of these is to present material which selectively will be appropriate for a quarter or semester course in time series analysis and which will cover both the finite parameter and spectral approach. The second object is the presentation of topics of current research interest and some open questions. I mention these now. In particular, there is a discussion in Chapter III of the types of limit theorems that will imply asymptotic normality for covariance estimates and smoothings of the periodogram. This discussion allows one to get results on the asymptotic distribution of finite parameter estimates that are broader than those usually given in the literature in Chapter IV. A derivation of the asymptotic distribution for spectral (second order) estimates is given under an assumption of strong mixing in Chapter V. A discussion of higher order cumulant spectra and their large sample properties under appropriate moment conditions follows in Chapter VI. Probability density, conditional probability density and regression estimates are considered in Chapter VII under conditions of short range dependence. Chapter VIII deals with a number of topics. At first estimates for the structure function of a large class of non-Gaussian linear processes are constructed. One can determine much more about this structure or transfer function in the non-Gaussian case than one can for Gaussian processes. In particular, one can determine almost all the phase information. It is in this sense that the Gaussian processes could be thought of as atypical rather than typical in the class of linear processes. The estimates make use of higher order moment or spectral information. The computational questions that arise when dealing with such estimates are considered. A number of prediction problems for stationary random fields are discussed. Random fields that are analogues of one dimensional autoregressive and moving average (ARMA) processes are introduced. There is a discussion of a one-dimensional deconvolution problem, a type of problem that arises often in seismic or geophysical applications. It is shown how a multi-dimensional version of one such problem can be dealt with. It will be clear to the reader that there are many open questions relating to non-Gaussian processes and random fields. There is discussion of applications and simulations in a number of chapters.

There are Monte Carlo simulations of some moving average and autoregressive processes that are considered in Chapter IV. Estimates of the parameters are made and they are compared with the parameter values for the schemes. The estimates used are the standard ones. A discussion of the homogeneous model of turbulence is given in Chapter V. An example of a spectral estimate in high Reynolds number turbulence is considered. Later on in Chapter VI the use of higher order spectral estimates to gauge the energy transfer spectrum is commented on. Finally in Chapter VIII the deconvolution procedure proposed is applied in the case of some Monte Carlo simulations of non-Gaussian processes.

Perhaps the first books on topics relating to time series were those of Grenander and Rosenblatt [1957] and Blackman and Tukey [1959]. Today there is an ever increasing literature in time series analysis. The object of this book is not that of a broad and detailed survey. As already noted one purpose is to present topics appropriate for a semester course in time series that will touch on both the finite parameter and spectral approach to time series analysis. This can be done by selecting topics from Chapters I to V. Chapter I presents basic results on the Fourier representation of the covariance function of a weakly stationary process and the harmonic analysis of the process itself. A variety of topics are mentioned in Chapter II. The relation between moments and cumulants is analyzed. The linear prediction problem is presented in the context of autoregressive and moving average (ARMA) processes. This is followed by comments on non-Gaussian linear processes. The chapter concludes with a discussion of the Kalman-Bucy filter. Various problems that could be given to students are mentioned at the end of the chapters.

A semester or quarter course on time series analysis on an advanced undergraduate level or on a beginning graduate level can be based on an appropriate choice of material. One convenient choice would consist of Chapters I, II, IV and V. This would give the student contact with statistical methods dealing with finite parameter models as well as spectral methods. Though it is true that some of the results of Chapter III are used in Chapters IV and V the material could simply be referred to and motivated on heuristic grounds. In fact, it will often be convenient to initially discuss most of the material in an informal intuitive manner and only then consider the formal derivation of results. Such a course would require elements of probability theory and Fourier analysis as a background though these elements could be introduced at an appropriate time in the course.

The literature on stationary sequences, time series analysis, and random fields is very large. This book discusses a selection of topics from these that are, it is hoped representative, coherent, and relate to each other. The references to the literature are not intended to be encyclopedic or historically motivated. They are supposed to be useful to the reader.

I appreciate the support of the Office of Naval Research which funded much of the research on which this book is partially based. The comments of S. Cambanis, K. Helland, Peter Lewis, Keh-shin Lii, E. Masry, and J. A. Rice were helpful. I thank C. Friehe, K. Helland and H. Nagib for permission to use various photographs and graphs.

La Jolla, California *Murray Rosenblatt*

Table of Contents

Chapter I

Stationary Processes

1. General Discussion

One of the basic models in much of time series analysis is that of a weakly stationary process, a process whose mean value function is constant and whose covariance function depends only on the difference in the times at which the observations are made. The importance of this assumption is due to the fact that it implies that a Fourier (or harmonic) analysis of both the covariance function and the process itself can be carried out. These results will be developed in this chapter. As we shall later see, these basic probabilistic or structural results for these processes will motivate some of the statistical methods used in spectral analysis of data. The development will be carried out for discrete time parameter weakly stationary processes. However, there will be problems and occasional discussion suggesting analogous results and representations for continuous and or multidimensional time parameter processes. The results will be obtained for complex-valued weakly stationary processes.

As a simple example of a weakly stationary process, we first consider a *model of random phases*. Let

$$X_n = \sum_{\nu=1}^{k} c_\nu \, e^{i(n\lambda_\nu + \varphi_\nu)}$$

where the λ_ν are real constants in the interval $(-\pi, \pi)$ and the random variables φ_ν are independent and uniformly distributed on $(-\pi, \pi)$. Clearly,

$$E\, X_n \equiv 0$$

and

$$\text{(1)} \qquad \operatorname{cov}(X_n, X_m) = r_{n-m} = E(X_n \overline{X}_m) = \sum_{\nu=1}^{k} | c_\nu |^2 e^{i(n-m)\lambda_\nu} .$$

This is a special case of the Fourier representation of the covariance r_n

$$r_n = \int_{-\pi}^{\pi} e^{in\lambda} \, dG(\lambda)$$

in terms of a bounded nondecreasing function G that will be derived in section 2. The function G is called the spectral distribution function of $\{X_n\}$. Equation (1) implies that the spectral distribution function G of the process $\{X_n\}$ is a pure jump function with jump $| c_\nu |^2$ at the point λ_ν.

A second example is given by a *weakly stationary real Gaussian process* $\{Y_n\}$. The weak stationarity noted above means that

$$E\, Y_n \equiv m$$

and

$$\operatorname{cov}(Y_n, Y_m) = r_{n-m} .$$

It should be noted that by a real Gaussian process one means a process with the property that every finite subcollection of random variables is jointly Gaussian (or multivariate Gaussian). Another equivalent way of stating this (see Rosenblatt [1974]) is that every linear combination of a finite number of random variables is a Gaussian random variable. It should be noted that the first and second order moments (covariances) of a Gaussian process determine its full probability structure. This implies that the probability structure of a weakly stationary real Gaussian process is invariant (does not change) with time shifts. A process whose probability structure is invariant under time shifts is called a *strictly stationary process*. Thus a weakly stationary real Gaussian process is strictly stationary. The joint probability structure of any fixed k-tuple $Y_{n_1}, \ldots, Y_{nk}$ is the same as that of any shift $Y_{n_1+m}, \ldots, Y_{nk+m}$ of the k-tuple. However, it should be noted that generally the notions of weak stationarity and strict stationarity overlap. A strictly stationary process may not be weakly stationary because it may not have finite first and second order moments. On the other hand a weakly stationary process may not be strictly stationary because its finite third order moments do not depend only on time differences. The covariances of a real stationary Gaussian process are such that

$$r_{-n} = r_n .$$

Of course, this is true of any real weakly stationary process. However, this does imply that the mass of the spectral distribution function G is symmetric about zero, that is, $dG(\lambda) = dG(-\lambda)$. This will generally not be the case for a complex-valued weakly stationary process. We also remark that one can define some nonGaussian stationary processes in terms of Gaussian stationary processes by taking a nonlinear function of a Gaussian stationary process and its shifts. If $X = (X_n)$ is a stationary Gaussian process, a simple example of such a nonGaussian stationary process is given by $Y_n = p(X_n, X_{n-1}, \ldots, X_{n-k})$ where p is a polynomial $p(u_0, \ldots, u_k)$ in the variables $u_0, \ldots, u_k$.

Still another example of a weakly stationary process is given by an orthogonal sequence of random variables $\{X_n\}$, $E\, X_n \equiv 0$, with covariance

$$E(X_n \overline{X_m}) = \sigma^2 \delta_{n-m}, \quad \sigma^2 > 0 .$$

Here it is clear that the spectral distribution G is differentiable with $G'(\lambda) = \sigma^2(2\pi)^{-1}$. Such a process is called a "white noise" process because the spectral mass $G'(\lambda)$ is uniformly distributed over the interval $(-\pi, \pi)$. The term white is used because white light is thought of as composed equally of light from the whole visible spectral range.

2. *Positive Definite Functions*

Let $\{X_n; n = \ldots, -1, 0, 1, \ldots\}$ be a *weakly stationary process*, that is, (i) $E\, X_n \equiv m$ and (ii) $\operatorname{cov}(X_s, X_t) = E\{(X_s - m)\overline{(X_t - m)}\} = r_{s-t}$. The covariance function r_s, $s = \ldots, -1, 0, 1, \ldots$, is a *positive definite function* since all quadratic forms

$$\sum_{|j|,|k| \leq N} c_j r_{j-k} \bar{c}_k = E \left| \sum_{|j| \leq N} c_j X_j \right|^2 \geq 0, \quad N = 0, 1, \ldots,$$

are positive semi-definite. We now prove the Herglotz theorem which establishes a Fourier representation for the covariance sequence of a weakly stationary process.

(Herglotz) **Theorem 1.** *Given a positive definite function* $\{r_s\}$, *there is an essentially unique monotone nondecreasing function* G *with* $G(-\pi) = 0$ *such that*

$$r_s = \int_{-\pi}^{\pi} e^{is\lambda}\, dG(\lambda) . \tag{2}$$

Proof. Let

$$\begin{aligned} g_N(\lambda) &= \frac{1}{2\pi N} \sum_{j,k=0}^{N-1} r_{j-k}\, e^{-i(j-k)\lambda} \\ &= \frac{1}{2\pi} \sum_{s=-N+1}^{N-1} r_s\, e^{-is\lambda} \left(1 - \frac{|s|}{N}\right) \end{aligned}$$

and

$$G_N(\lambda) = \int_{-\pi}^{\lambda} g_N(u)\, du .$$

The functions $G_N(\lambda)$ are monotone nondecreasing with $G_N(\pi) \equiv r_0$. A subsequence N_k can be found such that $G_{N_k}(\lambda)$ converges weakly to a limiting monotone nondecreasing function $G(\lambda)$ as $N_k \to \infty$ by the Helly convergence theorem (see section 1 of the Appendix). Since

$$r_s \left(1 - \frac{|s|}{N}\right) = \int_{-\pi}^{\pi} e^{is\lambda}\, dG_N(\lambda), \quad |s| \leq N ,$$

it follows on letting $N = N_k \to \infty$ that

$$r_s = \int_{-\pi}^{\pi} e^{is\lambda}\, dG(\lambda) .$$

On integrating by parts the relation

$$r_s = (-1)^s r_0 - i s \int_{-\pi}^{\pi} e^{is\lambda}\, G(\lambda)\, d\lambda$$

is obtained. The uniqueness of G up to an additive constant then follows from the uniqueness of the Fourier representation for a function of bounded variation (see Zygmund [1959], p. 60). G is called the spectral distribution function of the process $\{X_n\}$. If G is absolutely continuous (differentiable), its derivative $g(\lambda) = G'(\lambda)$ is called the spectral density of the process $\{X_n\}$.

3. *Fourier Representation of a Weakly Stationary Process*

We wish to obtain a Fourier representation for a weakly stationary process that corresponds to the Fourier representation for the covariance function of such a process. Since the Fourier representation of the process will involve a random integral we shall have to introduce certain auxiliary processes called processes of orthogonal increments and develop the notion of a random integral in terms of these processes.

A process $Z(\lambda)$ is one of orthogonal increments if

$$\text{(i)} \qquad E\, Z(\lambda) \equiv 0\,,$$

$$\text{(ii)} \qquad E[dZ(\lambda)\, \overline{dZ(\mu)}] = \delta_{\lambda,\,\mu}\, dH(\lambda)\,,$$

where H is a bounded nondecreasing function. By (ii) we mean that

$$E[\{Z(\lambda+h) - Z(\lambda)\}\, \overline{\{Z(\mu+h) - Z(\mu)\}}] = \begin{cases} 0 & \text{if the intervals } (\lambda, \lambda+h] \text{ and } (\mu, \mu+h] \text{ are disjoint} \\ H(\lambda+h) - H(\lambda) & \text{if } \lambda = \mu\,. \end{cases}$$

The definition of a random Stieltjes integral

$$\int g(\lambda)\, dZ(\lambda)$$

is introduced for an appropriate class of functions g. First the integral is defined for step functions

$$g(\lambda) = g_k \quad \text{if} \quad a_{k-1} < \lambda \le a_k\,, \quad k = 1, \ldots, i\,, \tag{3}$$

$-\pi \le a_0 < a_1 < \cdots < a_i \le \pi$. For such a function the random integral is defined as

$$\int g(\lambda)\, dZ(\lambda) = \sum_k g_k[Z(a_k) - Z(a_{k-1})]$$

with

$$E \int g(\lambda)\, dZ(\lambda) = 0$$

and

$$E\left|\int g(\lambda)\,dZ(\lambda)\right|^2 = \sum |g_k|^2 [H(a_k) - H(a_{k-1})] \tag{4}$$
$$= \int |g(\lambda)|^2 \,d\eta(\lambda)\,.$$

Here η is the measure generated by H (see Appendix 1).

The integral can be defined for any continuous function as follows. Given a continuous function g, there is a sequence of step functions g_n of the form (3) such that

$$\int |g(\lambda) - g_n(\lambda)|^2 \,d\eta \to 0$$

as $n \to \infty$. Set

$$J_n = \int g_n(\lambda)\,dZ(\lambda)\,.$$

Then

$$E|J_n - J_m|^2 = \int |g_n(\lambda) - g_m(\lambda)|^2 \,d\eta \to 0$$

as $m, n \to \infty$ by (4). The Riesz-Fischer theorem (see Appendix) implies that J_n converges in mean square to a random variable J with

$$E|J|^2 = \lim_{n\to\infty} E|J_n|^2 = \int |g(\lambda)|^2 \,d\eta(\lambda)\,.$$

We define $\int g(\lambda)\,dZ(\lambda)$ as J since J does not depend on the specific sequence g_n with which g is approximated. This integral can also be defined for the larger class of Borel functions g for which $\int |g(\lambda)|^2 \,d\eta(\lambda)$ is finite by a similar argument. This integral has the following properties of an integral:

1. $\int [ag(\lambda) + bh(\lambda)]\,dZ(\lambda) = a\int g(\lambda)\,dZ(\lambda) + b\int h(\lambda)\,dZ(\lambda)$ for any complex numbers a, b.

2. $\lim_{n\to\infty} \int g_n(\lambda)\,dZ(\lambda) = \int g(\lambda)\,dZ(\lambda)$ if and only if

$$\int |g_n(\lambda) - g(\lambda)|^2 \,d\eta(\lambda) \to 0$$

as $n \to \infty$.

3. $E\left[\int g(\lambda)\,dZ(\lambda)\,\overline{\int h(\lambda)\,dZ(\lambda)}\right] = \int g(\lambda)\,\overline{h(\lambda)}\,d\eta(\lambda).$

These properties can be obtained by using approximating step functions and going to the limit.

We shall show that a weakly stationary process itself has a random Fourier representation in terms of an appropriately defined process of orthogonal increments. Let $\{X_n\}$ be a weakly stationary process with $E\, X_n \equiv 0$ and spectral distribution function G. Let λ, μ, $\lambda < \mu$, be continuity points of G. Introduce the random variables

$$Z_N(\lambda, \mu) = \frac{1}{2\pi} \sum_{|j| \le N} X_j \int_\lambda^\mu e^{-ijy}\, dy\,. \tag{5}$$

The first object is to show that the random variables $Z_N(\lambda, \mu)$ are a Cauchy sequence as $N \to \infty$ in that

$$E\, |\, Z_N(\lambda, \mu) - Z_M(\lambda, \mu)\, |^2 \to 0 \tag{6}$$

as $M, N \to \infty$. Let us assume that $M < N$. Then

$$E\, |\, Z_N(\lambda, \mu) - Z_M(\lambda, \mu)\, |^2 = \frac{1}{4\pi^2} E \left| \sum_{M < |j| \le N} X_j \int_\lambda^\mu e^{-ijy}\, dy \right|^2$$

$$= \int_{-\pi}^{\pi} dG(u)\, |\, h_M(u) - h_N(u)\, |^2$$

where

$$h_N(u) = \frac{1}{2\pi} \sum_{|j| \le N} \int_\lambda^\mu e^{-ij(y-u)}\, dy \tag{7}$$

by making use of (2). The function $h_N(u)$ is the truncated (at N) Fourier series of the indicator function

$$I_{(\lambda, \mu]}(u) = \begin{cases} 1 & \text{if } \lambda < u \le \mu \\ 0 & \text{otherwise.} \end{cases}$$

By standard results on Fourier series one knows that the functions $h_N(u)$ are uniformly bounded in N and u and converge uniformly to $I_{(\lambda, \mu]}$ as $N \to \infty$ outside of ε-intervals ($\varepsilon > 0$) about λ, μ for any fixed $\varepsilon > 0$. From this it follows that (6) is valid as $N, M \to \infty$ since λ, μ are continuity points of G. By the Riesz-Fischer theorem there is a limiting random variable (in mean square) as $N \to \infty$.

$$Z(\lambda, \mu) = \lim_{N \to \infty} Z_N(\lambda, \mu)\,. \tag{8}$$

Let us call (7) $h_N(u; \lambda, \mu)$ since it is the truncated Fourier series corresponding to $I_{(\lambda, \mu]}$. Consider now λ, μ, $\lambda < \mu$, and λ', μ', $\lambda' < \mu'$, all continuity points of G. Then $E\, Z(\lambda, \mu) = 0$ and

$$E[Z(\lambda, \mu)\, \overline{Z(\lambda', \mu')}] = \lim_{N \to \infty} E[Z_N(\lambda, \mu)\, \overline{Z_N(\lambda', \mu')}]$$

(9)
$$= \lim_{N \to \infty} \int dG(u)\, h_N(u; \lambda, \mu)\, h_N(u; \lambda', \mu')$$

$$= \int_{(\lambda, \mu] \cap (\lambda', \mu']} dG(u)\,.$$

Notice that if $(\lambda, \mu]$, $(\lambda', \mu']$ are disjoint intervals then $Z(\lambda, \mu)$ and $Z(\lambda', \mu')$ are uncorrelated (orthogonal). Also (9) implies that

$$E\, |\, Z(\lambda, \mu)\, |^2 = G(\mu) - G(\lambda)\,.$$

Further let

(10)
$$Z(\mu) = Z(-\pi, \mu)\,.$$

From (10) it follows that

$$Z(\lambda, \mu) = Z(\mu) - Z(\lambda)\,.$$

The process $Z(\lambda)$ is defined at discontinuity points λ of G by setting

$$Z(\lambda) = \lim_{\mu \to \lambda +} Z(\mu)$$

where μ approaches λ from above through continuity points of G. For this reason, it is convenient to take G continuous to the right. Property (9) can be stated in terms of the process $Z(\mu)$ as

$$E[(Z(\mu) - Z(\lambda))\, \overline{(Z(\mu') - Z(\lambda'))}] = 0$$

if $(\lambda, \mu]$, $(\lambda', \mu']$ are disjoint and

$$E\, |\, Z(\mu) - Z(\lambda)\, |^2 = G(\mu) - G(\lambda)$$

or as

(11)
$$E[dZ(\lambda)\, \overline{dZ(\mu)}] = \delta_{\lambda, \mu}\, dG(\lambda)\,.$$

Thus, we have constructed a process $Z(\mu)$ of orthogonal increments from the process $\{X_n\}$ and will derive the following theorem.

Theorem 2. *Let $\{X_n\}$ be a weakly stationary process with $E\, X_n \equiv 0$ and spectral distribution function G. The process X_n then has the random Fourier representation*

(12)
$$X_n = \int_{-\pi}^{\pi} e^{in\lambda}\, dZ(\lambda)$$

in terms of the process $Z(\lambda)$ of orthogonal increments given by (5), (8) *and* (10). *Notice that the process $Z(\lambda)$ has property* (11).

The theorem is a consequence of the following simple computation. First

$$E\, X_n\, \overline{Z_N(\lambda, \mu)} = \frac{1}{2\pi} \sum_{|j| \le N} r_{n-j} \int_\lambda^\mu e^{ijy}\, dy$$

$$= \int_{-\pi}^{\pi} e^{inu} \frac{1}{2\pi} \int_\lambda^\mu \sum_{|j| \le N} e^{ij(y-u)}\, dy\, dG(u) \to \int_\lambda^\mu e^{inu}\, dG(u)$$

as $N \to \infty$ if λ, μ are continuity points of G. This implies that

$$E\, X_n \overline{\int g(\lambda)\, dZ(\lambda)} = \int e^{in\lambda}\, \overline{g(\lambda)}\, d\eta(\lambda) \tag{13}$$

for a step function g with η the measure generated by G. By an approximation argument one can see that (13) still holds for continuous g. Now take $g(\lambda) = e^{in\lambda}$ and compute

$$E \left| X_n - \int e^{in\lambda}\, dZ(\lambda) \right|^2 = r_0 - 2\, E \left\{ X_n \overline{\int e^{in\lambda}\, dZ(\lambda)} \right\} + r_0 = 0\,.$$

If the weakly stationary process $\{X_j\}$ is real-valued with spectral representation

$$X_j = \int_{-\pi}^{\pi} e^{ij\lambda}\, dZ(\lambda)$$

$$E[dZ(\lambda)\, \overline{dZ(\mu)}] = \delta_{\lambda,\mu}\, dF_x(\lambda)$$

it follows that

$$[\overline{Z(\mu) - Z(\lambda)}] = [Z(-\lambda) - Z(-\mu)]$$

for any continuity points μ, λ of F_x. If we then set

$$\begin{aligned} Z_1(\lambda) &= 2\, \mathrm{Re}\, \{Z(\lambda) - Z(0)\} \\ &= \{Z(\lambda) - Z(0) + Z(-\lambda) - Z(0)\} \\ Z_2(\lambda) &= -2\, \mathrm{Im}\, \{Z(\lambda) - Z(0)\} \\ &= i\, \{Z(\lambda) - Z(0) - Z(-\lambda) + Z(0)\} \end{aligned}$$

when λ and 0 are continuity points of F_x, the real representation

$$X_j = \int_0^\pi \cos j\lambda\, dZ_1(\lambda) + \int_0^\pi \sin j\lambda\, dZ_2(\lambda)$$

is obtained with

$$E[dZ_1(\lambda)\, dZ_2(\mu)] = 0$$

$$E[dZ_1(\lambda)\, dZ_1(\mu)] = E[dZ_2(\lambda)\, dZ_2(\mu)]$$

$$= 2\,\delta_{\lambda,\,\mu}\, dF(\lambda), \quad 0 \le \lambda, \mu \le \pi\,.$$

Theorem 2 can be given an interesting interpretation in the following manner. Consider the linear space $\mathscr{M}$ of random variables generated by finite linear combinations of the random variables X_j of the weakly stationary process $\{X_n\}$. Then consider $\overline{\mathscr{M}} = \mathscr{H}(X)$, the linear space obtained by adding to $\mathscr{M}$ all limits of Cauchy sequences of random variables in $\mathscr{M}$. This is a closed linear space in that all limits of Cauchy sequences are already in $\mathscr{H}(X)$. In fact, it is a Hilbert space with the inner product of two elements (random variables) U, V of $\mathscr{H}(X)$ given by the inner product

$$(U, V) = E(U\,\overline{V})\,.$$

Let us now consider the space of functions $l(u)$ square integrable with respect to the measure η generated by the spectral distribution function G of $\{X_n\}$

$$\int |\, l(u)\,|^2\, d\eta(u) < \infty\,.$$

Call this space of functions $L^2(d\eta)$. The inner product of two functions l_1, l_2 of the space is given by

$$(l_1, l_2) = \int l_1(u)\, \overline{l_2(u)}\, d\eta(u)$$

and the distance $d(l_1, l_2)$ between the two elements by

$$d(l_1, l_2) = \left\{ \int |\, l_1(u) - l_2(u)\,|^2\, d\eta(u) \right\}^{1/2}.$$

$L^2(d\eta)$ is a closed linear space that is also a Hilbert space. In effect Theorem 2 says that there is essentially a one-one mapping between $\mathscr{H}(X)$ and $L^2(d\eta)$ such that if $U \in \mathscr{H}(X)$ corresponds to $l \in L^2(d\eta)$ then

$$U = \int l(u)\, dZ(u)$$

and

$$E\,|\,U\,|^2 = E\left|\int l(u)\, dZ(u)\right|^2 = \int |\, l(u)\,|^2\, \eta(d\mu) = |\, l\,|^2$$

where $|\,l\,|$ is the length of l in $L^2(d\eta)$. Further, X_n corresponds to e^{inu} in this mapping. Such a one-one mapping is called an isometry because it preserves length.

In the spectral representation of the weakly stationary process X_n let

$$\Delta Z(\lambda) = \lim_{h \downarrow 0} \{Z(\lambda + h) - Z(\lambda - h)\}$$

be the possible jump in random spectral function Z at λ. We shall prove the following Theorem.

Theorem 3. *Let $\{X_n\}$ be a weakly stationary process. Then*

$$\lim_{n \to \infty} E \left| \frac{1}{n} \sum_{k=0}^{n-1} X_k e^{-ik\lambda} - \Delta Z(\lambda) \right|^2 = 0 .$$

Now

$$\frac{1}{n} \sum_{k=0}^{n-1} X_k e^{-ik\lambda} = \int_{-\pi}^{\pi} \frac{1}{n} \sum_{k=0}^{n-1} e^{ik(\mu - \lambda)} \, dZ(\mu)$$

$$= \int_{-\pi}^{\pi} \frac{e^{in(\mu - \lambda)} - 1}{e^{i(\mu - \lambda)} - 1} \, dZ(\mu) .$$

Notice that $E \mid \Delta Z(\lambda) \mid^2 = \Delta F(\lambda)$. Let

$$\tilde{Z}(\mu) = \begin{cases} Z(\mu) & \text{if } \mu < \lambda \\ Z(\mu) - \Delta Z(\lambda) & \text{if } \mu > \lambda \end{cases}$$

and

$$\tilde{F}(\mu) = \begin{cases} F(\mu) & \text{if } \mu < \lambda \\ F(\mu) - \Delta F(\lambda) & \text{if } \mu > \lambda . \end{cases}$$

Then

$$E \left| \frac{1}{n} \sum_{k=0}^{n-1} X_k e^{-ik\lambda} - \Delta Z(\lambda) \right|^2 \tag{14}$$

$$= E \left| \int_{-\pi}^{\pi} \frac{1}{n} \, \frac{e^{in(\mu - \lambda)} - 1}{e^{i(\mu - \lambda)} - 1} \, d\tilde{Z}(\mu) \right|^2$$

$$= \int_{-\pi}^{\pi} \frac{1}{n^2} \, \frac{\sin^2 \frac{n}{2} (\mu - \lambda)}{\sin^2 \frac{1}{2} (\mu - \lambda)} \, d\tilde{F}(\mu)$$

since

$$E[d\tilde{Z}(\mu) \, \overline{d\tilde{Z}(\mu')}] = \delta_{\mu - \mu'} \, d\tilde{F}(\mu) .$$

The sequence of Fejer kernels

$$F_n(\mu) = \frac{1}{2\pi n} \frac{\sin^2 \frac{n}{2}\mu}{\sin^2 \frac{1}{2}\mu}$$

have the following properties:

a) $F_n(\mu) \geq 0$,

b) $\int_{-\pi}^{\pi} F_n(\mu)\, d\mu \equiv 1$,

c) $\sup_{|\mu| \geq \varepsilon} F_n(\mu) \to 0$ as $n \to \infty$ given any fixed $\varepsilon > 0$.

It is clear that $\tilde{F}(\mu)$ is continuous at $\mu = \lambda$. This together with the properties of the Fejer kernel imply that (14) tends to zero as $n \to \infty$.

The principal results of this Chapter show that weak stationarity of a random process implies that both the covariance function and the process itself have representations as continuous sums or integrals of harmonics. This is explicitly indicated in formula (2) for the covariance function and formula (12) for the process itself. In the case of the representation of the process, the weights in the continuous sum are random and given by a process of orthogonal increments. There is a close relationship between the two Fourier representations for the covariance function and the process, as is indicated by relation (11). As we shall later see, much of spectral analysis of processes is motivated by these results.

Problems

1. A continuous parameter weakly stationary process $X(t)$, $-\infty < t < \infty$, is said to be continuous in mean square if $E\,|X(t) - X(s)|^2 \to 0$ if $t - s \to 0$. Let $r(t) = \operatorname{cov}\big(X(t), X(0)\big) = \operatorname{cov}\big(X(t+s), X(s)\big)$ be the covariance function of $X(t)$. Show that if $X(t)$ is continuous in mean square the covariance function $r(t)$ is a continuous function.

2. Show that the covariance function $r(t)$ of a continuous time weakly stationary process is positive definite.

3. A theorem of Bochner states that a continuous positive definite function $r(t)$ has a Fourier-Stieltjes representation

$$r(t) = \int_{-\infty}^{\infty} e^{it\lambda}\, dG(\lambda)$$

in terms of a bounded nondecreasing function G. Can you suggest how one might derive this result using the Herglotz theorem.

4. Suppose that the continuous time parameter weakly stationary process $X(t)$ is observed at the discrete time intervals $k\,h$, $k = \cdots, -1, 0, 1, \ldots$, $h > 0$. Given the representation of the covariance function $r(t)$ of $X(t)$ in terms of its spectral distribution function $G(\lambda)$, show that

$$r(k\,h) = \int_{-\pi/h}^{\pi/h} e^{ikh\lambda}\, dG_h(\lambda)$$

with

$$G_h(\lambda) = \sum_{k=-\infty}^{\infty} \left\{ G\left(\frac{2k\pi}{h} + \lambda\right) - G\left(\frac{2k-1}{h}\pi\right) \right\},$$

$|\lambda| \leq \pi/h$. It is clear that from $r(k h)$, k integral one can only determine $G_h(\lambda)$. $G_h(\lambda)$ is sometimes called an aliased spectral distribution function of the process $X(t)$.

5. Suppose $\{\mathbf{X}_n\}$ is a k-vector valued process $X_n = (X_n^{(s)};\ s = 1, \ldots, k)$ with the property that for each k-vector α, $\alpha \cdot \mathbf{X}_n$ is a weakly stationary process. Then $\{\mathbf{X}_n\}$ is referred to as a k-vector valued weakly stationary process. Show that there is a $k \times k$ Hermitian matrix valued function $G(\lambda)$ that is nondecreasing in the sense that $G(\lambda) - G(\mu)$ is positive semidefinite whenever $\lambda \geq \mu$ such that the $k \times k$ covariance matrices

$$\begin{aligned} r_j &= E\,X_j X_0' = E\,X_{j+s} X_s' \\ &= \int_{-\pi}^{\pi} e^{ij\lambda}\, dG(\lambda)\,. \end{aligned}$$

Here $\mathbf{X}_n$ is understood to be a column vector. Given a matrix M, the matrix M' denotes the conjugated transpose of M.

6. Show that if $X(t)$ is a continuous parameter weakly stationary process continuous in mean square, there is a process $Z(\lambda)$, $-\infty < \lambda < \infty$, of orthogonal increments such that

$$E[dZ(\lambda)\,\overline{dZ(\mu)}] = \delta_{\lambda,\mu}\, dG(\lambda)$$

[G is the spectral distribution function of $X(t)$] and

$$X(t) = \int_{-\infty}^{\infty} e^{it\lambda}\, dZ(\lambda)\,.$$

7. Let $\{\mathbf{X}_n\}$ be a k-vector valued weakly stationary process. Show that there is a k-vector valued process $\mathbf{Z}(\lambda)$ of orthogonal increments (each component a process of orthogonal increments) such that

$$E\,d\mathbf{Z}(\lambda)\, d\mathbf{Z}(\mu)' = \delta_{\lambda,\mu}\, dG(\lambda)$$

(G is the $k \times k$ matrix-valued spectral distribution function of $\{\mathbf{X}_n\}$) and

$$\mathbf{X}_n = \int_{-\pi}^{\pi} e^{in\lambda}\, d\mathbf{Z}(\lambda)\ .$$

8. Let the assumptions in problem 7 be satisfied. In addition let the components of $\mathbf{X}_n$ be real-valued. Show that then

$$dG(\lambda) = \overline{dG(-\lambda)}\ .$$

Further, if $\mathbf{Z}_1(\lambda) = 2\,\mathrm{Re}\,\mathbf{Z}(\lambda)$, $\mathbf{Z}_2(\lambda) = -2\,\mathrm{Im}\,\mathbf{Z}(\lambda)$, $0 < \lambda < \pi$, then

$$\mathbf{X}_n = \int_0^{\pi} \cos n\lambda\, d\mathbf{Z}_1(\lambda) + \int_0^{\pi} \sin n\lambda\, d\mathbf{Z}_2(\lambda)\ .$$

Use the fact that $d\mathbf{Z}(\lambda) = \overline{d\mathbf{Z}(-\lambda)}$. The processes $\mathbf{Z}_1(\lambda)$ and $\mathbf{Z}_2(\lambda)$ are of orthogonal increments. Further

$$E\, d\mathbf{Z}_1(\lambda)\, d\mathbf{Z}_2(\mu)' = 0 \text{ if } \lambda \neq \mu$$

$$E\, d\mathbf{Z}_1(\lambda)\, d\mathbf{Z}_1(\mu)' = \delta_{\lambda,\mu}\, 2\,\mathrm{Re}\, dG(\lambda)$$

$$= E\, dZ_2(\lambda)\, dZ_2(\mu)'$$

$$E\, d\mathbf{Z}_1(\lambda)\, d\mathbf{Z}_2(\mu)' = \delta_{\lambda,\mu}\, 2\,\mathrm{Im}\, dG(\lambda)\ .$$

9. Let n be a k-vector of integers. Consider a process $\{X_n\}$ parameterized by these lattice points. The process is then a random field on the lattice points in k-space. The process is called weakly stationary if it has constant mean value and covariance

$$\mathrm{cov}(X_{\mathbf{n}}, X_{\mathbf{m}}) = r_{\mathbf{n}-\mathbf{m}}$$

depending only on the difference $\mathbf{n} - \mathbf{m}$ of the k-vectors $\mathbf{n}$ and $\mathbf{m}$. Show that

$$r_{\mathbf{n}-\mathbf{m}} = \underbrace{\int_{-\pi}^{\pi} \cdots \int_{-\pi}^{\pi}}_{k} e^{i(\mathbf{n}-\mathbf{m})\cdot\lambda}\, dG(\lambda)$$

where G is bounded and nondecreasing in the sense that mixed kth order differences $\Delta_1 \cdots \Delta_k\, G$ are nonnegative.

Notes

1.1 It is clear that a corresponding concept of stationarity can be considered for any process $\{X_g\}$ with $E\, X_g \equiv 0$, $E\,|X_g|^2 < \infty$ and the index set G a group. In section 1 the index set is the group of integers under addition. In general, the covariance function of the process is a function on the group G.

1.2 If the index set is a group G, the covariance function of $\{X_g\}$ is a positive definite function on the group G. Problems 3 and 9 of this chapter deal with the case of G the real line and the set of lattice points in k-space under addition respectively. The covariance function

$$r(g - g') = \operatorname{cov}(X_g, X_{g'}), \; g, g' \in G$$

if G is commutative and the group operation is indicated by addition. The covariance function can be seen to be positive definite since

$$\sum_{n, m = 1}^{N} c_n \bar{c}_m \, r(g_n - g_m) \geq 0$$

for every choice of elements $g_1, \ldots, g_N$ of G and every set of complex numbers $c_1, \ldots, c_N$. An extensive generalization of the Fourier representation (3) exists for continuous positive definite functions r on a locally compact commutative group G. A complex-valued function γ on such a group G is called a *character* if $|\gamma(g)| \equiv 1$ for all $g \in G$ and if

$$\gamma(g + g') = \gamma(g)\,\gamma(g'), \quad g, g' \in G\,.$$

The set of all continuous characters of G forms a group Γ called the *dual group* of G with *addition* given by G

$$(\gamma_1 + \gamma_2)\,(g) = \gamma_1(g)\,\gamma_2(g), \quad g \in G, \quad \gamma_1, \gamma_2 \in \Gamma\,.$$

Γ is given a topology so that $\gamma(g) = (g, \gamma)$ is a continuous function on $G \times \Gamma$ and Γ is a locally compact commutative group. Further one is able to show that a continuous function r on G is positive definite if and only if

$$r(g) = \int_{\Gamma} (g, \gamma)\,\mu(d\gamma)$$

with μ a finite measure on the σ-algebra generated by the topology on Γ. A detailed derivation of this result and related concepts can be found in Rudin [1962]. Notice that in the case of the Herglotz theorem the characters are of the form $\gamma_\lambda(s) = e^{is\lambda}$, with the character indicator λ an element of the dual group, the circle group under addition.

1.3 Theorem 2 is due to H. Cramér [1940]. It is clear that a corresponding result can be obtained in the more general context alluded to in note 1.2.

At times processes W_n with weakly stationary increments $X_n = W_n - W_{n-1}$ are of interest. Assuming that $E\,W_n \equiv 0$ it is immediately clear from Theorem 2 that

$$X_n = \int_{-\pi}^{\pi} e^{in\lambda}\, dZ(\lambda)$$

with

$$E[dZ(\lambda)\, dZ(\mu)] = \delta_{\lambda,\mu}\, dG(\lambda)$$

with G a bounded nondecreasing function. This implies that

$$W_n - W_0 = \int_{-\pi}^{\pi} \frac{e^{i(n+1)\lambda} - 1}{e^{i\lambda} - 1}\, dZ(\lambda) .$$

It is of some interest to consider the continuous time parameter analogue of these simple remarks. We then consider a process $W(t)$ with weakly stationary increments $W(t+h) - W(t)$ that is continuous in mean square. The process is said to have weakly stationary increments if the derived processes $W(t+h) - W(t)$ are weakly stationary as processes in t for each h. It is continuous in mean square if $E \mid W(t+h) - W(t) \mid^2 \to 0$ as $h \to 0$ for each t. Again assume that $E\, W(t) \equiv 0$. One can then show that

$$W(t) - W(0) = \int_{-\infty}^{\infty} \frac{e^{it\lambda} - 1}{i\,\lambda}\, dZ(\lambda)$$

with

$$E[dZ(\lambda)\, \overline{dZ(\mu)}] = \delta_{\lambda,\mu}\, dG(\lambda)$$

and G a finite-valued nondecreasing function. Here G is no longer necessarily a bounded function. Notice that

$$E\,\{[W(t) - W(0)]\, \overline{[W(s) - W(0)]}\} = \int_{-\infty}^{\infty} \frac{e^{it\lambda} - 1}{i\,\lambda} \left\{\overline{\frac{e^{is\lambda} - 1}{i\,\lambda}}\right\} dG(\lambda) .$$

The cases of the Wiener process and Poisson process are natural examples to consider in this context. The Wiener process $W(t)$ is a Gaussian process with $E\, W(t) \equiv 0$, $E \mid W(t) - W(s) \mid^2 = \mid t - s \mid \sigma^2$, $\sigma^2 > 0$, and increments of disjoint intervals independent. The Poisson process $P(t)$ has $E\, P(t) = t\,\eta$, $\eta > 0$, with $E \mid P(t) - P(s) - (t-s)\,\eta \mid^2 = \mid t - s \mid \eta$ and increments of disjoint intervals independent. The distribution of $P(t) - P(s)$, $t > s$, is Poisson with mean $(t-s)\,\lambda$. The covariance properties of the Wiener and Poisson processes are essentially the same. In the case of the Wiener process

$$dG(\lambda) = \frac{\sigma^2}{2\pi}\, d\lambda$$

so that the total increase of the function G is infinite. For the Poisson process

$$P(t) - P(0) - t\,\eta = \int_{-\infty}^{\infty} \frac{e^{it\lambda} - 1}{i\,\lambda}\, dZ(\lambda)$$

with

$$E[dZ(\lambda)\, \overline{dZ(\mu)}] = \delta_{\lambda,\mu}\, dG(\lambda) = \delta_{\lambda,\mu} \frac{\eta}{2\pi}\, d\lambda .$$

In the case of a continuous parameter random field $W(t)$, $t = (t_1, \ldots, t_k)$, on k-space with weakly stationary increments that is continuous in mean square, one would be led to a corresponding representation. In the case $k = 2$ the increments would be

$$\Delta_{h_1} \Delta_{h_2} W(t_1, t_2) = W(t_1 + h_1, t_2 + h_2) - W(t_1 + h_1, t_2) - W(t_1, t_2 + h_2) + W(t_1, t_2) .$$

As usual assume $E\, W(t_1, t_2) \equiv 0$. We would then have

$$\Delta_{h_1} \Delta_{h_2} W(t_1, t_2) = \int_{-\infty}^{\infty} \int_{-\infty}^{\infty} e^{i(t_1 \lambda_1 + t_2 \lambda_2)} \frac{e^{ih_1\lambda_1} - 1}{i \lambda_1} \frac{e^{ih_2\lambda_2} - 1}{i \lambda_2} dZ(\lambda_1, \lambda_2)$$

with

$$E[dZ(\lambda_1, \lambda_2) \overline{dZ(\mu_1, \mu_2)}] = \delta_{\lambda_1, \mu_1} \delta_{\lambda_2, \mu_2} dG(\lambda_1, \lambda_2)$$

and G a finite (but not necessarily bounded) nondecreasing function of two variables. See Yaglom [1962] for a more detailed discussion of related questions.

Chapter II

Prediction and Moments

1. Prediction

Consider a random variable X, $E\, X^2 < \infty$, that we wish to approximate (or predict) by a random variable Y belonging to a linear space of random variables H with finite second moments that is closed in the sense of mean square convergence. The following result is useful.

Theorem 1. *The best predictor (in the sense of minimizing the mean square error of the predictor) is given by the projection of X on H*

$$X^* = \mathscr{P}_H X$$

and is characterized by the property that

$$E[(X - X^*)\, \overline{Y}] = 0$$

for all $Y \in H$.

We shall sketch the derivation of this result in the case that H is separable, that is, when there is a countable collection of random variables $\{U_j\}$ in H that is dense in H. This means that given any random variable V in H one can find a sequence out of $\{U_j\}$ that converges to V in mean square. First an orthonormal sequence of random variables W_j, $E(W_j\, \overline{W_k}) = \delta_{jk}$, is generated from the collection $\{U_j\}$ by the Gramm-Schmidt orthogonalization procedure. This is carried out recursively. Assume that $W_1, \ldots, W_j$ are already orthonormal and generated from $U_1, \ldots, U_j$ (assumed linearly independent). Given U_{j+1} linearly independent of $U_1, \ldots, U_j$ and hence of $W_1, \ldots, W_j$ let

$$Z = U_{j+1} - \sum_{k=1}^{j} E(U_{j+1}\, \overline{W_k})\, W_k\,.$$

Set

$$W_{j+1} = Z \,/\, \{E\, |\, Z\, |^2\}^{1/2}\,.$$

It is clear that

$$E(W_{j+1}\, \overline{W_k}) = 0 \quad \text{for} \quad k = 1, \ldots, j$$

and $E\, |\, W_{j+1}\, |^2 = 1$.

Given the random variable X, it follows that the best predictor of X in terms of random variables out of H is

$$\mathscr{P}_H X = \sum_{j=1}^{\infty} E(X\, \overline{W_j})\, W_j\,. \tag{1}$$

One can see that this is so since any linear predictor out of H can be written

$$\sum c_j\, W_j, \quad \sum |\, c_j\, |^2 < \infty\,.$$

The coefficients that minimize

$$E \,|\, X - \sum c_j W_j \,|^2$$

are given by $c_j = E(X \overline{W}_j)$. Notice that $\mathscr{P}_H X$ as given by (1) is orthogonal to all the W_j's and hence to all of H.

The characterization of the best linear predictor given above can be used in a nonlinear prediction problem. Suppose that a random variable X with $E\,X^2 < \infty$ is to be approximated in mean square by a function (generally nonlinear) of the random variables Y_α, $\alpha \in I$. Here I is simply a set of indices. The random variables Y_α are the possible or potential observables. The set of possible predictors is simply the linear space H of random variables Z with $E\,Z^2 < \infty$ that are measurable with respect to the σ-field generated by the random variables Y_α, $\alpha \in I$. H is just the linear space of (generally nonlinear) decent functions of the Y_α's. As in the case discussed above, the best predictor X^* of X is the projection of X on H and is characterized by the property that

$$E[(X - X^*)\,\overline{Z}] = 0 \tag{2}$$

for all $Z \in H$. In this problem, the condition (2) implies that

$$X^* = E[X \mid Y_\alpha, \quad \alpha \in I]$$

the conditional expectation of X given the Y_α, $\alpha \in I$.

In the case of a linear prediction problem H would be determined as the linear space of random variables Z with $E\,Z^2 < \infty$ that are linear in the random variables Y_α, $\alpha \in I$.

A k-vector $\mathbf{X} = (X_j)$ of real-valued random variables is said to be a normal vector (or a vector of jointly normal random variables) if every inner product $\mathbf{t} \cdot \mathbf{X}$ of $\mathbf{X}$ with a real k-vector $\mathbf{t}$ is one-dimensional normal. This implies that the distribution of $\mathbf{X}$ is determined by its mean vector $\boldsymbol{\mu} = E\,\mathbf{X} = (\mu_i)$, $\mu_i = E\,X_i$, and its covariance matrix

$$R = E(\mathbf{X} - \boldsymbol{\mu})\,(\mathbf{X} - \boldsymbol{\mu})' = (r_{i,j})$$

$$r_{i,j} = \operatorname{cov}(X_i, X_j)\,.$$

Normal (or Gaussian) random variables have all moments finite. If R is nonsingular (R is automatically positive semidefinite) the distribution of $\mathbf{X}$ has the probability density

$$\varphi(\mathbf{x}) = (2\,\pi)^{-k/2}\,|\,R\,|^{-1/2} \exp\left\{-\frac{1}{2}(\mathbf{x} - \boldsymbol{\mu})'\,R^{-1}(\mathbf{x} - \boldsymbol{\mu})'\right\}.$$

Whether R is nonsingular or not, the characteristic function of $\mathbf{X}$ is given by

$$\psi(\mathbf{t}) = E\,\exp(i\,\mathbf{t} \cdot \mathbf{X}) = \int \exp(i\,\mathbf{t} \cdot \mathbf{x})\,d\Phi(\mathbf{x})\,.$$

with Φ the distribution function of $\mathbf{X}$.

The k-vector $\mathbf{X} = (X_j)$ of complex-valued random variables is called complex normal with mean vector $\boldsymbol{\mu}$ and covariance matrix R if the $2k$ vector of real-valued random variables

$$\begin{pmatrix} \operatorname{Re} \mathbf{X} \\ \operatorname{Im} \mathbf{X} \end{pmatrix}$$

is a normal vector with mean vector $\begin{pmatrix} \operatorname{Re} \boldsymbol{\mu} \\ \operatorname{Im} \boldsymbol{\mu} \end{pmatrix}$ and covariance matrix

$$\frac{1}{2} \begin{pmatrix} \operatorname{Re} R & -\operatorname{Im} R \\ \operatorname{Im} R & \operatorname{Re} R \end{pmatrix}.$$

One should note that R is self-adjoint (Hermitian) and so $(\operatorname{Im} R)' = -\operatorname{Im} R$. Consider a normal k-vector valued stationary process $\{X_n\}$ (see problem 8 of Chapter I). If the process has a pure jump spectrum, one can see that the spectral jumps $\Delta Z(\lambda_j)$ are complex normal if $\lambda_j \neq 0$.

As a simple example consider the case of two jointly normal random variables X, Y with means $E\,X = E\,Y = 0$, variances $E\,X^2 = E\,Y^2 = 1$ and covariance ϱ, $|\varrho| < 1$. The joint probability density is

$$(2\pi)^{-1} (1 - \varrho^2)^{-1/2} \exp\left\{-\frac{1}{2(1-\varrho^2)} [x^2 - 2\varrho x y + y^2]\right\}.$$

This implies that the conditional density of x given y is

$$f(x \mid y) = (2\pi)^{-1/2} (1 - \varrho^2)^{-1/2} \exp\left\{-\frac{1}{2(1-\varrho^2)} (x - \varrho y)^2\right\},$$

a normal density function with mean ϱy and variance $1 - \varrho^2$. Notice that the conditional mean of X given Y

$$\begin{aligned} E(X \mid Y = y) &= \int x f(x \mid y)\, dx \\ &= \varrho y \end{aligned}$$

is linear in y. Given jointly normal random variables, the conditional mean of one given the others can be shown to be linear in the others by a similar argument. It is simple to compute

$$\begin{aligned} E(X^2 \mid Y = y) &= \int x^2 f(x \mid y)\, dx \\ &= (\varrho y)^2 + (1 - \varrho^2) \end{aligned}$$

and one can show that

$$E[\{X^2 - E(X^2 \mid Y)\}\, g(Y)] = 0$$

for any function g with $E\,g(Y)^2 < \infty$.

2. *Moments and Cumulants*

Let $\varphi_{\mathbf{X}}(t) = \varphi_{\mathbf{X}}(t_1, \ldots, t_k)$ be the joint characteristic function of the k-vector $\mathbf{X} = \begin{pmatrix} X_1 \\ \vdots \\ X_k \end{pmatrix}$. Then

$$\varphi_{\mathbf{X}}(t_1, \ldots, t_k) = E \exp(i\,\mathbf{t} \cdot \mathbf{X}) \tag{3}$$

$$= \sum_{\nu_1 + \cdots + \nu_k \leq n} \frac{i^{\nu_1 + \cdots + \nu_k}}{\nu_1! \cdots \nu_k!} m_X^{(\nu_1, \cdots, \nu_k)} t_1^{\nu_1} \ldots t_k^{\nu_k} + o(|\mathbf{t}|^n)$$

if all moments (absolute) up to order n exist with

$$m_{\mathbf{X}}^{(\nu_1, \cdots, \nu_k)} = E(X_1^{\nu_1} \cdots X_k^{\nu_k}) .$$

The logarithm of the characteristic function can be expanded about the origin in a Taylor series truncated at order n under the same conditions

$$\log \varphi_{\mathbf{X}}(t_1, \ldots, t_k) = \sum_{\nu_1 + \cdots + \nu_k \leq n} \frac{i^{\nu_1 + \cdots + \nu_k}}{\nu_1! \cdots \nu_k!} c_X^{(\nu_1, \cdots, \nu_k)} t_1^{\nu_1} \cdots t_k^{\nu_k} + o(|\mathbf{t}|^n) \tag{4}$$

with the coefficients $c_{\mathbf{X}}^{(\nu_1, \cdots, \nu_k)}$ the cumulants of $X_1^{\nu_1}, \ldots, X_k^{\nu_k}$ respectively. If we abbreviate $(\nu_1, \ldots, \nu_k)$ by $\boldsymbol{\nu}$, $(\alpha_1, \ldots, \alpha_k)$ by $\boldsymbol{\alpha}$ and let

$$\boldsymbol{\beta}^{\boldsymbol{\mu}} = \beta_1^{\mu_1} \cdots \beta_k^{\mu_k}, \ \boldsymbol{\mu}! = \mu_1! \cdots \mu_k! , \ |\boldsymbol{\mu}| = \mu_1 + \mu_2 + \cdots + \mu_k$$

relations (3) and (4) can be rewritten

$$\varphi_{\mathbf{X}}(\mathbf{t}) = \sum_{|\boldsymbol{\nu}| \leq n} \frac{i^{|\boldsymbol{\nu}|}}{\boldsymbol{\nu}!} m_{\mathbf{X}}^{(\boldsymbol{\nu})} \mathbf{t}^{\boldsymbol{\nu}} + o(|\mathbf{t}|^n) \tag{5}$$

and

$$\log \varphi_{\mathbf{X}}(\mathbf{t}) = \sum_{|\boldsymbol{\nu}| \leq n} \frac{i^{|\boldsymbol{\nu}|}}{\boldsymbol{\nu}!} c_{\mathbf{X}}^{(\boldsymbol{\nu})} \mathbf{t}^{\boldsymbol{\nu}} + o(|\mathbf{t}|^n) . \tag{6}$$

By expanding $\exp\big(\log \varphi_X(t)\big)$ using (6) and comparing coefficients with (5) the relation

$$m_{\mathbf{X}}^{(\boldsymbol{\nu})} = \sum_{\boldsymbol{\lambda}^{(1)} + \cdots + \boldsymbol{\lambda}^{(q)} = \boldsymbol{\nu}} \frac{1}{q!} \frac{\boldsymbol{\nu}!}{\boldsymbol{\lambda}^{(1)}! \cdots \boldsymbol{\lambda}^{(q)}!} \prod_{p=1}^{q} c_{\mathbf{X}}^{(\boldsymbol{\lambda}^{(p)})}$$

is obtained where it is understood that one sums over all possible q. Also, if one expands $\log \varphi_{\mathbf{X}}(t)$ using (5) and compares coefficients with (6) the relation

$$c_{\mathbf{X}}^{(\boldsymbol{\nu})} = \sum_{\boldsymbol{\lambda}^{(1)} + \cdots + \boldsymbol{\lambda}^{(q)} = \boldsymbol{\nu}} \frac{(-1)^{q-1}}{q} \frac{\boldsymbol{\nu}!}{\boldsymbol{\lambda}^{(1)}! \cdots \boldsymbol{\lambda}^{(q)}!} \prod_{p=1}^{q} m_{\mathbf{X}}^{(\boldsymbol{\lambda}^{(p)})}$$

is obtained. As particular cases, we note the following relationships. First

$$\operatorname{cum}(Y_1, \ldots, Y_k) = \sum (-1)^{p-1}(p-1)!\, E\left(\prod_{j \in \nu_1} Y_j\right) \ldots E\left(\prod_{j \in \nu_p} Y_j\right)$$

where $\nu_1, \ldots, \nu_p$ is a partition of $(1, 2, \ldots, k)$ and one sums over all these partitions. Also

$$E(Y_1 \cdots Y_k) = \sum D_{\nu_1} \cdots D_{\nu_p}$$

where $\nu_1, \ldots, \nu_p$ is a partition of $(1, 2, \ldots, k)$,

$$D_{\nu_s} = \operatorname{cum}(Y_{\alpha_1}, \ldots, Y_{\alpha_m})$$

where the α_j are the elements of ν_s, and one sums over all partitions ν of $(1, 2, \ldots, k)$.

Consider a table

$$\begin{matrix} (1, 1), \ldots, (1, J_1) \\ \vdots \qquad\quad \vdots \\ (I, 1), \ldots, (I, J_I) \end{matrix}$$

A partition of the table $P_1, \ldots, P_M$ is said to be indecomposable if there are no sets $P_{m_1}, \ldots, P_{m_N}$ $(N < M)$ and rows $R_{i_1}, \ldots, R_{i_s}$ $(s < I)$ such that

$$P_{m_1} \cup \cdots \cup P_{m_N} = R_{i_1} \cup \cdots \cup R_{i_s}.$$

Theorem 2. *Consider a two way array of random variables* X_{ij}, $j = 1, \ldots, J_i$, $i = 1, \ldots, I$. *Let*

$$Y_i = \prod_{j=1}^{J_i} X_{ij}, \quad i = 1, \ldots, I.$$

The joint cumulant $\operatorname{cum}(Y_1, \ldots, Y_I)$ *is given by*

$$\sum_{\nu} \operatorname{cum}(X_{ij};\, ij \in \nu_1) \cdots \operatorname{cum}(X_{ij};\, ij \in \nu_p)$$

where the summation is over all indecomposable partitions $\nu_1 \cup \cdots \cup \nu_p = \nu$ *of the two-way table.*

As already noted

$$E(Y_1 \cdots Y_I) = \sum D_{\mu_1} \cdots D_{\mu_p}$$

with

$$D_{\mu_s} = \operatorname{cum}(Y_{\alpha_1}, \ldots, Y_{\alpha_m}),$$

$\mu_s = (a_1, \ldots; a_m)$, $\mu_1, \ldots, \mu_p$ a partition of $(1, \ldots, I)$ and one sums over all partitions of $(1, \ldots, I)$. One also has

$$E(Y_1 \cdots Y_I) = E\left(\prod_{i=1}^{I} \prod_{j=1}^{J_i} X_{ij}\right) = \quad c_{\nu_1} \cdots c_{\nu_p}$$

where $\nu_1, \ldots, \nu_p$ is a partition of the two-way table of X_{ij}'s and $c_{\nu_s} = \text{cum}\,(X_{ij}, ij \in \nu_s)$. Then

$$\text{cum}(Y_1, \ldots, Y_I) = \sum c_{\nu_1} \cdots c_{\nu_p} - \quad {}' D_{\mu_1} \cdots D_{\mu_p}$$

with $\sum'$ a summation over all partitions with $p \geq 2$. Theorem 2 follows on using this last identity with an induction argument on the number of rows I.

The following are simple properties of cumulants which follow almost immediately from the definition:

(i) $\text{cum}(a_1 Y_1, \ldots, a_r Y_r) = a_1 \cdots a_r \,\text{cum}(Y_1, \ldots, Y_r)$ for $a_1, \ldots, a_r$ constants.

(ii) If a nontrivial proper subgroup of the Y's are independent of the remaining Y's then $\text{cum}\,(Y_1, \ldots, Y_r) = 0$.

(iii) Given random variables $Z_1, Y_1, \ldots, Y_r$ $\text{cum}(Y_1 + Z_1, Y_2, \ldots, Y_r) = \text{cum}(Y_1, Y_2, \ldots, Y_r) + \text{cum}(Z_1, Y_2, \ldots, Y_r)$.

$$\text{cum}(X_1, X_2) = E\,X_1 X_2 - E\,X_1\,E\,X_2\,.$$

If $X_1 = X_2$ we have $\sigma^2 = E\,X^2 - (E\,X)^2$ the variance of X.

$$\begin{aligned}\text{cum}(X_1, X_2, X_3) = {} & E(X_1 X_2 X_3) - E\,X_1\,E(X_2 X_3) - E\,X_2\,E(X_1 X_3) \\ & - E\,X_3\,E(X_1 X_2) + 2\,E\,X_1\,E\,X_2\,E\,X_3\,.\end{aligned}$$

If $X_1 = X_2 = X_3$ we obtain

$$E\,X^3 - 3\,E\,X\,E\,X^2 + 2\,(E\,X)^3 = E(X - E\,X)^3$$

the third central moment of X. The coefficient of skewness of X is

$$E(X - E\,X)^3 \,/\, \sigma^3\,.$$

$$\begin{aligned}\text{cum}(X_1, X_2, X_3, X_4) = {} & E(X_1 X_2 X_3 X_4) - \{E\,X_1\,E(X_1 X_2 X_3) \\ & + E\,X_2\,E(X_1 X_3 X_4) + \cdots\} \\ & - \{E(X_1 X_2)\,E(X_3 X_4) + E(X_1 X_3)\,E(X_2 X_4) + \cdots\} \\ & + 2\{E\,X_1\,E\,X_2\,E(X_3 X_4) + E\,X_1\,E\,X_3\,E(X_2 X_4) + \cdots\} \\ & - 6\,E\,X_1\,E\,X_2\,E\,X_3\,E\,X_4\,.\end{aligned}$$

Notice that we have

$$E\,X^4 - 4\,E\,X\,E\,X^3 + 6(E\,X)^2\,E\,X^2 - 3(E\,X)^4 = E(X - E\,X)^4$$

the fourth central moment of X.

The coefficient of kurtosis of X is

$$E(X - E\,X)^4 / \sigma^4 - 3\,.$$

Assuming $E\,X_j \equiv 0$ we obtain

$$\begin{aligned}
\operatorname{cum}(X_1, X_2) &= E\,X_1 X_2 \\
\operatorname{cum}(X_1, X_2, X_3) &= E\,X_1 X_2 X_3 \\
\operatorname{cum}(X_1, X_2, X_3, X_4) &= E(X_1 X_2 X_3 X_4) - \{E(X_1 X_2)\,E(X_3 X_4) \\
&\quad + E(X_1 X_3)\,E(X_2 X_4) + E(X_1 X_4)\,E(X_2 X_3)\}\,.
\end{aligned}$$

Let us note that if one has jointly Gaussian random variables all cumulants of order higher than second are zero. This follows from the fact that the characteristic function of jointly Gaussian random variables with mean vector μ and covariance matrix R is given by

$$\varphi(t) = \exp\left\{i\,t \cdot \mu - \frac{1}{2}\,t'\,R\,t\right\}.$$

The first order and second order cumulants are just the means and covariances. We can now look at a simple illustration of Theorem 2. Let $X = \{X_n\}$ be a Gaussian stationary sequence with mean zero and covariance function r_n. Consider the derived stationary sequence $Y = \{Y_n\}$ with $Y_n = X_n X_{n-1}$. We wish to compute the covariance $\operatorname{cov}(Y_n, Y_{n-k}) = \operatorname{cum}(Y_n, Y_{n-k})$. This computation requires that one look at the two-way table

$$\begin{matrix} X_n & X_{n-1} \\ X_{n-k} & X_{n-k-1} \end{matrix}$$

One only needs to look at indecomposable partitions consisting of sets of pairs because only covariances or second order cumulants of the X random variables can be nonzero. The two indecomposable partitions of this type are

$$\{(X_n, X_{n-k}), (X_{n-1}, X_{n-k-1})\}$$

$$\{(X_n, X_{n-k-1}), (X_{n-1}, X_{n-k})\}$$

indicating that

$$\operatorname{cov}(Y_n, Y_{n-k}) = r_k^2 + r_{k+1}\,r_{k-1}\,.$$

Consider a random variable X with distribution function $F(x) = P[X \leq x]$. If $F(-x) + 1 - F(x)$ decreases to zero sufficiently rapidly as $x \to \infty$, all the moments

$$m_X^{(\nu)} = \int x^\nu\,dF(x), \quad \nu = 1, 2, \ldots$$

of the distribution F will exist. Assuming that all the moments exist, one can ask whether one has a determinate moment problem in the sense that there is a unique distribution corresponding to the given moments. Carleman (see Akhiezer [1965]) gave the following sufficient condition for the moment problem to be determinate

$$\sum_{1}^{\infty} \{m^{(2\nu)}\}^{1/2\nu} = \infty .$$

From this one can see, for example, that the moments of the exponential distribution and the moments of the normal distribution lead to a determinate moment problem.

Related ideas can be used to prove certain types of limit theorems.

Theorem 3. *Consider a sequence of distribution functions $F_n(x)$ with corresponding moments $m_n^{(\nu)}$, $\nu = 1, 2, \ldots, n = 1, 2, \ldots$. Suppose*

$$\lim_{n \to \infty} m_n^{(\nu)} = m^{(\nu)}, \quad n = 1, 2, \ldots$$

and the numbers $m^{(\nu)}$ are the moments of a determinate moment problem with corresponding distribution function $F(x)$. Then F_n converges to F weakly as $n \to \infty$.

This result can be obtained by using the Helly convergence theorem (see Loève [1963]). In particular since the normal distribution has a determinate moment problem, one way of proving convergence to a normal distribution is by showing that $m_n^{(\nu)}$ converges as $n \to \infty$ to the νth moment m_ν of a normal distribution, $\nu = 1, 2, \ldots$.

3. Autoregressive and Moving Average Processes

An autoregressive moving average process (ARMA process) $\{Y_t\}$ satisfies

$$\sum_{k=0}^{p} \beta_k Y_{t-k} = \sum_{g=0}^{q} a_g V_{t-g}, \tag{7}$$

$a_0, \beta_0 \neq 0$,

$t = \ldots, -1, 0, 1, \ldots$ with the sequence $\{V_t\}$ consisting of independent, identically distributed random variables. Assume that $E\, V_t \equiv 0$, $E\, V_t^2 = \sigma^2 > 0$. For the moment, let us assume only that the V_t's are orthogonal so that $\{V_t\}$ is a "white noise" process. Notice that if $a_g = 0$ for $g \neq 0$ we have an autoregressive process while if $\beta_k = 0$ for $k \neq 0$ we have a moving average process. Our object is to determine circumstances under which there is a weakly stationary solution to (7). Let

$$\beta(z) = \sum_{k=0}^{p} \beta_k z^k, \; a(z) = \sum_{g=0}^{q} a_g z^g .$$

If $\{Y_t\}$ is a weakly stationary solution of (7), let $Z_Y(\lambda)$ be the corresponding process of orthogonal increments

$$Y_t = \int_{-\pi}^{\pi} e^{it\lambda}\, dZ_Y(\lambda) .$$

Also

$$V_t = \int_{-\pi}^{\pi} e^{it\lambda}\, dZ_V(\lambda)$$

$$E[dZ_V(\lambda)\, \overline{dZ_V(\mu)}] = \delta_{\lambda,\mu} \frac{\sigma^2}{2\pi}\, d\lambda .$$

Equation (7) can be rewritten as

$$\int_{-\pi}^{\pi} e^{it\lambda}\, \beta(e^{-i\lambda})\, dZ_Y(\lambda) = \int_{-\pi}^{\pi} e^{it\lambda}\, \alpha(e^{-i\lambda})\, dZ_V(\lambda) . \tag{8}$$

This implies that the spectral density $g(\lambda)$ of $\{Y_t\}$ has the form

$$g(\lambda) = \frac{\sigma^2}{2\pi} \left| \frac{\alpha(e^{-i\lambda})}{\beta(e^{-i\lambda})} \right|^2 . \tag{9}$$

If $\alpha(z)$, $\beta(z)$ both have zeros at $z = e^{-i\xi}$ (ξ real) the order of the zero for α must be greater than or equal to the order of the zero for β. For otherwise, $g(\lambda)$ would not be integrable. Relation (8) also implies that

$$\int_{-\pi}^{\mu} \beta(e^{-i\lambda})\, dZ_Y(\lambda) = \int_{-\pi}^{\mu} \alpha(e^{-i\lambda})\, dZ_V(\lambda) . \tag{10}$$

If $Z_Y(\lambda)$ has no jump at $\lambda = \xi$, the order of the zero at $z = e^{-i\xi}$ for β can be taken to be zero. If $Z_Y(\lambda)$ has a jump at $\lambda = \xi$ the order of the zero at $z = e^{-i\xi}$ for β can be taken to be one. This follows from (10). The zeros of α and β properly inside or outside the unit circle can be taken to be distinct for the purpose of our argument.

Let us assume that $Z_Y(\lambda)$ has no jumps. The spectral distribution function G of $\{Y_t\}$ can have no continuous singular part and hence must be absolutely continuous. β can then be assumed to have no zeros on $|z| = 1$. The process $\{Y_t\}$ will then have a one-sided representation

$$Y_t = \sum_{j=0}^{\infty} \gamma_j V_{t-j} \tag{11}$$

in terms of the process $\{V_t\}$ if and only if all zeros of $\beta(z)$ have modulus greater than one. Similarly, $\{V_t\}$ will have a one-sided representation

$$V_t = \sum_{j=0}^{\infty} \eta_j Y_{t-j} \tag{12}$$

in terms of $\{Y_t\}$ if and only if $\alpha(z)$ has all its zeros of modulus greater than one. This can be derived from the fact that Y_t has the representation

$$Y_t = \int_{-\pi}^{\pi} e^{it\lambda}\, dZ_Y(\lambda) = \int_{-\pi}^{\pi} e^{it\lambda} \frac{\alpha(e^{-i\lambda})}{\beta(e^{-i\lambda})}\, dZ_V(\lambda) \tag{13}$$

if β has no zeros of modulus one. One should note that in all the cases of ARMA schemes, whatever the location of the zeros of $\alpha(z)$ and $\beta(z)$, there is a representation of the process $\{Y_t\}$ in terms of the independent, identically distributed random variables $\{V_t\}$ of the form

$$Y_t = \sum_{j=-\infty}^{\infty} \gamma_j V_{t-j}\,. \tag{14}$$

The remarks made above can be seen to hold by using an argument of the following character. Consider a partial fraction expansion of $\alpha(z) / \beta(z)$

$$\frac{\alpha(z)}{\beta(z)} = p(z) + \sum_j \frac{A_j}{z - z_j} \tag{15}$$

with $p(z)$ a polynomial and the z_j's the roots (assumed simple) of $\beta(z)$. It is naturally assumed that $\alpha(z)$ and $\beta(z)$ have no zeros in common. Then

$$\begin{aligned}(e^{-i\lambda} - z_j)^{-1} &= (-z_j)^{-1}(1 - z_j^{-1} e^{-i\lambda})^{-1} \\ &= (-z_j)^{-1} \sum_{k=0}^{\infty} z_j^{-k} e^{-ik\lambda}\end{aligned} \tag{16}$$

if $|z_j| > 1$ while

$$\begin{aligned}(e^{-i\lambda} - z_j)^{-1} &= e^{i\lambda}(1 - e^{i\lambda} z_j)^{-1} \\ &= \sum_{k=0}^{\infty} z_j^{k} e^{i(k+1)\lambda}\end{aligned} \tag{17}$$

if $|z_j| < 1$. The representations (11) and (14) of Y_t are obtained by making use of (15), (16), (17) and formula (13). The representation of V_t as given by (12) is obtained by interchanging the role of $\alpha(z)$ and $\beta(z)$. In the case of ARMA schemes the weights γ_j decay to zero as $|j| \to \infty$ at least at an exponential rate. A process of the form (14) with Y_t given in terms of independent, identically distributed random variables and with

$$\sum_j |\gamma_j|^2 < \infty$$

is called a *linear process*. Notice that if the roots of $\alpha(z)$, $\beta(z)$ are outside the unit disc, the equation (7) has an interesting interpretation in terms of the linear prediction problem. Let $\mathcal{M}_n = \mathcal{M}(Y_t, t \leq n)$ denote the closed linear space (closed in the sense of mean square convergence) generated by linear forms in the Y_t's, $t \leq n$. Now $V_t \in \mathcal{M}_t$ since the zeros of $\alpha(z)$ have modulus greater than one. Y_t has a one-sided representation (11) in terms of the V's since the zeros of $\beta(z)$ have modulus greater than one. Therefore V_t is orthogonal to $\mathcal{M}_{t-1}$. It follows that

$$Y_t^* = -\sum_{k=1}^{p} \frac{\beta_k}{\beta_0} Y_{t-k} + \sum_{g=1}^{q} \frac{\alpha_g}{\beta_0} V_{t-g}$$

is in $\mathcal{M}_{t-1}$ and $Y_t - Y_t^* = (\alpha_0/\beta_0) V_t$ is orthogonal to $\mathcal{M}_{t-1}$. Thus Y_t^* is the projection of Y_t on $\mathcal{M}_{t-1}$ and consequently the best linear predictor of Y_t in terms of the past Y_j, $j \leq t-1$. The variance of the prediction error is

$$E \mid Y_t - Y_t^* \mid^2 = \left| \frac{\alpha_0}{\beta_0} \right|^2 \sigma^2 .$$

We wish to show that this equals

$$2\pi \exp\left\{ \frac{1}{2\pi} \int_{-\pi}^{\pi} \log g(\lambda)\, d\lambda \right\} \tag{18}$$

with g the spectral density

$$g(\lambda) = \frac{\sigma^2}{2\pi} \left| \frac{\alpha(e^{-i\lambda})}{\beta(e^{-i\lambda})} \right|^2 .$$

If the z_v's are the zeros of $\alpha(z)$

$$\begin{aligned} \mid \alpha(e^{-i\lambda}) \mid^2 &= \left| \prod_{v=1}^{q} (e^{-i\lambda} - z_v) \right|^2 \cdot \mid \alpha_q \mid^2 \\ &= \prod_{v=1}^{q} \mid z_v \mid^2 \prod_{v=1}^{q} \mid 1 - e^{-i\lambda} z_v^{-1} \mid^2 \mid \alpha_q \mid^2 . \end{aligned}$$

If $\mid w \mid < 1$

$$\log(1 - w) = -\sum_{j=1}^{\infty} \frac{w^j}{j} .$$

Now

$$\log \mid \alpha(e^{-i\lambda}) \mid^2 = \log \mid \alpha_0 \mid^2 + \sum_{v=1}^{q} \log \mid 1 - e^{-i\lambda} z_v^{-1} \mid^2 .$$

Since $\mid z_v \mid > 1$

$$\int_{-\pi}^{\pi} \log(1 - e^{-i\lambda} z_v^{-1})\, d\lambda = -\sum_{j=1}^{\infty} \int_{-\pi}^{\pi} \frac{e^{-ij\lambda} z_v^{-j}}{j}\, d\lambda = 0 .$$

By assumption all the zeros of $\alpha(z)$ and $\beta(z)$ are outside the unit disc. Thus

$$\frac{1}{2\pi} \int_{-\pi}^{\pi} \log g(\lambda)\, d\lambda = \log \frac{\sigma^2}{2\pi} + \log \mid \alpha_0 \mid^2 - \log \mid \beta_0 \mid^2$$

and so

$$\sigma^2 \left| \frac{\alpha_0}{\beta_0} \right|^2 = 2\pi \exp\left\{ \frac{1}{2\pi} \int_{-\pi}^{\pi} \log g(\lambda)\, d\lambda \right\}. \tag{19}$$

Let us now assume that $\{Y_t\}$ is a weakly stationary process that satisfies the system of equations

$$\sum_{k=0}^{p} b_k Y_{t-k} = \sum_{j=0}^{q} a_j V_{t-j}\,, \tag{20}$$

with $a_0, b_0 \neq 0$. Assume that $a(z)$ has no roots of modulus one. It may be that some of the roots of $b(z)$ and $a(z)$ are inside the unit disc in the complex plane. In that case the system of equations (20) does not correspond to the prediction problem for $\{Y_t\}$. However, one can show how the system (20) can be replaced by another one that does correspond to the prediction problem for $\{Y_t\}$ in which all the roots have absolute value greater than one. Let the z_v and w_v be the zeros of $a(z)$ and $b(z)$. Set

$$\alpha(z) = \prod_{|z_v|<1} \frac{(z\bar{z}_v - 1)}{(z - z_v)}$$

$$\beta(z) = \prod_{|w_v|<1} \frac{(z\bar{w}_v - 1)}{(z - w_v)}$$

and

$$\tilde{a}(z) = a(z)\,\alpha(z)$$

$$\tilde{b}(z) = b(z)\,\beta(z)\,.$$

In constructing $\tilde{a}(z)$ from $a(z)$ each root z_v with $|z_v| < 1$ is replaced by the root $\bar{z}_v^{-1}$ (outside the unit disc) while roots outside the unit disc are retained. Then

$$\int_{-\pi}^{\pi} e^{it\lambda}\, \tilde{b}(e^{-i\lambda})\, \beta(e^{-i\lambda})^{-1}\, dZ_Y(\lambda)$$

$$= \int_{-\pi}^{\pi} e^{it\lambda}\, \tilde{a}(e^{-i\lambda})\, \alpha(e^{-i\lambda})^{-1}\, dZ_V(\lambda)$$

and this implies that

$$\int_{-\pi}^{\pi} e^{it\lambda}\, \tilde{b}(e^{-i\lambda})\, dZ_Y(\lambda) = \int_{-\pi}^{\pi} e^{it\lambda}\, \tilde{a}(e^{-i\lambda})\, dZ_\eta(\lambda)$$

where

$$dZ_\eta(\lambda) = \beta(e^{-i\lambda})\, \alpha(e^{-i\lambda})^{-1}\, dZ_V(\lambda)\,.$$

Notice that

$$\eta_t = \int_{-\pi}^{\pi} e^{it\lambda}\, dZ_{\eta}(\lambda)$$

is a white noise process with $E[\eta_t \overline{\eta_\tau}] = \delta_{t,\tau}\, \sigma^2$. The representation (20) can be rewritten in the form

$$\sum_{j=0}^{p} \tilde{b}_j\, Y_{t-j} = \sum_{k=0}^{q} \tilde{a}_k\, \eta_{t-k}$$

where the $\tilde{b}_j$'s and $\tilde{a}_k$'s are the coefficients of the polynomials $\tilde{b}(z)$ and $\tilde{a}(z)$. This representation of $\{Y_t\}$ as an ARMA process corresponds to the linear prediction problem for that process. If the process Y_t is realvalued the polynomials $b(z)$ and $a(z)$ will have real coefficients and the roots will be real or will appear in conjugate pairs. We have the following theorem.

Theorem 3. *The system of equations* (7) *have a weakly stationary solution* $\{Y_t\}$ *if and only if the function* $g(\lambda)$ *given in* (9) *is integrable. If* $\beta(z)$ *has no zeros on* $|z| = 1$, *there is a unique weakly stationary solution* $\{Y_t\}$. *The spectral distribution function G of* $\{Y_t\}$ *is absolutely continuous with spectral density* $g(\lambda)$. *The one step prediction error variance is given by* (19).

Consider

$$\sum_{k=0}^{p} b_k\, Y_{t-k} = \sum_{j=0}^{q} a_j\, \xi_{t-j}, \quad a_0, b_0 \neq 0 \,.$$

If the ξ_j's are independent identically distributed with finite second moment and the roots of $b(z)$, $a(z)$ are outside the unit circle then the best predictor (one-step) of Y_{t+1} in terms of Y_s, $s \leq t$, in the sense of minimizing mean square error of prediction is

$$(21) \qquad -\frac{1}{b_0} \sum_{k=1}^{p} b_k\, Y_{t+1-k} + \frac{1}{b_0} \sum_{j=1}^{q} a_j\, \xi_{t+1-j}$$

with variance of the prediction error

$$E\left|\frac{a_0}{b_0}\, \xi_{t+1}\right|^2 = \left|\frac{a_0}{b_0}\right|^2 \sigma^2 \,.$$

This follows from the fact that the predictor given above is in the closed linear space $\mathcal{M}_t(Y)$ generated by Y_s, $s \leq t$, since ξ_s, $s \leq t$, is in $\mathcal{M}_t(Y)$ (because the roots of $a(z)$ are outside the unit disc). On the other hand, Y_s, $s \leq t$, is in the closed linear space $\mathcal{M}_t(\xi)$ generated by ξ_s, $s \leq t$, because the roots of $b(z)$ are outside the unit disc in the complex plane. Since ξ_{t+1} is independent statistically of $\mathcal{M}_t(\xi)$ it follows that ξ_{t+1} is independent of $\mathcal{M}_t(Y)$ (which is equal to

$\mathscr{M}_t(\xi)$ by the argument just given). This implies that the expression (21) given above is

$$E[Y_{t+1} \mid Y_t, Y_{t-1}, \ldots]$$

and hence the best predictor of Y_{t+1} given Y_s, $s \le t$.

We shall make a few additional remarks about the autoregressive equations

$$\sum_{k=0}^{p} \beta_k Y_{t-k} = V_t \tag{22}$$

with the V_t's a white noise process. It has already been noted that there is a weakly stationary solution $\{\tilde{Y}_t\}$ of the system of equations if and only if the polynomial $\beta(z) = \sum_{k=0}^{p} \beta_k z^k$ has no zeros of modulus one. Further the weakly stationary solution is uniquely determined. One can also consider the equations a recursive system for the determination of Y_t in terms of Y_s, $s < t$, and V_s, $s \le t$. Assume that one is given initial data $Y_{-p+1}, \ldots, Y_0$ and generates Y_t, $t > 0$, recursively from this initial data by using the system (22) for $t \ge 0$. If all the roots of the polynomial $\beta(z)$ have modulus greater than one, the behavior of the solution Y_t is asymptotically stable as $t \to \infty$ (does not diverge) and

$$\lim_{t \to \infty} E \mid Y_t - \tilde{Y}_t \mid^2 = 0 .$$

On the other hand, if some roots of the polynomial $\beta(z)$ have modulus less than one, usually the solution Y_t will be asymptotically unstable as $t \to \infty$. These results will not be derived. We only illustrate them by considering the simple case of a first order autoregressive system

$$Y_t - a\, Y_{t-1} = V_t .$$

Here $\beta(z) = 1 - a\, z$. Assume Y_0 given. Then

$$Y_t = \sum_{j=0}^{t-1} a^j V_{t-j} + a^t Y_0 . \tag{23}$$

When $\mid a \mid < 1$, the unique weakly stationary solution is

$$\tilde{Y}_t = \sum_{j=0}^{\infty} a^j V_{t-j}$$

and it is clear that

$$E \mid Y_t - \tilde{Y}_t \mid^2 \to 0$$

as $t \to \infty$. This is the stable case. However, when $\mid a \mid > 1$ the unique weakly stationary solution is

$$\tilde{Y}_t = \sum_{j=1}^{\infty} (-a)^j V_{t+j} .$$

If $Y_0 \neq \tilde{Y}_0$, Y_t [as given by (23)] will typically diverge as $t \to \infty$. Thus $|a| > 1$ is from this recursive perspective essentially unstable. Of course, if we reverse the time direction, the roles of $|a| < 1$ and $|a| > 1$ are interchanged relative to this notion of stability.

One might say that since an autoregressive scheme will have a divergent solution given most initial conditions if $\beta(z)$ has some zeros of absolute value less than one, this can be detected by observing the solution as $t \to \infty$. This ought to be true even if the process is Gaussian. However, this leads us out of the domain of stationary processes. But as is clear, any moving average

$$Y_t = \sum_{j=0}^{q} a_j V_{t-j}$$

with the $\{V_t\}$ sequence white noise is weakly stationary. If the V_t variables are independent Gaussian, one cannot distinguish in terms of observations on $\{Y_t\}$ alone between schemes with different polynomials $a(z)$ which are equal in modulus for $z = e^{-i\lambda}$. However, we shall later see in section 4 that such a distinction is generally possible when the V_t variables are independent, identically distributed and non-Gaussian.

A simple approximation argument using the Weierstrass approximation theorem (see problem II. 5) allows us to show that the one-step prediction error is given by the expression (19) for any weakly stationary process with spectral density g positive and continuous on $[-\pi, \pi]$ with $g(-\pi) = g(\pi)$. In fact, a corresponding result holds much more generally. Let $\{X_n\}$ be a weakly stationary process with spectral distribution function G. Then the one step linear prediction error for the process is given by

$$2\pi \exp\left\{ \frac{1}{2\pi} \int_{-\pi}^{\pi} \log G'(\lambda)\, d\lambda \right\}.$$

Notice that this implies that X_n can be predicted perfectly (with zero error) in terms of X_s, $s \leq n-1$, linearly if and only if

$$\int_{-\pi}^{\pi} \log G'(\lambda)\, d\lambda = -\infty .$$

There is an analogous result for continuous time parameter weakly stationary processes $X(t)$ that are continuous in mean square, i.e. $E|X(t) - X(s)|^2 \to 0$ as $t \to s$. As before, let G denote the spectral distribution function of $\{X(t)\}$. Then $X(t)$ can be predicted perfectly in terms of $X(s)$, $s \leq \tau < t$, linearly if and only if

$$\int_{-\infty}^{\infty} \frac{\log G'(\lambda)}{1+\lambda^2}\, d\lambda = -\infty . \tag{24}$$

In Chapter I, one considered the Hilbert space $H(X)$ generated by a weakly stationary sequence $\{X_n\}$ and $L^2(d\eta)$ where η is the measure generated by the spectral distribution function G of $\{X_n\}$. There is then an isometry between $H(X)$ and $L^2(d\eta)$ in which X_n maps onto $e^{in\lambda}$. There is a corresponding result in the case of a weakly stationary process $\{X(t)\}$ continuous in mean square. Let $H(X)$ as before be the Hilbert space generated by $\{X(t)\}$, i.e., generated by all finite linear combinations $\sum a_j X(t_j)$. $L^2(d\eta)$ is now the space of all functions f square integrable

$$\int_{-\infty}^{\infty} | f(u) |^2 \, d\eta(u) < \infty$$

with η the measure generated by the spectral distribution G of $\{X(t)\}$ (see problem I. 3). Then as before there is an isometry between $H(X)$ and $L^2(d\eta)$ where now $X(t)$ maps onto $e^{it\lambda}$. In view of this we can reinterpret (24) as a necessary and sufficient condition that the finite linear combinations generated by $e^{it\lambda}$, $t \leq 0$, be dense in $L^2(d\eta)$. We now sketch the argument leading to the following result.

Theorem 4. *Let η possess finite moments of all orders*

$$s_k = \int_{-\infty}^{\infty} u^k \, d\eta(u), \quad k = 0, 1, 2, \ldots .$$

Then if

$$\int_{-\infty}^{\infty} \frac{\log G'(u)}{1 + u^2} \, du > -\infty \tag{25}$$

the sequence $\{s_k, k = 0, 1, \ldots\}$ generates an indeterminate moment problem.

By (25) there is a function $f \in L^2(d\eta)$ such that

$$\int_{-\infty}^{\infty} | f(u) |^2 \, d\eta(u) \neq 0, \quad \int_{-\infty}^{\infty} f(u) \, e^{itu} \, d\eta(u) = 0 \tag{26}$$

for $t \leq 0$. Differentiating the equation on the right of (26) one finds that

$$\int_{-\infty}^{\infty} f(u) \, u^k \, d\eta(u) = 0, \quad k = 0, 1, 2, \ldots .$$

This implies that the polynomials are not dense in $L^2(d\eta)$. However, the polynomials not being dense in $L^2(d\eta)$ implies that the moment problem for η is indeterminate (see Akhiezer [1965], p. 42).

4. Non-Gaussian Linear Processes

Let $\{a_j\}$ be a sequence of real constants with

$$\sum_{j=-\infty}^{\infty} a_j^2 < \infty .$$

Assume that the random variables V_t, $t = \ldots, -1, 0, 1, \ldots$ are independent and identically distributed with mean zero $E\, V_t \equiv 0$ and variance one $E\, V_t^2 \equiv 1$. As already remarked, the process

$$X_t = \sum_{j=-\infty}^{\infty} a_j V_{t-j}$$

is called a linear process. We introduce the z-transform $a(z) = \sum_j a_j z^j$ corresponding to the process. The function

$$a(e^{-i\lambda}) = \sum_j a_j e^{-ij\lambda}$$

is called the frequency response function or transfer function. The spectral density of the process $\{X_t\}$

$$f(\lambda) = \frac{1}{2\pi} \mid a(e^{-i\lambda}) \mid^2 .$$

In the Gaussian case the complete probability structure of $\{X_t\}$ is determined by $f(\lambda)$ (or equivalently by the modulus $\mid a(e^{-i\lambda}) \mid$) since the probability distribution of jointly Gaussian variables is specified by their first and second order moments. The phase of $a(e^{-i\lambda})$ is not identifiable (cannot be estimated) in the Gaussian case in terms of observations on $\{X_t\}$ alone. We shall show that under reasonably broad conditions the phase information in $a(e^{-i\lambda})$ can be estimated in the case of non-Gaussian linear processes. In this sense Gaussian linear processes are atypical. A detailed discussion of phase estimates will be given in Chapter VIII but a qualitative treatment of how they can be constructed will be given in this section. It is a consequence of the following theorem.

Theorem 5. *Consider a non-Gaussian linear process with the independent random variables $\{V_t\}$ having all moments finite. Let*

$$\sum |j| \quad |a_j| < \infty$$

and $a(e^{-i\lambda}) \neq 0$ for all λ. The function $a(e^{-i\lambda})$ can then be identified in terms of observations on $\{X_t\}$ alone up to an indeterminate integer a in a factor $e^{ia\lambda}$ and the sign of $a(1) = \sum a_j$. In fact, for this result it is sufficient to have some moment of order $k > 2$ finite with cumulant $\gamma_k \neq 0$.

Because the V_t's are non-Gaussian with all moments finite, there will be a cumulant of V_t, $\gamma_k \neq 0$ of smallest subscript $k > 2$. The kth joint cumulant of the random variables $X_t, X_{t+j_1}, \ldots, X_{t+j_{k-1}}$ is given by

$$\operatorname{cum}(X_t, X_{t+j_1}, \ldots, X_{t+j_{k-1}}) = \gamma_k \sum_u a_u a_{u+j_1} \cdots a_{u+j_{k-1}} .$$

In section I. 2 the (second order) spectral density or Fourier transform of second order moments was first introduced. Here we introduce the kth order cumulant spectral density or Fourier transform of kth order cumulants

$$f_k(\lambda_1, \ldots, \lambda_{k-1}) = (2\pi)^{-k+1} \sum_{j_1, \ldots, j_{k-1}} \operatorname{cum}(X_t, X_{t+j_1}, \ldots, X_{t+j_{k-1}}) \exp\left(-\sum_{s=1}^{k-1} i j_s \lambda_s\right)$$

$$= \gamma_k (2\pi)^{-k+1} \alpha(e^{-i\lambda_1}) \cdots \alpha(e^{-i\lambda_{k-1}}) \alpha(e^{i(\lambda_1 + \cdots + \lambda_{k-1})}) .$$

Set

$$h(\lambda) = \arg\left\{\alpha(e^{-i\lambda}) \frac{\alpha(1)}{|\alpha(1)|}\right\}.$$

It then follows that

$$\{\alpha(1) / |\alpha(1)|\}^k \gamma_k = (2\pi)^{(k/2)-1} f_k(0, \ldots, 0) \{f(0)\}^{-k/2}$$

and also that

$$h(\lambda_1) + \cdots + h(\lambda_{k-1}) - h(\lambda_1 + \cdots + \lambda_{k-1})$$

$$= \arg\left[\left\{\frac{\alpha(1)}{|\alpha(1)|}\right\}^k \gamma_k^{-1} f_k(\lambda_1, \ldots, \lambda_{k-1})\right]$$

since $h(-\lambda) = -h(\lambda)$. Further

$$h'(0) - h'(\lambda) = \lim_{\Delta \to 0} \frac{1}{(k-2)\Delta} \{h(\lambda) + (k-2) h(\Delta) - h(\lambda + (k-2)\Delta)\}.$$

Let

$$h_1(\lambda) = \int_0^\lambda \{h'(u) - h'(0)\} du .$$

Then

$$h(\lambda) = h_1(\lambda) + c\lambda$$

with $c = h'(0)$. Now

$$h(\pi) = h_1(\pi) + c\pi .$$

The a_j's real imply that $h(\pi) = a\pi$ for some integer a. Let $h_1(\pi)/\pi = \delta$. Since $h(\pi) = a\pi = (\delta + c)\pi$ it follows that

$$c = a - \delta .$$

The integer a cannot be specified without additional information since it corresponds to reindexing or subscripting the V_t's. The sign of $\alpha(1)$ is also indeterminate because one can multiply the a_j's and V_t's by (-1) without changing the observed process $\{X_t\}$. Notice that up to sign

$$\alpha(e^{-i\lambda}) = |\, 2\,\pi\, f(\lambda)\,|^{1/2} \exp\{i\, h(\lambda)\}\,.$$

Consistent (convergent) estimates of the (second order) spectral density will be discussed in section V. 2. Consistent estimates of kth order cumulant spectra will be considered in section VI. 4. Consistent estimates of the kth order cumulants exist by the ergodic theorem. Under the conditions assumed this implies that consistent estimates of the kth order cumulant spectra exist. This in turn implies that one can get consistent estimates of $h_1(\lambda)$ and δ.

5. Kalman-Bucy Filter

In the case of the Kalman-Bucy filter one basically deals with a prediction problem for a Markovian process (conditional independence of past and future given precise knowledge of the present). The process is Gaussian. The novelty and interest lies in the recursive character of the estimate.

Consider the system described by

$$\mathbf{X}(t+1) = \Phi\,\mathbf{X}(t) + \mathbf{V}(t)$$

$$\mathbf{Y}(t) = \theta\,\mathbf{X}(t) + \mathbf{e}(t)$$

where $\mathbf{X}$ is an n-dimensional state vector, $\mathbf{Y}$ a p-dimensional observed output vector, and $\{\mathbf{V}(t)\}$, $\{\mathbf{e}(t)\}$ sequences of independent normal vectors with zero mean values and covariances

$$E\,\mathbf{V}(t)\,\mathbf{V}(t)' = R_1$$

$$E\,\mathbf{V}(t)\;\mathbf{e}(t)' = 0$$

$$E\,\mathbf{e}(t)\;\mathbf{e}(t)' = R_2\,.$$

The initial state $\mathbf{X}(t_0)$ is assumed independent of $\mathbf{V}$ and $\mathbf{e}$ and normal with mean $\mathbf{m}$ and covariance R_0. The object is to get a best predictor (estimator) of $\mathbf{X}(t+1)$ based on the observations $\mathbf{Y}(t)$, $\mathbf{Y}(t-1)$, $\ldots$, $\mathbf{Y}(t_0)$ in the sense of minimizing the mean square error of prediction (approximation). Let $\mathscr{Y}_t$ be the Borel field generated by $\mathbf{Y}(t_0)$, $\ldots$, $\mathbf{Y}(t)$ and

$$\hat{\mathbf{X}}(t+1\,|\,t) = E[\mathbf{X}(t+1)\,|\,\mathscr{Y}_t]\,.$$

Now

$$\hat{\mathbf{X}}(t+1\,|\,t) = E[\mathbf{X}(t+1)\,|\,\mathscr{Y}_{t-1}, \mathbf{Y}(t)]\,.$$

Further $\mathscr{Y}_{t-1}$ and

$$\begin{aligned}\tilde{\mathbf{Y}}(t) &= \mathbf{Y}(t) - E[\mathbf{Y}(t) \mid \mathscr{Y}_{t-1}] \\ &= \mathbf{Y}(t) - E[\theta\,\mathbf{X}(t) + \mathbf{e}(t) \mid \mathscr{Y}_{t-1}] \\ &= \mathbf{Y}(t) - \theta\,\hat{\mathbf{X}}(t) = \theta\,\tilde{\mathbf{X}}(t) + \mathbf{e}(t)\end{aligned}$$

are independent. Here $\tilde{\mathbf{X}} = \mathbf{X} - \hat{\mathbf{X}}$. Now

$$\begin{aligned}\hat{\mathbf{X}}(t+1) &= E[\mathbf{X}(t+1) \mid \mathscr{Y}_{t-1}, \tilde{\mathbf{Y}}(t)] \\ &= E[\mathbf{X}(t+1) \mid \mathscr{Y}_{t-1}] + E[\mathbf{X}(t+1) \mid \tilde{\mathbf{Y}}(t)] - E\,\mathbf{X}(t+1)\,.\end{aligned}$$

Notice that

$$\begin{aligned}E[\mathbf{X}(t+1) \mid \mathscr{Y}_{t-1}] &= E[\Phi\,\mathbf{X}(t) + \mathbf{V}(t) \mid \mathscr{Y}_{t-1}] = E[\Phi\,\mathbf{X}(t) \mid \mathscr{Y}_{t-1}] \\ &= E[\mathbf{X}(t) \mid \mathscr{Y}_{t-1}] = \Phi\,\hat{\mathbf{X}}(t \mid t-1)\,.\end{aligned}$$

We have only to evaluate $E[\mathbf{X}(t+1) \mid \tilde{\mathbf{Y}}(t)]$ in order to get the best predictor. By a simple computation we see that

$$\begin{aligned}\operatorname{cov}[\mathbf{X}(t+1), \tilde{\mathbf{Y}}(t)] &= \operatorname{cov}[\Phi\,\mathbf{X}(t) + \mathbf{V}(t), \theta\,\tilde{\mathbf{X}}(t) + \mathbf{e}(t)] \\ &= E[\Phi\,\mathbf{X}(t) + \mathbf{V}(t) - \Phi\,E\,\mathbf{X}(t)]\,[\theta\,\tilde{\mathbf{X}}(t) + \mathbf{e}(t)]' \\ &= E[\Phi(\hat{\mathbf{X}}(t) + \tilde{\mathbf{X}}(t))\,\tilde{\mathbf{X}}(t)'\,\theta'] = \Phi\,E\,\tilde{\mathbf{X}}(t)\,\tilde{\mathbf{X}}(t)'\,\theta'\end{aligned}$$

where the next to last step follows from the independence of $\mathbf{e}(t)$, $\mathbf{V}(t)$ and $\mathbf{X}(t)$. The covariance matrix of $\mathbf{Y}(t)$

$$\begin{aligned}\sigma^2(\tilde{\mathbf{Y}}(t)) &= E[\theta\,\tilde{\mathbf{X}}(t) + \mathbf{e}(t)]\,[\theta\,\tilde{\mathbf{X}}(t) + \mathbf{e}(t)]' \\ &= \theta\,E[\tilde{\mathbf{X}}(t)\,\tilde{\mathbf{X}}(t)']\,\theta' + R_2\end{aligned}$$

since $\mathbf{e}(t)$ and $\tilde{\mathbf{X}}(t)$ are independent. Let

$$P(t) = E[\tilde{\mathbf{X}}(t)\,\tilde{\mathbf{X}}(t)']\,.$$

Then

$$E[\mathbf{X}(t+1) \mid \tilde{\mathbf{Y}}(t)] = E\,\mathbf{X}(t+1) + k(t)\,\tilde{\mathbf{Y}}(t)$$

with

$$\begin{aligned}k(t) &= \operatorname{cov}[\mathbf{X}(t+1), \tilde{\mathbf{Y}}(t)]\,\sigma^{-2}(\tilde{\mathbf{Y}}(t)) \\ &= \Phi\,P(t)\,\theta'[\theta\,P(t)\,\theta' + R_2]^{-1}\end{aligned}$$

by standard linear or normal theory. The estimate $\hat{\mathbf{X}}(t+1 \mid t)$ is then given by the recursive set of equations

$$\hat{\mathbf{X}}(t+1 \mid t) = \Phi\,\hat{\mathbf{X}}(t \mid t-1) + k(t)\,\tilde{\mathbf{Y}}(t)$$

$$\tilde{\mathbf{Y}}(t) = \mathbf{Y}(t) - \theta\,\hat{\mathbf{X}}(t \mid t-1)\,.$$

Also notice that $\mathbf{X}(t+1) - \hat{\mathbf{X}}(t+1 \mid t)$

$$= \tilde{\mathbf{X}}(t+1 \mid t) = \Phi\,\tilde{\mathbf{X}}(t \mid t-1) + V(t) - k(t)\,\tilde{\mathbf{Y}}(t)$$

$$= [\Phi - k(t)\,\theta]\,\tilde{\mathbf{X}}(t \mid t-1) + \mathbf{V}(t) - k(t)\,\mathbf{e}(t)$$

and

$$E\,\tilde{\mathbf{X}}(t+1 \mid t) = 0\,.$$

$P(t)$ is the covariance matrix of the estimation error and by direct computation

$$\begin{aligned} P(t+1) &= [\Phi - k(t)\,\theta]\,P(t)\,[\Phi - k(t)\,\theta]' + R_1 + k(t)\,R_2\,k(t)' \\ &= \Phi\,P(t)\,\Phi' + R_1 - k(t)\,\theta\,P(t)\,\Phi' - \Phi\,P(t)\,\theta'\,k(t)' + k(t)\,\theta\,P(t)\,\theta'\,k(t)' \\ &\quad + k(t)\,R_2\,k(t)' \\ &= [\Phi - k(t)\,\theta]\,P(t)\,\Phi(t)' + R_1\,. \end{aligned}$$

Problems

1. If $E\,X^2 < \infty$ show that $E\,X$ is the constant that minimizes $E(X-c)^2$.

2. If X is a random variable with distribution symmetric about zero and $E\,X^{2m} < \infty$ for some integer $m > 1$, show that $E\,X$ is the constant that minimizes $E(X-c)^{2m}$.

3. Assume that X and Y have a continuous positive joint density function. Give the best predictor of X in terms of Y (in the sense of minimizing mean square error of prediction) in terms of the joint density function.

4. Determine the density function of X^m, $m = 1, 2, \ldots$, for X a standard normal variable with mean zero and variance one.

5. Use the Weierstrass theorem for approximating continuous functions f on $[-\pi, \pi]$, $f(\pi) = f(-\pi)$, uniformly by trigonometric polynomials (see Rosenblatt [1974] p. 21) to show that (18) still gives the one step linear prediction error variance when $\{X_n\}$ is a weakly stationary process with spectral density positive and continuous.

6. Show that the gamma and normal distributions have a determinate moment problem by using the Carleman criterion.

7. By using (25) and problem 4 show that with X a standard normal random variable, the distribution of X^2 has a determinate moment problem, while the distribution of X^m for $m \geq 3$ has an indeterminate moment problem.

8. Show that if $\{X_j\}$ is a Gaussian stationary sequence, the probability structure of $\{X_j\}$ is the same as that of $\{X_{-j}\}$. This means that the process looks the same with time running forward or time reversed.

9. Consider a Gaussian stationary sequence of two-vectors $\begin{pmatrix} X_j \\ Y_j \end{pmatrix}$. Here the probability structure may not be the same with time running forward or reversed. Such an example is given when $Y_j = X_{j+1}$.

10. Let $X_j = \frac{1}{2} X_{j-1} + V_j$ with the V_j's independent, identically distributed random variables with

$$V_j = \begin{cases} 0 & \\ & \text{with probability } 1/2 \\ 1 & \end{cases}$$

and V_{j+s}, $s \geq 0$, independent of X_{j-t}, $t > 0$. Show that if $\{X_j\}$ is a stationary sequence satisfying this system of equations X_j is uniformly distributed on $[0, 1]$. Further, prediction from the future to the past can be carried out perfectly with

$$X_j = 2\,X_{j+1} \quad \text{modulo one}\ .$$

Prediction from the past to the future is with positive mean square error of prediction.

11. Let $\{X_j\}$ be a Gaussian stationary sequence with a discrete harmonic $\Delta Z(\lambda) \neq 0$ at frequency λ, $0 < \lambda < \pi$. Show that $\arg\{\Delta Z(\lambda)\}$ is uniformly distributed on $[0, 2\,\pi]$.

12. Let $X_1, \ldots, X_n$ be jointly Gaussian random variables with nonsingular covariance matrix R_n. Show that the mean square error of prediction for X_n in terms of $X_1, \ldots, X_{n-1}$ is $|\,R_n\,| \,/\, |\,R_{n-1}\,|$. Use formula (18) and this remark to show that if $\{X_j\}$ is a stationary Gaussian process with positive continuous spectral density $f(\lambda)$

$$\frac{1}{n} \log |\,R_n\,| \to \frac{1}{2\pi} \int_{-\pi}^{\pi} \log f(\lambda)\, d\lambda$$

as $n \to \infty$.

Notes

2.1 One should notice that a k-vector of complex-valued random variables with real and imaginary parts of the components, jointly normal need not be complex normal in the sense specified in section II. 1.

2.2 A discussion of the computation and usefulness of cumulant functions in some problems of statistical physics can be found in the paper of Abdulla-Zadeh, Minlos and Pogosian [1980].

2.3 A full discussion of the linear prediction problem for weakly stationary processes can be found in Rozanov [1967a]. An analysis can be found there also of the linear prediction problem when the process is a k-vector valued weakly stationary problem. A full counterpart of the univariate results does not exist. However one can show that the determinant of the covariance matrix of the one-step prediction error is given by

$$(2\pi)^k \exp\left\{ \frac{1}{2\pi} \int_{-\pi}^{\pi} \log \det \{G'(\lambda)\}\, d\lambda \right\}$$

where $G(\lambda)$ is the $k \times k$ matrix-valued spectral distribution function of the process, $G'(\lambda)$ is the derivative of $G(\lambda)$ (which exists almost everywhere) and $\det(A)$ denotes the determinant of the matrix A. A development of this result can also be found in Rozanov's book together with related references.

2.4 A simple example of a non-Gaussian linear process is given by

$$X_t = \sum_{j=0}^{\infty} 2^{-j+1} V_{t-j}$$

with the random variables V_t independent identically distributed random variables with $P(V_t = 1) = P(V_t = -1) = \frac{1}{2}$. Notice that

$$X_t = \frac{1}{2} X_{t-1} + 2 V_t .$$

The spectral density of the stationary process $\{X_t\}$ is

$$g(\lambda) = \frac{1}{2\pi} \left| 1 - e^{-i\lambda} / 2 \right|^{-2} .$$

From the discussion in section 3 it is clear that the best predictor of X_t given the past $X_{t-1}, X_{t-2}, \ldots$ (in the sense of minimal mean square error of prediction) is the linear predictor $X_t^* = \frac{1}{2} X_{t-1}$. This is because the zero of the polynomial $b(z) = 1 - z/2$ is outside the unit circle. Notice that if one reverses time and considers the process $Y_t = X_{-t}$, the best one step predictor for Y_t is nonlinear and predicts perfectly. The best predictor of Y_t given $Y_{t-1}, Y_{t-2}, \ldots$ is

$$Y_t^* = 2(Y_{t-1} + 4) \text{ modulo } 4 - 4 = Y_t .$$

Interesting comments on the prediction problem for non-Gaussian linear processes can be found in the paper of Shepp, Slepian, and Wyner [1980].

Chapter III

Quadratic Forms, Limit Theorems and Mixing Conditions

1. Introduction

There are a number of related topics that are discussed in this chapter. One object is to obtain results on the large sample distribution of covariance estimates under appropriate conditions when sampling from a stationary process. Related quadratic forms are also considered. For this, we need to derive appropriate types of central limit theorems that will be employed in this chapter as well as in derivations in later chapters. However, in most central limit theorems for random processes, some type of effective "mixing" condition expressing asymptotic independence of random variables with large time separation is required. It is for this reason that the apparently distinct topics are joined in this chapter.

2. Quadratic Forms

We first introduce some useful terminology and some relevant concepts. Let $\{X_n\}$ be a strictly stationary process with $E|X_j| < \infty$. Consider any integrable function $f(X_1, \ldots, X_k)$ of a finite number of random variables $X_1, \ldots, X_k$. Suppose the time averages of f and its shifts

$$\frac{1}{n} \sum_{t=1}^{n} f(X_{1+t}, \ldots, X_{k+t})$$

converge with probability one to the mean of f

$$E\, f(X_1, \ldots, X_k) .$$

If this is valid for every such f, the process $\{X_n\}$ is said to be an *ergodic* process. This property of ergodicity is often spoken of loosely as the interchangeability of "time averages" and "space averages".

Let $\mathscr{B}_n$ be the σ-field of events generated by the random variables X_k, $k \leq n$. $\mathscr{B}_n$ then essentially corresponds to the information carried by the random variables X_k, $k \leq n$. Suppose we consider the best predictor (possibly nonlinear) of X_j in terms of the past $X_{j-1}, X_{j-2}, \ldots$, relative to j that minimizes the mean square error of prediction. From earlier discussion it is clear that the best predictor is

$$E(X_j \mid \mathscr{B}_{j-1}) = E(X_j \mid X_{j-1}, X_{j-2}, \ldots)$$

with prediction error

$$Y_j = X_j - E(X_j \mid \mathscr{B}_{j-1}) .$$

Notice that

$$E(Y_j \mid \mathscr{B}_{-j}) \equiv 0$$

for all j and this in turn implies that

$$E(Y_j \mid \mathscr{B}_{j-s}) \equiv 0 \tag{1}$$

for all j and all $s \geq 1$. A process $\{Y_j\}$ satisfying (1) is called a *martingale difference process*. It is interesting to note that such processes can arise as prediction error processes derived from strictly stationary processes. As one might anticipate processes Z_k obtained as partial sums of Y variables

$$Z_k = \sum_{j=1}^{k} Y_j$$

have the property that

$$E(Z_k \mid Z_{k-j}, Z_{k-j-1}, \ldots) = Z_{k-j} \tag{2}$$

$k = 1, 2, \ldots,$ for $k - j \geq 1$. Property (2) is the martingale property. This property had been used informally by S. Bernstein [1927] and P. Lévy [1937] in deriving limit theorems. In its explicit form it has been extensively investigated by Doob [1953] and others.

Generally in the case of martingale difference processes, one does not require that $\mathscr{B}_n$ be the σ-field generated by X_k, $k \leq n$. It is enough if X_n is measurable with respect to $\mathscr{B}_n$ and $\mathscr{B}_n = \tau^{-n} \mathscr{B}_0$ where τ is the measure-preserving shift (time) transformation with $\mathscr{B}_n$ increasing in n. Such an example is given if one considers independent, identically distributed random variables ξ_j, $j = \ldots, -1, 0, 1, \ldots$ with $X_n = f(\xi_n, \xi_{n-1}, \ldots)$ and f a Borel function. Here the shift transformation takes ξ_n into ξ_{n+1} and $\mathscr{B}_n$ is the σ-field generated by the random variables ξ_j, $j \leq n$. Clearly X_n is measurable with respect to $\mathscr{B}_n$ but $\mathscr{B}_n$ is not the σ-field generated by X_j, $j \leq n$, unless the relation $X_n = f(\xi_n, \xi_{n-1}, \ldots)$ can be inverted with ξ_n as a function of X_k, $k \leq n$.

At this point a central limit theorem for martingale differences will be stated without derivation. Proofs of this result can be found in Rosenblatt [1974] or Billingsley [1961].

Theorem 1. *Let $X = (X_n)$ be an ergodic strictly stationary martingale difference sequence with finite second moment $E\, X_n^2 = 1$. It then follows that*

$$\frac{1}{\sqrt{n}} \sum_{j=1}^{n} X_j$$

is asymptotically normally distributed with mean zero and variance one as $n \to \infty$.

By using a device of Gordin [1969] together with this result for martingale differences, a central limit theorem for ergodic stationary sequences useful in a number of applications will be obtained.

Theorem 2. *Let $Y = (Y_n)$ be an ergodic strictly stationary sequence with $E\,Y_j \equiv 0$, $E\,Y_j^2 < \infty$. Assume that*

$$\sum_{j=1}^{\infty} \{E(E(Y_0 \mid \mathscr{B}_{-j}))^2\}^{1/2} < \infty \tag{4}$$

and

$$E\,Y_0^2 + 2 \sum_{k=1}^{\infty} E(Y_k\,Y_0) = \sigma^2 > 0\,. \tag{5}$$

It then follows that

$$\frac{1}{\sqrt{n}\,\sigma} \sum_{j=1}^{n} Y_j$$

is asymptotically normal with mean zero and variance one.

Set $S_n = \Sigma_{j=1}^n Y_j$ and notice that

$$E\,S_n^2 = n\,E\,Y_0^2 + 2 \sum_{k=1}^{n} (n-k)\,E(Y_k\,Y_0)\,. \tag{6}$$

The Schwartz inequality implies that

$$|\,E(Y_k\,Y_0)\,| = |\,E(E(Y_k \mid \mathscr{B}_0)\,Y_0)\,| \le \{E(Y_0^2)\,E(E(Y_k \mid \mathscr{B}_0)^2)\}^{1/2}\,. \tag{7}$$

From (4), (5), (6) and (7) it is clear that

$$\frac{1}{n}\,E\,S_n^2 \to \sigma^2 > 0$$

as $n \to \infty$. Let

$$u_r = E(Y_r \mid \mathscr{B}_0) - E(Y_r \mid \mathscr{B}_{-1})\,, \quad 0 \le r < \infty\,.$$

Assumption (4) implies that $\{E\,|\,\Sigma_{r=0}^{\infty} u_r\,|^2\}^{1/2} \le \Sigma_{r=0}^{\infty} \{E\,|\,u_r\,|^2\}^{1/2} < \infty$. Set

$$X_0 = \sum_{r=0}^{\infty} u_r$$

and

$$X_r = \tau^r\,X_0$$

where it is understood that τ is the one-step shift operator. It is clear that $X = (X_r)$ is a martingale difference sequence because

$$E(X_r \mid \mathscr{B}_{r-1}) \equiv 0\,.$$

The properties of conditional expectations imply

$$\begin{aligned} E(u_r u_{r+k}) &= E\big(E(Y_r \mid \mathscr{B}_0)\, E(Y_{r+k} \mid \mathscr{B}_0)\big) - E\big(E(Y_r \mid \mathscr{B}_{-1})\, E(Y_{r+k} \mid \mathscr{B}_{-1})\big) \\ &= E\big(Y_{r+k}\, E(Y_r \mid \mathscr{B}_0)\big) - E\big(Y_{r+k}\, E(Y_r \mid \mathscr{B}_{-1})\big) \\ &= E\big(Y_k\, E(Y_0 \mid \mathscr{B}_{-r})\big) - E\big(Y_k\, E(Y_0 \mid \mathscr{B}_{-r-1})\big) . \end{aligned}$$

Also

$$\begin{aligned} E \mid X_0 \mid^2 &= \lim_{n\to\infty} E \left| \sum_{r=0}^{n} u_r \right|^2 \\ &= \lim_{n\to\infty} \left[\sum_{r=0}^{n} E\, u_r^2 + 2 \sum_{k=1}^{n} \sum_{r=0}^{n-k} E(u_r u_{r+k}) \right] \\ &= \lim_{n\to\infty} \Big[E\, Y_0^2 - E\big(Y_0\, E(Y_0 \mid \mathscr{B}_{-n-1})\big) \\ &\quad + 2 \sum_{k=1}^{n} \{ E(Y_k Y_0) - E\big(Y_k\, E(Y_0 \mid \mathscr{B}_{-n+k-1})\big) \} \Big] \\ &= E\, Y_0^2 + 2 \sum_{k=1}^{n} E(Y_k Y_0) = \sigma^2 > 0 . \end{aligned}$$

Let $T_n = \Sigma_{j=1}^{n} X_j$. Now X is a martingale difference sequence. Theorem 1 therefore implies that $(1/\sqrt{n}\,\sigma)\, T_n$ is asymptotically normal with mean zero and variance one. Further, $E\, T_n^2 = n\,\sigma^2$. It is now enough to show that $\frac{1}{n} E\, S_n T_n \to \sigma^2$ as $n \to \infty$ since that implies $\frac{1}{n} E(S_n - T_n)^2 \to 0$ as $n \to \infty$ and therefore the conclusion of Theorem 2. But notice that

$$\begin{aligned} E\, S_n T_n &= \sum_{i=1}^{n} \sum_{j=1}^{n} E(Y_i X_j) = n\, E(X_0 Y_0) + \sum_{j=1}^{n} (n-j) \{ E(Y_0 X_j) + E(Y_0 X_{-j}) \} \\ &= \sum_{j=0}^{n-1} (n-j)\, E(Y_0 X_{-j}) \\ &= \sum_{j=0}^{n-1} (n-j)\, E \left\{ Y_0 \sum_{r=0}^{\infty} \big(E(Y_{-j+r} \mid \mathscr{B}_{-j}) - E(Y_{-j+r} \mid \mathscr{B}_{-j-1})\big) \right\} \\ &= \left(\sum_{j=0}^{n-1} \sum_{r=0}^{\infty} (n-j) - \sum_{j=1}^{n} \sum_{r=1}^{\infty} (n-j+1) \right) E\big(Y_0\, E(Y_{-j+r} \mid \mathscr{B}_{-j})\big) \\ &= \sum_{j=0}^{n-1} (n-j)\, E(Y_0 Y_{-j}) - \sum_{j=1}^{n} \sum_{r=1}^{\infty} E\big(Y_0\, E(Y_{-j+r} \mid \mathscr{B}_{-j})\big) \\ &\quad + \sum_{r=1}^{\infty} n\, E\big(Y_0\, E(Y_r \mid \mathscr{B}_0)\big) \\ &= \sum_{j=0}^{n-1} (n-j)\, E(Y_0 Y_j) - \sum_{j=1}^{n} \sum_{r=1}^{\infty} E\big(Y_j\, E(Y_r \mid \mathscr{B}_0)\big) + n \sum_{r=1}^{\infty} E(Y_0 Y_r) . \end{aligned}$$

Therefore $n^{-1} E(S_n T_n) \to \sigma^2$ as $n \to \infty$.

Our object is to apply Theorem 2 so as to obtain a result on the asymptotic distribution of covariance estimates. Let $X = (X_n)$ be a strictly stationary sequence with $E X_n \equiv 0$ and $E X_n^4 < \infty$. Let

$$r_k = E X_n X_{n+k}$$

and $c_{a,b,d}$ be the fourth cumulant

$$c_{a,b,d} = \operatorname{cum}(X_n, X_{n+a}, X_{n+b}, X_{n+d}) .$$

The following result will be derived as a consequence of Theorem 2.

Theorem 3. *Let $X = (X_n)$ be an ergodic strictly stationary sequence with mean zero and $E X_n^4 < \infty$. Set*

$$r_u(N) = \frac{1}{N} \sum_{j=1}^{N-u} X_j X_{j+u}$$

$u = 0, 1, \ldots, s$ and let X_j, $j \leq n$, be measurable with respect to $\mathcal{B}_n$. Assume that

$$\sum_k | r_k | < \infty ,$$

$$\sum_{a,b,d} | c_{a,b,d} | < \infty ,$$

and also that

$$\sum_{n=1}^{\infty} \{E | E(X_0 X_u | \mathcal{B}_{-n}) - r_u |^2\}^{1/2} < \infty , \tag{8}$$

$$E(X_0 X_u - r_u)^2 + 2 \sum_{k=1}^{\infty} E\{(X_k X_{k+u} - r_u)(X_0 X_u - r_u)\} = \sigma_u^2 > 0 , \quad u = 0, 1, \ldots, s . \tag{9}$$

It then follows that $\sqrt{N}(r_u(N) - r_u)$, $u = 0, 1, \ldots, s$ are asymptotically normal with mean zero and covariances

$$m_{u,v} = \sum_a \{r_a r_{a+v-u} + r_{a+v} r_{a-u} + c_{u,a,a+v}\} ,$$

$u, v = 0, 1, \ldots, s$. Notice that $\sigma_u^2 = m_{u,u}$.

First consider asymptotic normality for just one of the estimates $r_u^{(N)}$. Set

$$Y_j = X_j X_{j+u} - r_u , \quad j = \ldots, -1, 0, 1, \ldots . \tag{10}$$

The assumptions (8) and (9) then imply that (4) and (5) are satisfied for the sequence Y_j as defined in (10). Theorem 2 then directly implies the asymptotic normality of $\sqrt{N}(r_u^{(N)} - r_u)$. For joint normality one simply applies the same argument to any linear combination of $X_j X_{j+u} - r_u$, $u = 0, 1, \ldots, s$. The characterization of the asymptotic covariance structure follows from the observation that

$$\operatorname{cov}\bigl(r_u(N), r_v(N)\bigr) \cong \frac{1}{N^2} \sum_{j,k=1}^{N} \operatorname{cov}(X_j X_{j+u}, X_k X_{k+v})$$

$$= \frac{1}{N^2} \sum_{j,k=1}^{N} \left[r_{k-j} r_{k-j+v-u} + r_{k-j+v} r_{k-j-u} + c_{u,k-j,k-j+v}\right] \cong \frac{1}{N} m_{u,v} .$$

Corollary 1. The conclusion of Theorem 3 still holds with all assumptions the same except for $\Sigma \mid r_k \mid < \infty$ which is replaced by $\Sigma r_k^2 < \infty$.

It is of interest to see what the conditions of the Corollary amount to in the case of a one-sided linear process

$$X_t = \sum_{k=0}^{\infty} a_k V_{t-k}$$

with the V_t's independent, identically distributed and $V_t \equiv 0$, $E V_t^2 \equiv 1$. We assume as usual that $\Sigma a_k^2 < \infty$ so that the series representation for X_t in terms of the V sequence is convergent. Let $\mathscr{B}_n = \mathscr{B}(V_t, t \leq n)$. Notice that then

$$(E X_0 X_u \mid \mathscr{B}_{-n}) = E\left(\sum_k a_k V_{-k} \sum_j a_j V_{u-j} \mid \mathscr{B}_{-n}\right)$$

$$= \sum_{k \geq n} a_k V_{-k} \sum_{j \geq u+n} a_j V_{u-j} + \sum_{k=0}^{n-1} a_k a_{u+k}$$

and so

$$E(X_0 X_u \mid \mathscr{B}_{-n}) - r_u = \sum_{k \geq n} a_k V_{-k} \sum_{j \geq u+n} a_j V_{u-j} - \sum_{k=n}^{\infty} a_k a_{u+k} .$$

Thus

$$E \mid E(X_0 X_u \mid \mathscr{B}_{-n}) - r_u \mid^2$$

$$= \sum_{k \geq n} a_k^2 \sum_{j \geq u+n} a_j^2 - \left(\sum_{k=n}^{\infty} a_k a_{u+k}\right)^2$$

$$+ \sum_{k \geq n} a_k^2 a_{u+k}^2 (m^{(4)} - 3) .$$

This implies that

$$\sum_{n=1}^{\infty} \sum_{k \geq n} a_k^2 = \sum_{k=1}^{\infty} k a_k^2 < \infty$$

will imply that condition (8) is satisfied. Also $\Sigma_k r_k^2 < \infty$ and $\Sigma_k | c_{u,k,k+u} | < \infty$ implies that expression (9) is absolutely convergent. Thus

$$\sum_{k=1}^{\infty} k\, a_k^2 < \infty , \quad \sum_k r_k^2 < \infty$$

imply that covariance estimates are asymptotically normal in the case of a one-sided linear process.

An old condition under which one has asymptotic normality for covariance estimates of a linear process is

$$\sum | a_k | < \infty$$

(see T. W. Anderson [1971]). Notice that if one has

$$a_k \simeq c\, k^{-1} (\log k)^{-\frac{1}{2}+\eta} , \quad 0 < \eta < \frac{1}{2} ,$$

as $k \to \infty$ then $\Sigma\, k\, a_k^2 < \infty$, $\Sigma\, r_k^2 < \infty$ so that the conditions of the Corollary are satisfied but $\Sigma | a_k | = \infty$.

Hannan [1970] and Hall and Heyde [1980] have considered a generalization of a one-sided linear process in which the $\{ V_t \}$ sequence is not necessarily one of independent, identically distributed random variables but rather a martingale difference sequence

$$E[V_t \mid \mathscr{B}_{t-1}] \equiv 0$$

with $E(V_t\, V_\tau) = \delta_{t-\tau}$. Our results not only deal with a large class of these processes but also many in which $\{ V_t \}$ is a white noise process but not a martingale difference sequence. An example of such a sequence V_t is given by

$$V_t = \varepsilon_{t-1}\, \varepsilon_t^2$$

where the ε_t's are independent, identically distributed symmetric random variables with $E\, \varepsilon_t^4 < \infty$, $E\, \varepsilon_t^2 = 1$. Then $E\, V_t\, V_\tau = \delta_{t-\tau}$ but if $\mathscr{B}_n = \mathscr{B}(\xi_j, j \le n)$

$$E(V_t \mid \mathscr{B}_{t-1}) = \varepsilon_{t-1} .$$

We now would like to consider a larger class of quadratic forms and determine their limiting distribution. Let $A(\lambda)$ be a weight function symmetric about zero that is square integrable with Fourier coefficients

$$\alpha_t = \frac{1}{2\pi} \int_{-\pi}^{\pi} A(\lambda)\, e^{-it\lambda}\, d\lambda .$$

The quadratic form (with $r_s(N) = r_{-s}(N)$)

$$\sum_{|s| \le N-1} \alpha_s\, r_s(N) = \int_{-\pi}^{\pi} \frac{1}{2\pi} \sum_{|s| \le N-1} r_s(N)\, e^{-is\lambda}\, A(\lambda)\, d\lambda \tag{11}$$

is of interest. Notice that

$$\frac{1}{2\pi} \sum_{|s| \leq N-1} r_s(N)\, e^{-is\lambda} = \frac{1}{2\pi N} \left| \sum_{t=1}^{N} X_t\, e^{-it\lambda} \right|^2 = I_N(\lambda) .$$

The expression $I_N(\lambda)$ is commonly called the *periodogram* and is up to a scalar multiple the modulus squared of a finite Fourier transform of the data $X_1, \ldots, X_N$. The expression (11) can therefore be written as

$$\int_{-\pi}^{\pi} I_N(\lambda)\, A(\lambda)\, d\lambda .$$

It is also useful to introduce the fourth order cumulant spectral density

$$f_4(\lambda, \mu, \eta) = \frac{1}{(2\pi)^3} \sum_{a, b, d} c_{a, b, d}\, e^{-i(a\lambda + b\mu + d\eta)} .$$

The following result is a Corollary of Theorem 3.

Corollary 2. Let $X = (X_n)$ be an ergodic strictly stationary process satisfying the assumptions of Theorem 3. Let

$$\int_{-\pi}^{\pi} I_N(\lambda)\, A_1(\lambda)\, d\lambda, \quad \int_{-\pi}^{\pi} I_N(\lambda)\, A_2(\lambda)\, d\lambda$$

be two quadratic forms in X with weight functions $A_1(\lambda)$, $A_2(\lambda)$ symmetric about zero and square integrable. The quadratic forms are then asymptotically normal with means

$$\int_{-\pi}^{\pi} f(\lambda)\, A_i(\lambda)\, d\lambda , \quad i = 1, 2,$$

and covariance

$$\cong \frac{2\pi}{N} \left\{ 2 \int_{-\pi}^{\pi} A_i(\lambda)\, A_j(\lambda)\, f^2(\lambda)\, d\lambda + \iint f_4(\lambda, -\mu, \mu)\, A_i(\lambda)\, A_j(\mu)\, d\lambda\, d\mu \right\},$$
$$i, j = 1, 2 .$$

Clearly

$$E \int_{-\pi}^{\pi} I_N(\lambda)\, A(\lambda)\, d\lambda = \sum_{|s| \leq N-1} a_s\, r_s \left(1 - \frac{|s|}{N}\right) \cong \sum_s a_s\, r_s \tag{12}$$
$$= \int_{-\pi}^{\pi} A(\lambda)\, f(\lambda)\, d\lambda .$$

Under the assumptions made on the process X and the weight functions A, one can approximate the quadratic forms (12) by a finite sum of the type

$$\sum_{|s| \leq k} a_s r_s(N)$$

with k fixed but large in mean square. The asymptotic distribution of such a form as $N \to \infty$ is seen to be asymptotically normal by Theorem 3. Under the assumptions made the asymptotic estimate for the covariance would be

$$(13) \qquad \cong \frac{1}{N} \left\{ \sum_{a,t,s} r_a r_{a+t-s} \alpha_s^{(i)} \alpha_t^{(j)} + \sum_{a,t,s} r_{a+t} r_{a-s} \alpha_s^{(i)} \alpha_t^{(j)} + \sum_{a,t,s} c_{s,a,a+t} \alpha_s^{(i)} \alpha_t^{(j)} \right\}, \quad i,j = 1,2.$$

The first sum of (13) is equal to

$$\int_{-\pi}^{\pi} 2\pi f^2(\lambda) \sum_{t,s} \alpha_s^{(j)} \alpha_t^{(k)} e^{i(t-s)\lambda} \, d\lambda = 2\pi \int_{-\pi}^{\pi} A_j(-\lambda) A_k(\lambda) f^2(\lambda) \, d\lambda, \quad j,k = 1,2.$$

The second sum of (13) can similarly be shown to be

$$2\pi \int_{-\pi}^{\pi} A_j(\lambda) A_k(\lambda) f^2(\lambda) \, d\lambda.$$

The last sum of that expression is

$$\frac{1}{(2\pi)^2} \iint A_j(\lambda) A_k(\mu) \sum_{a,s,t} c_{s,a,a+t} e^{-it\lambda} e^{-is\mu} \, d\lambda \, d\mu$$
$$= 2\pi \iint A_j(\lambda) A_k(\mu) f_4(\mu, -\lambda, \lambda) \, d\lambda \, d\mu.$$

3. A Limit Theorem

In this section we will derive a limit theorem for dependent triangular sequences that will later be used to derive asymptotic normality for estimates of the spectral density function under appropriate conditions in Chapter V.

Let $X = \{X_n\}$ be a strictly stationary process. Just as in section 2, let $\mathscr{B}_n = \mathscr{B}(X_k, k \leq n)$ be the σ-field generated by the random variables X_k, $k \leq n$. Also, let $\mathscr{F}_n = \mathscr{B}(X_k, k \geq m)$ be the σ-field generated by X_k, $k \geq m$. It is natural to think of $\mathscr{B}_n$ and $\mathscr{F}_m$ as past and future σ-fields relative to n and m respectively. At this point we shall introduce a *strong mixing* condition that specifies a form of asymptotic independence of the past and future of the process $\{X_n\}$. This condition was originally introduced in Rosenblatt [1956a] to obtain a central limit theorem and has been considered at length in a number of papers since then. We mention the papers of Kolmogorov and Rozanov [1960] and Ibragimov [1962] in particular. The process $X = (X_n)$ is said to be *strongly mixing* if

$$\sup_{\substack{B \in \mathscr{B}_0 \\ F \in \mathscr{F}_n}} |P(B\,F) - P(B)\,P(F)| = \alpha(n) \to 0 \tag{14}$$

as $n \to \infty$. In a certain sense, the strong mixing condition says that the dependence of the process X is short range.

The proof of the central limit theorem stated below follows the lines of that given in Rosenblatt [1956a]. This central limit theorem will be useful in demonstrating the asymptotic normality of a large class of spectral density estimates in section 3 of Chapter V.

Theorem 4. *Let* $\{Y_j^{(n)}, j = \ldots, -1, 0, 1, \ldots\}$, $E\,Y_n^{(n)} \equiv 0$, $n = 1, 2, \ldots,$ *be a sequence of strictly stationary processes defined on the probability space of the strongly mixing stationary process* $X = (X_n)$. *Further assume that* $Y_j^{(n)}$ *is measurable with respect to* $\mathscr{F}_{j-c(n)} \cap \mathscr{B}_{j+c(n)}$ *with* $c(n) = o(n)$, $c(n) \uparrow \infty$ *as* $n \to \infty$. *Let*

$$h_n(b-a) = E\left|\sum_{j=a}^{b} Y_j^{(n)}\right|^2 .$$

Assume that for any two sequences $s(n)$, $m(n)$ *with* $c(n) = o\big(m(n)\big)$, $m(n) \le n$ *and* $s(n)/m(n) \to 0$ *one has*

$$h_n\big(s(n)\big)/h_n\big(m(n)\big) \to 0 . \tag{15}$$

Further let

$$\{h_n(m)\}^{-(2+\delta)/2}\, E\left|\sum_{k=1}^{m} Y_k^{(n)}\right|^{2+\delta} = O(1) \tag{16}$$

for $m = m(n)$ *and some* $\delta > 0$. *There are then sequences* $k(n)$, $p(n) \to \infty$ *as* $n \to \infty$ *with* $k(n)\,p(n) \cong n$ *such that*

$$\sum_{j=1}^{n} Y_j^{(n)} \Big/ \sqrt{k(n)\,h_n\big(p(n)\big)} \tag{17}$$

is asymptotically normally distributed with mean zero and variance one. Also if $k(n)\,h_n\big(p(n)\big) \cong h_n(n)$ *the normalization in* (17) *can be replaced by* $\sqrt{h_n(n)}$.

First let us note that the measurability of $Y_j^{(n)}$ with respect to $\mathscr{F}_{j-c(n)} \cap \mathscr{B}_{j+c(n)}$ just means that $Y_j^{(n)}$ can be taken as a function of $X_{j-c(n)}, X_{j-c(n)+1}, \ldots, X_{j+c(n)}$. Let

$$S_n = \sum_{j=1}^{n} Y_j^{(n)} .$$

The proof proceeds essentially by means of a decomposition of S_n into big blocks separated by small blocks. The sum of the small blocks is shown to be negligible and the big blocks almost independent by virtue of their separation by small blocks and the strong mixing condition. In this way by means of a set of estimates, the proof of asymptotic normality is reduced to an application of a central limit theorem for sums of independent random variables. Set

$$U_r(n) = \sum_{j=(r-1)(p(n)+q(n))+1}^{r\,p(n)+(r-1)\,q(n)} Y_j^{(n)} ,$$

$$V_r(n) = \sum_{j=r\,p(n)+(r-1)\,q(n)+1}^{r(p(n)+q(n))} Y_j^{(n)} ,$$

$r = 1, \ldots, k(n)$, with $k(n)\,\big(p(n) + q(n)\big) = n$. The sequences $k(n)$, $p(n)$, $q(n)$ will be chosen so that $p(n), q(n), k(n) \to \infty$ and $q(n)/p(n) \to \infty$. Thus the random variables U_r are the big blocks and the V_r the small blocks. First observe that

$$E^{1/2}\left|\sum_{r=1}^{k} \frac{V_r(n)}{\sqrt{k(n)\,h_n(p(n))}}\right|^2 \le \sum_{r=1}^{k} \frac{E^{1/2}\,|\,V_r(n)\,|^2}{\sqrt{k(n)\,h_n(p(n))}} \sim \{k(n)\,h_n\big(q(n)\big)/h_n\big(p(n)\big)\}^{1/2} .$$

By the assumption (15) the sequences $k(n)$, $p(n)$, $q(n)$ can be chosen so that

$$k(n)\,h_n\big(q(n)\big)/h_n\big(p(n)\big) \to 0 \tag{17}$$

as $n \to \infty$. If (17) is satisfied it then follows that

$$\sum_{r=1}^{k(n)} V_r \Big/ \sqrt{k(n)\,h_n\big(p(n)\big)} \to 0$$

in probability as $n \to \infty$. Additional assumptions on the sequences $k(n)$, $p(n)$, $q(n)$ will be made but we will later see they can all be satisfied. The sum of the big blocks will now be considered. Let

$$G_{r,n}(x) = P[U_r(n)\,\{k(n)\,h_n\big(p(n)\big)\}^{-1/2} \le x]$$

and $A(r, n, l_r, \delta)$ be the event

$$\left\{l_r\,\delta < \frac{U_r(n)}{\sqrt{k(n)\,h_n(p(n))}} \le (l_r + 1)\,\delta\right\} = A(r, n, l_r, \delta)$$

with l_r an integer. The inequality

$$\sum_{(l_1+\cdots+l_k+k)\,\delta \le x} P\left(\bigcap_{r=1}^{k} A(r, n, l_r, \delta)\right)$$

$$\le P\left(\sum_{r=1}^{k} \frac{U_r(n)}{\sqrt{k(n)\,h_n(p(n))}} \le x\right)$$

$$\le \sum_{(l_1+\cdots+l_k)\,\delta \le x} P\left(\bigcap_{r=1}^{k} A(r, n, l_r, \delta)\right)$$

is clearly valid. The simple observation that

$$E\left|\max_{r=1,\ldots,k}\left|\frac{U_r(n)}{\sqrt{k(n)}\,h_n(p(n))}\right|\right|^2 \le E\left|\sum_{r=1}^{k}\left|\frac{U_r(n)}{\sqrt{k(n)}\,h_n(p(n))}\right|\right|^2 \le k(n)$$

implies that

$$P\left(\max_{r=1,\ldots,k}\left|\frac{U_r(n)}{\sqrt{k(n)}\,h_n(p(n))}\right| > t_k\right) < \varepsilon \tag{18}$$

if $t_k = (k(n)/\varepsilon)^{1/2}$. We make a few remarks leading to the following lemma.

Lemma 1. If $c(n) = o(q(n))$ *then*

$$\left|\sum_{(l_1+\cdots+l_k)\delta\le x} P\left(\bigcap_{r=1}^{k} A(r,n,l_r,\delta)\right) - \sum_{(l_1+\cdots+l_k)\delta\le x}\prod_{r=1}^{k} P(A(r,n,l_r,\delta))\right| \le k\left(\frac{2t_k}{\delta}\right)^k \alpha(q(n)-c(n)) + \varepsilon\,. \tag{19}$$

The probability of the union of all events $\cap_{r=1}^{k} A(r,n,l_r,\delta)$ for which $\max|U_r/\sqrt{k}\,h_n(p(n))| > t_k$ is at most ε by (18). Consider an event $\cap_{r=1}^{k} A(r,n,l_r,\delta)$ for which $\max|U_r/\sqrt{k}\,h_n(p(n))| \le t_k$. Repeated application of condition (14) leads to

$$\left|P\left(\bigcap_{r=1}^{k} A(r,n,l_r,\delta)\right) - \prod_{r=1}^{k} P(A(r,n,l_r,\delta))\right| \le k\,\alpha(q(n)-c(n))\,.$$

Since there are $(2t_k/\delta)^k$ events of this type the lemma follows.

Consider the convolution

$$G_{1,n} * \cdots * G_{k,n}(x) \tag{20}$$

of $G_{1,n}(x), \ldots, G_{k,n}(x)$. We have the inequalities

$$G_{1,n} * \cdots * G_{k,n}(x) \le \sum_{(l_1+\cdots+l_k)\delta\le x}\prod_{r=1}^{k} P(A(r,n,l_r,\delta)) \le G_{1,n} * \cdots * G_{k,n}(x+k\delta) \tag{21}$$

and

$$G_{1,n} * \cdots * G_{k,n}(x-k\delta) \le \sum_{(l_1+\cdots+l_k+k\delta)\le x}\prod_{r=1}^{k} P(A(r,n,l_r,\delta)) \le G_{1,n} * \cdots * G_{k,n}(x)\,. \tag{22}$$

Now the distribution (20) tends to the standard normal distribution as $n \to \infty$ by (16) and the Liapounov form of the central limit theorem. We wish to let $k(n)$, $p(n)$, $q(n) \to \infty$, $k(n)\,p(n) \cong n$ and $k(n)\,\delta(n) \to 0$ in such a way that $c(n) = o(p(n))$, $c(n) = o(q(n))$

$$k(n)\left(\frac{2\,t_k}{\delta}\right)^k \alpha(q(n) - c(n)) \to 0 \tag{23}$$

$$\frac{k(n)\,h_n(q(n))}{h_n(p(n))} \to 0\,. \tag{24}$$

Let $\delta = k^{-2}$ so that $k\,\delta \to 0$. Notice that

$$k\left(\frac{2\,t_k}{\delta}\right)^k \leq k^{5k}\,C^k$$

with $C = 2/\varepsilon^{1/2}$. We can assume that $\alpha(n) > 1/n$ for all n. If $k(n)$ is chosen so that

$$k \leq \left[-\log \alpha\left(\frac{1}{2}\,q(n)\right)\right]^{1/2} \tag{25}$$

then condition (23) will be satisfied. Since

$$h_n(n) = E\left|\sum_{j=1}^{n} Y_j^{(n)}\right|^2 \leq k^2\,h_n(n/k)$$

we have $k^2/h_n(n) \geq h_n(p(n))$. This implies that

$$\frac{k\,h_n(q(n))}{h_n(p(n))} \leq \frac{k^3\,h_n(q(n))}{h_n(n)}\,.$$

Thus (24) will be satisfied if

$$h_n(q(n)) = o\left(\frac{h_n(n)}{k(n)^3}\right). \tag{26}$$

Sequences $k(n)$, $p(n)$, $q(n)$, with $c(n) = o(p(n))$, $c(n) = o(q(n))$, $k(n)\,p(n) \cong n$ satisfying (25) and (26) can be determined. But then conditions (23) and (24) are satisfied. Inequalities (23), (24) and (18) and the lemma imply that

$$\sum_{r=1}^{k} \frac{U_r(n)}{\sqrt{k(n)\,h_n(p(n))}} \tag{27}$$

asymptotically has a standard normal distribution. Since (27) is asymptotically normal, it follows that the same is true of

$$S_n/\sqrt{k(n)\,h_n(p(n))}\,.$$

Kolmogorov and Rozanov [1960] have obtained a convenient sufficient condition for a Gaussian stationary sequence to be strongly mixing. It is that the spectral distribution function be absolutely continuous with a spectral density that is continuous and bounded away from zero. Helson and Sarason

[1967] derived an interesting necessary and sufficient condition for a stationary Gaussian sequence to be strongly mixing. However, it does not have an immediate intuitive interpretation. The following result of Ibragimov and Rozanov [1978] gives estimates on the rate at which the strong mixing coefficient $\alpha(n)$ tends to zero as $n \to \infty$.

Theorem 5. *A necessary and sufficient condition for* $\alpha(n) = O(n^{-r-\beta})$ *when the random sequence is stationary Gaussian and* r *is a nonnegative integer and* $0 < \beta < 1$ *runs as follows. The spectral distribution function is absolutely continuous and the spectral density* $f(\lambda)$ *of the form* $|P(e^{i\lambda})|^2 w(\lambda)$ *with* $P(z)$ *a polynomial with zeros on* $|z| = 1$ *and* $w(\lambda)$ *bounded away from zero,* r *times differentiable with the* rth *derivative satisfying a Hölder condition of order* β.

This theorem is derived by making use of results in approximation theory. Notice that if a process is strongly mixing, any process derived from it by a nonlinear operation of finite range and its shifts is also strongly mixing (see problem 8).

4. *Summability of Cumulants*

Summability conditions on cumulants often turn out to be convenient to assume. Suppose $X = (X_k)$ is a strictly stationary process with mean zero and $E X^2 < \infty$. Then summability of the covariances

$$\sum_k |r_k| < \infty$$

implies that one has an absolutely continuous spectrum with a continuous spectral density

$$f(\lambda) = \frac{1}{2\pi} \sum_k r_k e^{-ik\lambda} .$$

In Theorem 3 of this Chapter summability of fourth order cumulants is also assumed in a result on the asymptotic distribution of covariance estimates. If $E |X_j|^k < \infty$, let

$$c(u_1, \ldots, u_{k-1}) = \operatorname{cum}(X_t, X_{t+u_1}, \ldots, X_{t+u_{k-1}})$$

be the kth order cumulant of the random variables $X_t, X_{t+u_1}, \ldots, X_{t+u_{k-1}}$. Summability of this cumulant as a function of the u's implies that the kth order cumulant spectral density

$$f_k(\lambda_1, \ldots, \lambda_{k-1}) = (2\pi)^{-k+1} \sum_{u_1, \ldots, u_{k-1} = -\infty}^{\infty} c(u_1, \ldots, u_{k-1}) \exp\left\{ -i \sum_{j=1}^{k-1} u_j \lambda_j \right\}$$

exists and is continuous. Such cumulant spectra were introduced earlier in section 4 of Chapter II. Later on in the book conditions like

$$(28) \qquad \sum_{u_1, \ldots, u_{k-1}} | c(u_1, \ldots, u_{k-1}) | < \infty$$

for $k = 2, 3, \ldots$ or

$$(29) \qquad \sum (1 + | u_j |) \; | c(u_1, \ldots, u_{k-1}) | < \infty$$

for $j = 1, \ldots, k - 1$ and $k = 2, 3, \ldots$ will be assumed in a discussion of the asymptotic properties of a class of spectral estimates. Either of these conditions can be viewed as a mixing condition given in terms of moments or cumulants. Since the cumulants of order higher than the second are zero in the case of a Gaussian process, summability of these cumulants implies that the process in some sense is not too far from a Gaussian process. Nonetheless it is interesting to construct a class of non-Gaussian process satisfying some of the cumulant summability conditions.

In the course of constructing some examples of such non-Gaussian processes we shall derive an interesting result due to Slepian [1972] that allows us to compute moments of functions of jointly Gaussian random variables. Let $X_1, \ldots, X_m$ be independent (jointly) Gaussian randon variables with mean zero and covariance matrix R. Let us first assume that R is nonsingular. The joint density function of the random variables is given by

$$(30) \qquad \varphi(\mathbf{x}; R) = (2\pi)^{-m/2} \int \overset{\infty}{\underset{-\infty}{\cdots}} \int \exp\{i\, \mathbf{t}' \mathbf{x}\} \exp\left\{ - \frac{1}{2} \mathbf{t}' R\, \mathbf{t} \right\} dt$$

where

$$\mathbf{x} = \begin{pmatrix} x_1 \\ \vdots \\ x_m \end{pmatrix}, \mathbf{t} = \begin{pmatrix} t_1 \\ \vdots \\ t_m \end{pmatrix}.$$

Notice that for $j \neq k$

$$\frac{\partial \varphi(\mathbf{x}; R)}{\partial r_{jk}} = \frac{\partial^2 \varphi(\mathbf{x}; R)}{\partial x_j \, \partial x_k}.$$

If we expand (30) in the off-diagonal elements of R, the following expansion is obtained

$$(31) \qquad \varphi(\mathbf{x}; R) = \sum_{\nu_{12}=0}^{\infty} \cdots \sum_{\nu_{m-1,m}=0}^{\infty} \prod_{i<j} \frac{r_{ij}^{\nu_{ij}}}{\nu_{ij}!} \prod_{k=1}^{m} \frac{\partial^{s_k}}{\partial x_k^{s_k}} \varphi\left(\frac{x_k}{\sqrt{r_{kk}}} \right)$$

where $\nu = (\nu_{jk})$ is an $m \times m$ symmetric matrix with nonnegative integer entries, $s_k = \Sigma_{j \neq k} \nu_{jk}$, and $\varphi(\cdot)$ is the standard univariate Gaussian density function. Let $\varphi_s(z) = \partial^s / \partial z^s \, [\varphi(z)]$. If $r_{ii} = 1$ equation (31) can be written as

$$\varphi(\mathbf{x}; R) = \sum_{\nu} \frac{r^{\nu}}{\nu!} \prod_{1}^{m} \varphi_{s_i}(x_i)$$

where it is understood that $r^\nu = \Pi_{j<k} r_{jk}^{\nu_{jk}}$, $\nu! = \Pi_{j<k} \nu_{jk}!$. The Hermite polynomials

$$H_j(z) = (-1)^\nu \varphi_j(z)/\varphi(z)\ , \quad j = 0, 1, \ldots$$

are orthogonal with respect to the standard Gaussian density

$$\int Hj(z)\, H_k(z)\, \varphi(z)\, dz = \delta_{jk}\, j!\ .$$

In terms of the Hermite polynomials one can rewrite (31) as

$$\varphi(\mathbf{x}; R) = \sum \frac{r^\nu}{\nu!} H_{s_1}(x_1) \ldots H_{s_m}(x_m)\, \varphi(x_1) \ldots \varphi(x_m)\ . \tag{32}$$

The case $m = 2$ can be recognized as the classical Mehler's formula (see problem 2). This is an interesting generalization of Mehler's formula and can be used in some computations that we shall make. Notice that (32) implies that

$$E\{H_{s_1}(X_1) \ldots H_{s_m}(X_m)\} = s_1! \ldots s_m! \sum_{\{s_1, \cdots, s_m\}} \frac{r^\nu}{\nu!} \tag{33}$$

where $\Sigma_{\{s_1, \ldots, s_m\}}$ indicates that we are to sum over all symmetric matrices ν with nonnegative integer entries, $\nu_{ii} = 0$ and the row sums equal to $s_1, \ldots, s_m$. The result (33) was derived under the assumption that the covariance matrix R is nonsingular. By continuity, it is clear that it still must be valid even with R singular. The entries ν_{jk}, $j \neq k$, of ν correspond to a partition of the table

$$\begin{matrix} x_1 & \ldots & x_1 \\ x_2 & \ldots & x_2 \\ & \ldots & \\ x_m & \ldots & x_m \end{matrix} \tag{34}$$

with s_k x_k's in the kth row, $k = 1, \ldots, m$. The partition is one into pairs and ν_{jk} indicates there are ν_{jk} pairs of elements, each of these pairs consisting of one element from the jth row and one element from the kth row. This implies that the cumulant

$$\text{cum}\,\{H_{s_1}(X_1), \ldots, H_{s_m}(X_m)\} = s_1! \ldots s_m! \sum_{\{s_1, \ldots, s_m\}}' \frac{r^\nu}{\nu!} \tag{35}$$

where $\Sigma'_{\{s_1, \ldots, s_m\}}$ denotes summing over all matrices ν with row sums $s_1, \ldots, s_m$ that correspond to irreducible partitions of the table (34). We therefore have the following theorem

Theorem 6. *Let $X_1, \ldots, X_m$ be jointly Gaussian random variables with covariance matrix R and $r_{ii} = 1$. Then the moment $E\{H_{s_1}(X_1) \ldots H_{s_m}(X_m)\}$ and the cumulant* cum$\{H_{s_1}(X_1), \ldots, H_{s_m}(X_m)\}$ *are given by* (33) *and* (35) *respectively.*

Let us now consider a function $\psi(x)$ such that

$$E\mid \psi(X)\mid^{2k} < \infty$$

for some fixed positive integer k with X a standard Gaussian variable. Further let the Fourier-Hermite expansion of ψ be

$$\psi(z) = \sum_j a_j H_j(z)$$

with the coefficients $a_j \geq 0$ for all j. Then

$$E[\psi(X)^j] = \sum_{k_1, \ldots, k_j} a_{k_1} \cdots a_{k_j} s_1! \ldots s_j! \sum_{\{s_1, \ldots, s_m\}} \frac{1}{\nu!} < \infty, \tag{36}$$

$j = 2, 3, \ldots, 2k$. The nonnegativity of the a_j's implies that the series in (36) is absolutely convergent. Suppose ψ is a nonlinear function. Consider then the non-Gaussian strictly stationary process

$$Y_n = \psi(X_n), \quad n = \ldots, -1, 0, 1, \ldots$$

derived from the stationary Gaussian process $\{X_n\}$ with covariances r_n. If the covariance sequence of the Gaussian process X_n is summable, $r_0 = 1$, and

$$\sum_n \mid r_n \mid < \infty,$$

then (36) implies that the cumulant functions of the process $Y = (Y_n)$ up to order $2k$ will be summable. Essentially the same argument tells us that if

$$\sum_m (1 + \mid m \mid) \mid r_m \mid < \infty$$

then summability conditions of the type (29) will hold for $Y = (Y_n)$ up to order $2k$.

One can easily generate functions ψ with nonnegative coefficients a_j by making use of the identity

$$e^{tz} e^{-t^2/2} = \sum_{j=0}^{\infty} H_j(z) t^j/j!.$$

Given any positive c

$$\begin{aligned}
\psi(z) &= \left(\frac{2\pi}{1+c}\right)^{1/2} \exp\left\{\frac{z^2}{2(1+c)}\right\} \\
&= \int_{-\infty}^{\infty} e^{tz} e^{-(1+c)z^2/2} dz \\
&= \sum_{j=0}^{\infty} \frac{H_j(z)}{j!} \int_{-\infty}^{\infty} t^j e^{-ct^2/2} dt \\
&= \sum_{j=0}^{\infty} \frac{H_{2j}(z)}{2^j j!} c^{-j-1/2}.
\end{aligned}$$

With c large enough the $(2k)$th moment of $\psi(X)$ will be finite but higher moments will be infinite.

5. Long-range Dependence

In section 2 we considered conditions under which covariance estimates are asymptotically normal. It is of some inerest to indicate circumstances under which asymptotic normality is not obtained even though all moments of the random variables $\{X_k\}$ are finite. The existence of another more "exotic" limit distribution for covariance estimates can be regarded as a sign of long range dependence by the process $X = \{X_k\}$. Our analysis will be carried out under special assumptions though the effect described holds under much more general circumstances.

Let $X = \{X_k\}$ be a normal stationary process with mean zero and covariance function r_k. Assume that

$$r_k \cong a\, k^{-\beta}$$

with $a, \beta > 0$ as $|k| \to \infty$. For convenience, assume that $r_0 = 1$. Let us assume initially that $\beta < 1$. If the spectral distribution function of the process $X = \{X_k\}$ is $F(\lambda)$, then

$$F(\lambda) - F(-\lambda) = 2\lambda + \sum_{k=1}^{\infty} r_k\, 2\, \frac{\sin k\lambda}{k}$$

$$\cong 2a \sum_{k=1}^{\infty} k^{-\beta} \frac{\sin k\lambda}{k} \cong 2a\lambda^{\beta} \int_0^{\infty} \frac{\sin u}{u^{1+\beta}}\, du$$

as $\lambda \to 0$. A simple example of such a covariance sequence is given by

$$r_k = (1 + k^2)^{-\beta/2}\,. \tag{37}$$

One can show that the spectral distribution function corresponding to (37) is absolutely continuous and has a spectral density with a singularity of the form $|\lambda|^{\beta-1}$ in the neighborhood of zero (see Rosenblatt [1961]).

We shall only consider the asymptotic distribution of the variance estimator. Let

$$Y_k = X_k^2 - 1\,.$$

The covariance function of the process $Y = \{Y_k\}$ is $2\, r_k^2$. Let $0 < \beta < 1/2$. By the argument given above, if $G(\lambda) = F^{(2)}(\lambda)$ is the spectral distribution function of Y then

$$G(\lambda) - G(-\lambda) \cong 4\, a^2 \lambda^{2\beta} \int_0^{\infty} \frac{\sin u}{u^{1+2\beta}}\, du$$

as $\lambda \to 0$ and the spectral density corresponding to covariance $2\,r_k^2 = 2(1+k^2)^{-\beta}$ has a singularity of the form $|\lambda|^{2\beta-1}$ in the neighborhood of zero. It will be seen that $N^{-1+\beta} \sum_{k=1}^{N} Y_k$ has a non-normal limiting distribution as $N \to \infty$. Let R denote the covariance matrix of $X_1, \ldots, X_N$. The characteristic function of $N^{-1+\beta} \sum_{k=1}^{N} Y_k$ is

$$|I - 2\,i\,t\,N^{-1+\beta} R|^{-1/2} \exp\{-i\,t\,N^{\beta}\} = \exp\left\{\frac{1}{2} \sum_{k=2}^{\infty} (2\,i\,t\,N^{-1+\beta})^k\, s\,p(R^k)/k\right\}$$

where $s\,p(M)$ denotes the trace of the matrix M. Now

$$(N^{-1+\beta})^k\, s\,p(R^k) = (N^{-1+\beta})^k \sum_{i_j=1}^{N} r_{i_1-i_2} r_{i_2-i_3} \cdots r_{i_k-i_1}$$

$$\to \underbrace{\int_0^1 \cdots \int}_{k} |x_1 - x_2|^{-\beta} |x_2 - x_3|^{-\beta} \cdots |x_k - x_1|^{-\beta}\, dx_1 \cdots dx_k = c_k > 0$$

as $N \to \infty$. The characteristic function of the limiting distribution is

$$\exp\left\{\frac{1}{2} \sum_{k=2}^{\infty} (2\,i\,t)^k\, c_k/k\right\}$$

the characteristic function of a non-normal distribution. The usual normalization for partial sums of a stationary sequence in the case of asymptotic normality is $N^{1/2}$. Notice that the normalization here is $N^{-1-\beta}$ with $0 < \beta < 1/2$. This grows at a faster rate than $N^{1/2}$ and is an aspect of the long range dependence. Of course, the normalization required is determined by the behavior of the spectrum in the neighborhood of zero.

6. Strong Mixing and Random Fields

We have already mentioned in section 3 that Kolmogorov and Rozanov had shown that a sufficient condition for a Gaussian stationary sequence to be strongly mixing is that it have an absolutely continuous spectral distribution function with a strictly positive continuous spectral density. In this section a corresponding result will be obtained for stationary random fields by a similar argument. Suppose $X_{\mathbf{n}}$, $\mathbf{n} = (n_1, \ldots, n_k)$, $n_i = \ldots, -1, 0, 1, \ldots$, is a k-dimensional weakly stationary random field with $E\,X_{\mathbf{n}} \equiv 0$. From problem 9 of Chapter I it is clear that the covariance

$$r_{\mathbf{n}-\mathbf{m}} = E(X_{\mathbf{n}} \overline{X}_{\mathbf{m}}) = \int_{-\pi}^{\pi} \cdots \int_{-\pi}^{\pi} e^{i(\mathbf{n}-\mathbf{m})\cdot\boldsymbol{\lambda}}\, dF(\lambda)$$

with the function F of k variables $\boldsymbol{\lambda} = (\lambda_1, \ldots, \lambda_k)$ the spectral distribution function of the process $\{X_n\}$, a nondecreasing function of $\boldsymbol{\lambda}$ If F is absolutely continuous, the derivative

$$f(\boldsymbol{\lambda}) = \frac{\partial}{\partial \lambda_1} \cdots \frac{\partial}{\partial \lambda_k} F(\boldsymbol{\lambda})$$

is the spectral density of the process.

Let us now assume that $\{X_{\mathbf{n}}\}$ is strictly stationary. In the case of a multidimensional index the following concept of *strong mixing* is introduced. Let S and S' be two sets of indices. The Borel fields $\mathscr{B}(S) = \mathscr{B}(X_{\mathbf{n}}, \mathbf{n} \in S)$ and $\mathscr{B}(S') = \mathscr{B}(X_{\mathbf{n}}, \mathbf{n} \in S')$ as usual are the sigma-fields generated by the random variables $X_{\mathbf{n}}$ with subscript elements of S and S' respectively. Consider the distance $d(S, S')$ between the sets of indices S and S'. The process $\{X_{\mathbf{n}}\}$ is said to be strongly mixing if

$$\sup_{\substack{A \in \mathscr{B}(S) \\ B \in \mathscr{B}(S')}} |P(A\,B) - P(A)\,P(B)| \leq \varphi(d(S, S'))$$

for any two sets of indices S and S' with φ a function such that $\varphi(d) \to 0$ as $d \to \infty$.

Our object in this section is to prove the following theorem.

Theorem 7. *Let $\{X_{\mathbf{n}}\}$ be a Gaussian stationary random field with an absolutely spectral distribution function and positive continuous spectral density*

$$f(\boldsymbol{\lambda}) = \frac{\partial}{\partial \lambda_1} \cdots \frac{\partial}{\partial \lambda_k} F(\boldsymbol{\lambda})$$

(considered as a function on the compact k-torus $(-\pi, \pi]^k$, $k > 1$). *The process is then strongly mixing.*

Let

$$\alpha(\mathscr{B}(S), \mathscr{B}(S')) = \sup_{\substack{A \in \mathscr{B}(S) \\ B \in \mathscr{B}(S')}} |P(A\,B) - P(A)\,P(B)|\,.$$

Consider the spaces $L^2(S)$, $L^2(S')$ of functions measurable with respect to $\mathscr{B}(S)$ and $\mathscr{B}(S')$ respectively with finite second moments. Set

$$\varrho(L^2(S), L^2(S')) = \sup_{\substack{g \in L^2(S) \\ h \in L^2(S')}} |\operatorname{corr}(g, h)|$$

with corr (g, h) the correlation of the random variables g, h. $\varrho(L^2(S), L^2(S'))$ is sometimes referred to as the maximal correlation coefficient between the sigma-algebras $\mathscr{B}(S)$ and $\mathscr{B}(S')$. It is clear that

$$\alpha(\mathscr{B}(S), \mathscr{B}(S')) \leq \varrho(L^2(S), L^2(S'))\,.$$

Let $H(S)$ and $H(S')$ be the closure in mean square of the vector spaces obtained by finite linear combinations of random variables X_n, $n \in S$, and X_n, $n \in S'$, respectively. Then set

$$\varrho(H(S), H(S')) = \sup_{\substack{g \in H(S) \\ h \in H(S')}} |\operatorname{corr}(g, h)| .$$

Notice that

$$\varrho(H(S), H(S')) \leq \varrho(L^2(S), L^2(S'))$$

since $H(S)$ is the linear Hilbert space generated by X_n, $n \in S$, while $L^2(S)$ is the nonlinear Hilbert space generated by the same collection of random variables and $H(S) \subset L^2(S)$. Nonetheless, in the case of a Gaussian random field one has the following result.

Lemma 2. Let X_n be a Gaussian random field. Then

$$\varrho(H(S), H(S')) = \varrho(L^2(S), L^2(S')) . \tag{38}$$

Further

$$\alpha(\mathscr{B}(S), \mathscr{B}(S')) \leq \varrho(L^2(S), L^2(S')) \leq 2\pi\, \alpha(\mathscr{B}(S), \mathscr{B}(S')) . \tag{39}$$

In proving this result, it is enough to consider finite sets S, S'. Also, in the spaces $H(S)$, $H(S')$ one can take the corresponding generating sets of random variables ξ_j, $j = 1, \ldots, n$, and η_k, $k = 1, \ldots, m$, respectively so that only the Gaussian random variables with identical subscripts are dependent with

$$E\,\xi_j = E\,\eta_k = 0 , \quad E\,\xi_j^2 = E\,\eta_k^2 = 1 ,$$

$j = 1, \ldots, n$, and $k = 1, \ldots, m$. This can be seen in the following manner. Suppose the generating random variables of $H(S)$ and $H(S')$ are initially $\{u_j\}$ and $\{v_k\}$ with covariance matrices of rank n and m respectively. The sets $\{u_j\}$ and $\{v_k\}$ can be assumed to have n and m elements respectively with covariance matrices the $n \times n$ and $m \times m$ identity matrices I_n and I_m since this can be effected by the orthogonal transformations reducing the original covariance matrices to diagonal form followed by appropriate rescaling operations. The joint covariance matrix is

$$\begin{pmatrix} I_n & F \\ F^T & I_m \end{pmatrix}$$

with F an $n \times m$ matrix. The reduction to sets of random variables $\{\xi_j\}$ and $\{\eta_k\}$ of the desired form can be carried out if we can find orthogonal matrices $U_1(n \times n)$ and $U_2(m \times m)$ such that

$$U_1^T F\, U_2 = \Lambda$$

with Λ an $n \times m$ diagonal matrix. This can indeed be done with U_1 the matrix diagonalizing $F\,F^T$ and U_2 the matrix diagonalizing $F^T F$.

Now consider any random variables $f \in L^2(S)$, $g \in L^2(S')$ with finite second moments. Since $f = f(\xi_1, \ldots, \xi_n)$ and $g = g(\eta_1, \ldots, \eta_m)$ one can write

$$f = \sum_1^n f_j(\xi_1, \ldots, \xi_j), \quad g = \sum_1^m g_k(\eta_1, \ldots, \eta_k)$$

with

$$f_j = E(f \mid \xi_1, \ldots, \xi_j) - E(f \mid \xi_1, \ldots, \xi_{j-1}),$$
$$g_k = E(g \mid \eta_1, \ldots, \eta_k) - E(g \mid \eta_1, \ldots, \eta_{k-1}).$$

Notice that if $j \geq k$

$$E\big(E(f \mid \xi_1, \ldots, \xi_j) \mid \eta_1, \ldots, \eta_k\big) = E\big(E(f \mid \xi_1, \ldots, \xi_k) \mid \eta_1, \ldots, \eta_k\big)$$

while if $k \geq j$

$$E\big(E(g \mid \eta_1, \ldots, \eta_k) \mid \xi_1, \ldots, \xi_j\big) = E\big(E(g \mid \eta_1, \ldots, \eta_j) \mid \xi_1, \ldots, \xi_j\big),$$

and this implies that $E(f_j g_k) = 0$ if $j \neq k$. Therefore

$$E(fg) = \sum_{k=1}^{\min(m,n)} E(f_k g_k) = \sum_{k=1}^{\min(m,n)} E\big(E(f_k g_k \mid \xi_1, \ldots, \xi_{k-1}, \eta_1, \ldots, \eta_{k-1})\big).$$

It is clear that ξ_k and η_k are jointly Gaussian and independent of $\xi_1, \ldots, \xi_{k-1}$, $\eta_1, \ldots, \eta_{k-1}$. An application of Mehler's formula (see problem 5) implies that

$$| E(f_k g_k \mid \xi_1, \ldots, \xi_{k-1}, \eta_1, \ldots, \eta_{k-1}) | \leq \varrho\, a_k b_k$$

with

$$\varrho = \varrho\big(H(S), H(S')\big),\ a_k^2 = E(f_k^2 \mid \xi_1, \ldots, \xi_{k-1}, \eta_1, \ldots, \eta_{k-1}),$$
$$b_k^2 = E(g_k^2 \mid \xi_1, \ldots, \xi_{k-1}, \eta_1, \ldots, \eta_{k-1}).$$

From this it is clear that

$$| E(fg) | \leq \varrho \sum_1^{\min(m,n)} E(a_k b_k) \leq \varrho$$

and (38) follows.

The second part (39) of the Lemma can be shown to be true in the following manner. Given any $\varepsilon > 0$ choose $\xi_\varepsilon \in H(S)$, $\eta_\varepsilon \in H(S')$ so that $E\,\xi_\varepsilon = E\,\eta_\varepsilon = 0$, $E\,\xi_\varepsilon^2 = E\,\eta_\varepsilon^2 = 1$ and $r = E\,\xi_\varepsilon\,\eta_\varepsilon > \varrho - \varepsilon$. Let $A_\varepsilon = \{\xi_\varepsilon > 0\} \in \mathscr{B}(S)$, $B_\varepsilon = \{\eta_\varepsilon > 0\} \in \mathscr{B}(S')$. Then (see Cramér [1946], p. 290)

$$P(A_\varepsilon B_\varepsilon) = \frac{1}{4} + \frac{1}{2\pi} \arcsin r,\ P(A_\varepsilon)\, P(B_\varepsilon) = \frac{1}{4}$$

and so

$$\frac{1}{2\pi} \arcsin r = P(A_\varepsilon B_\varepsilon) - P(A_\varepsilon)\, P(B_\varepsilon) \leq \alpha\big(\mathscr{B}(S), \mathscr{B}(S')\big).$$

If $\alpha > 1/4$ the inequality $\varrho \leq 2\pi\alpha$ is trivially satisfied. If $\alpha \leq 1/4$ we have $\varrho - \varepsilon \leq r \leq \sin 2\pi\alpha$ and consequently $\varrho \leq 2\pi\alpha + \varepsilon$. Since this holds for any $\varepsilon > 0$ it is clear that we must have

$$\varrho \leq 2\pi\alpha .$$

The proof of Lemma 2 is complete.

We shall make use of the following Lemma of a functional analytic character. The brief derivation of the Lemma will be given in the Appendix.

Lemma 3. Let L be a Banach space and L^* the conjugate space (of linear functionals on L). *Consider a subspace H of L. Let H^0 be the set of linear functionals on L that reduce to zero on H. Then for any $h^* \in L^*$ one has*

$$\sup_{h \in H,\ \|h\| = 1} h^*(h) = \inf_{h^0 \in H^0} \| h^* - h^0 \| .$$

We now return to the proof of Theorem 7. By virtue of Lemma 2 it is clear that to prove strong mixing for a Gaussian stationary random field it is enough to show that

$$\varrho\big(H(S), H(S')\big) \leq \varphi\big(d(S, S')\big)$$

for some function φ such that $\varphi(d) \to 0$ as $d \to \infty$. Let us set

$$\varphi(d) = \sup_{d(S, S') = d} \varrho\big(H(S), H(S')\big) .$$

Then it is clear that $\varphi(d)$ is the supremum of

$$\int p_1(\boldsymbol{\lambda})\, \overline{p_2(\boldsymbol{\lambda})}\, f(\boldsymbol{\lambda})\, d\boldsymbol{\lambda}$$

with $p_1(\boldsymbol{\lambda})$, $p_2(\boldsymbol{\lambda})$ trigonometric polynomials

$$p_1(\boldsymbol{\lambda}) = \sum_{t_s} a_k\, e^{i\boldsymbol{\lambda}\cdot\mathbf{t}_s} ,$$

$$p_2(\boldsymbol{\lambda}) = \sum_{\tau_j} b_j\, e^{i\boldsymbol{\lambda}\cdot\boldsymbol{\tau}_j}$$

with $\int | p_i(\boldsymbol{\lambda}) |^2 f(\boldsymbol{\lambda})\, d\boldsymbol{\lambda} \leq 1$ and $| \mathbf{t}_s - \boldsymbol{\tau}_j | \geq d$ for all $\mathbf{t}_s$, $\boldsymbol{\tau}_j$ one sums over in $p_1(\boldsymbol{\lambda})$, $p_2(\boldsymbol{\lambda})$ respectively. But this implies that

$$\varphi(d) \leq \sup_p \int p(\boldsymbol{\eta})\, f(\boldsymbol{\eta})\, d\boldsymbol{\eta}$$

where the polynomials $p(\boldsymbol{\eta})$ are of the form

$$p(\boldsymbol{\eta}) = \sum_{|\boldsymbol{\tau}_j| \geq d} a_j \exp\{i\,\boldsymbol{\eta}\cdot\boldsymbol{\tau}_j\} \tag{40}$$

and satisfy

$$\int | p(\boldsymbol{\eta}) | f(\boldsymbol{\eta}) \, d\boldsymbol{\eta} \leq 1 .$$

Let us now consider applying Lemma 3 with L the Banach space of functions p with $\int | p(\boldsymbol{\eta}) | f(\boldsymbol{\eta}) \, d\boldsymbol{\eta} < \infty$. H is the linear closure in L of trigonometric polynomials of the form (40). Then H^0 is the subspace of linear functionals h^0 (on L) with corresponding function $h_0(\boldsymbol{\lambda})$ (ess sup $| h_0(\boldsymbol{\lambda}) | < \infty$) such that

$$\int e^{i\eta \cdot \tau} h_0(\boldsymbol{\eta}) f(\boldsymbol{\eta}) \, d\boldsymbol{\eta} = 0$$

for $| \boldsymbol{\tau} | \geq d$. Thus

$$h_0(\boldsymbol{\lambda}) f(\boldsymbol{\lambda}) = \sum_{|\tau_j| < d} \beta_j e^{i\boldsymbol{\lambda} \cdot \boldsymbol{\tau}_j} .$$

Lemma 3 implies that

$$\text{(41)} \qquad \sup_{p \in H} \int p(\boldsymbol{\eta}) f(\boldsymbol{\eta}) \, d\boldsymbol{\eta} = \inf_{h_0 \in H^0} | 1 - h_0(\boldsymbol{\lambda}) | = \inf_{\psi} | f(\boldsymbol{\lambda}) - \psi(\boldsymbol{\lambda}) | / f(\boldsymbol{\lambda})$$

where the infimum is taken over ψ of the form

$$\psi = \sum_{|\tau_j| < d} \beta \, e^{i\boldsymbol{\lambda} \cdot \boldsymbol{\tau}_j} .$$

If f is a positive continuous spectral density function, it is bounded away from zero and the Weierstrass approximation theorem implies that (41) tends to zero as $d \to \infty$.

In problem 4 of this chapter a version of a central limit theorem for a strongly mixing stationary sequence is stated under a Lindeberg like condition. This is basically a central limit theorem of a type derived in Rosenblatt [1956a]. At this point we state a corresponding result for a strongly mixing random field $X = (X_{\mathbf{n}})$, $\mathbf{n} = (n_1, \ldots, n_k)$. This proof parallels that given in the one-dimensional case and requires only occasional modifications.

Theorem 8. *Let $X = (X_{\mathbf{n}})$, $E X_{\mathbf{n}} \equiv 0$, be a strictly stationary random field. Assume that*

$$E \left| \sum_{\substack{n_i = a_i \\ i = 1, \ldots, k}}^{b_i} X_{n_1, \ldots, n_k} \right|^2 = h(b_1 - a_1, \ldots, b_k - a_k) \to \infty$$

as $b_1 - a_1, \ldots, b_k - a_k \to \infty$. Further let

$$h(\alpha_1, \ldots, \alpha_k) = o\big(h(\beta_1, \ldots, \beta_k)\big)$$

if $\alpha_1, \ldots, \alpha_k \to \infty$ with $\alpha_i = O(\beta_i)$, $i = 1, \ldots, k$ but for some j, $\alpha_j = o(\beta_j)$. Then if

$$E \left| \sum_{\substack{n_i = a_i \\ i = 1, \ldots, k}}^{b_i} X_n \right|^{2+\delta} = O\big(h(b_1 - a_1, \ldots, b_k - a_k)\big)^{1+\delta/2}$$

as $b_1 - a_1, \ldots, b_k - a_k \to \infty$ *for some* $\delta > 0$, $\sum_{\substack{n_i=1\\ i=1,\ldots,k}}^{N} X_n$ *properly normed is asymptotically* $N(0, 1)$ *as* $N \to \infty$.

In the case of random fields, the following version of a periodogram

$$I_N(\lambda) = (2\pi N)^{-k} \left| \sum_{\substack{n_i=1\\ i=1,\ldots,k}}^{N} X_{n_1,\ldots,n_k} e^{-in_1\lambda_1 - \cdots - in_k\lambda_k} \right|^2$$

is a direct generalization of the one-dimensional periodogram. An application of Theorem 8 yields the following result which provides us with a proposition similar to Theorem 2 that holds for random fields.

Theorem 9. *Let* $X = (X_n)$ *be a strongly mixing strictly stationary random field whose cumulant functions up to order eight are absolutely summable. Let*

$$\int_{-\pi}^{\pi} \cdots \int I_N(\boldsymbol{\lambda}) A_1(\boldsymbol{\lambda})\, d\boldsymbol{\lambda}, \quad \int_{-\pi}^{\pi} \cdots \int I_N(\boldsymbol{\lambda}) A_2(\boldsymbol{\lambda})\, d\boldsymbol{\lambda}$$

be two quadratic forms with weight functions $A_i(\boldsymbol{\lambda})$, $i = 1, 2$, *symmetric about zero* $(A_i(\boldsymbol{\lambda}) = A_i(-\boldsymbol{\lambda}))$ *and square integrable. The quadratic forms are then asymptotically normal with means*

$$\int_{-\pi}^{\pi} \cdots \int f(\boldsymbol{\lambda}) A_i(\boldsymbol{\lambda})\, d\boldsymbol{\lambda}, \quad i = 1, 2$$

and covariance

$$\cong \left(\frac{2\pi}{N}\right)^k \left\{ 2 \int_{-\pi}^{\pi} \cdots \int A_i(\boldsymbol{\lambda}) A_j(\boldsymbol{\lambda}) f^2(\boldsymbol{\lambda})\, d\boldsymbol{\lambda} + \int_{-\pi}^{\pi} \cdots \int \int_{-\pi}^{\pi} \cdots \int f_4(\boldsymbol{\lambda}, -\boldsymbol{\mu}, \boldsymbol{\mu}) A_i(\boldsymbol{\lambda}) A_j(\boldsymbol{\mu})\, d\boldsymbol{\lambda}\, d\boldsymbol{\mu} \right\},$$

$i, j = 1, 2$.

Problems

1. Consider a smoothed periodogram $H_N = \int_{-\pi}^{\pi} I_N(\lambda) A(\lambda)\, d\lambda$ with A a piecewise continuous bounded even function. Show that if the spectral density $f(\lambda)$ of the stationary process is continuously differentiable that then the mean

$$E\{H_N\} = \int_{-\pi}^{\pi} f(\lambda) A(\lambda)\, d\lambda + O(\log N/N)\,.$$

2. Show that the variance of the smoothed periodogram H_N is

$$N^{-2} \sum_{a,a',b,b'=1}^{N} \{r_{a-a'}\, r_{b-b'} + r_{a-b'}\, r_{a'-b} + c_{b-a,\, a'-a,\, b'-a}\}\, \alpha_{a-b}\, \alpha_{a'-b'} \,.$$

Indicate how the absolute summability of the covariances r_k and fourth order cumulants $c_{a,\,b,\,d}$ implies that H_N can be approximated in mean square by a finite sum $\Sigma_{|s| \leq k}\ \alpha_s\, r_s(N)$ if k is sufficiently large.

3. Let $\{X_t\}$ be a stationary linear process with the independent random variables V_t having a finite fourth moment. Let γ_4 be the fourth cumulant of V_t. Show that the asymptotic behavior of the covariance of covariance estimates is given by

$$\lim_{N \to \infty} N \operatorname{cov}(r_u(N), r_v(N)) = 4\pi \int \cos u\lambda \cos v\lambda\, f^2(\lambda)\, d\lambda + \gamma_4\, r_u\, r_v \,.$$

4. By using the ideas employed in the proof of Theorem 4, show that if $\{X_j\}$ is a strongly mixing process with mean zero, $E\,|\,\Sigma_a^b X_j\,|^2 \simeq h(b-a)$ as $b - a \to \infty$ with $h(m) \uparrow \infty$, $E\,|\,\Sigma_{j=a}^{b} X_j\,|^{2+\delta} = O\,(h(b-a))^{1+\delta/2}$ for some $\delta > 0$, that then $\Sigma_{j=1}^{n} X_j$ properly normalized is asymptotically normal $N(0, 1)$ as $n \to \infty$ (also see Rosenblatt [1956a]).

5. If $X = \{X_j\}$ is a normal stationary process with mean zero and covariance function r_k, and φ is a function with $\int \varphi(x)^2 (2\pi)^{-1/2} \exp(-x^2/2)\, dx < \infty$ and Fourier-Hermite coefficients a_j, show that

$$E\,\{\varphi(X_0)\, \varphi(X_k)\} = \sum_{j=0}^{\infty} j!\, a_j^2\, r_k^j \,.$$

6. Under the assumptions of the previous example show that if F is the spectral distribution of the process $X = \{X_k\}$ that then the spectral distribution function of $\{\varphi(X_k)\}$ is

$$\sum_{j=1}^{\infty} j!\, a_j^2\, F^{(j)}(\lambda) \,.$$

7. Determine the asymptotic behavior of a covariance estimate $r_u(N)$ for a Gaussian stationary process with covariance function (39).

8. Let $\{X_j\}$ be a strongly mixing strictly stationary process. Consider the process $\{Y_j\}$ generated by $f(X_1, \ldots, X_m)$ (m finite) and its shifts. Show that $\{Y_j\}$ is strongly mixing.

Notes

3.1 A discussion of various types of mixing conditions useful in deriving a number of limit theorems for stationary sequences can be found in the book of Hall and Heyde [1980].

3.2 Notice that the martingale difference condition $E[V_t \mid \mathscr{B}_{t-1}] \equiv 0$ considered by Hannan and Hall and Heyde in the representation of the process

$$X_t = \sum_{k=0}^{\infty} a_k V_{t-k}$$

implies that the best predictor (in the sense of minimum mean square error of prediction) is the best linear predictor. Under our assumptions this needn't be the case.

R. Dahlhaus [1984] obtains asymptotic normality of smoothed periodograms under certain mixing conditions with rates of decay on the mixing coefficients. In the course of his derivation, he obtains a generalization of a result of Grenander and Rosenblatt [1957] on the maximal deviation of an estimate of the spectral distribution function from its mean.

3.3 If $X_1, X_2, \ldots$ are independent, identically distributed random variables with mean $E X_t \equiv 0$ and variance $E X_i^2 \equiv 1$, it is well known that $\frac{1}{\sqrt{n}} \Sigma_{k=1}^{n} X_j$ is asymptotically normally distributed with mean zero and variance one as $n \to \infty$. This is a classical result that lies close to the boundary of the domain of the central limit theorem. The question as to where this boundary lies when one deals with partial sums of strictly stationary sequences is still open. The following result making use of the strong mixing condition is due to Ibragimov [1962]. Let $\{X_k\}$ be a strictly stationary condition satisfying the strong mixing condition with $E X_0 \equiv 0$ and $E X_0^{2+\delta} < \infty$ for some δ, $0 < \delta < \infty$. Let $\Sigma_{n=1}^{\infty} [\alpha(n)]^{\delta/(2+\delta)} < \infty$ and set $S_n = \Sigma_{j=1}^{n} X_j$. Then $\lim_{n\to\infty} n^{-1} E S_n^2 = \sigma^2$ with $0 \le \sigma^2 < \infty$. If $\sigma^2 > 0$, $S_n \sigma^{-1} n^{-1/2}$ converges in distribution to the normal distribution with mean zero and variance one.

An interesting construction due to Herrndorf [1983] indicates that the moment conditions cannot be relaxed so that one has only existence of second order moments, strong mixing and a central limit theorem. He shows that given any positive sequence ε_n, one can construct a strictly stationary strongly mixing sequence $\{X_k\}$ with $E X_0 = 0$, $E(X_j X_k) = \delta_{jk}$ and $\alpha(n) \le \varepsilon_n$ such that $\inf_{n \ge 1} P(S_n = 0) > 0$, the family of distributions of S_n is tight, $S_n/b_n \to 0$ in probability as $n \to \infty$ for every sequence $b_n \to \infty$ as $n \to \infty$.

3.4 The representation of cumulants or moments of polynomials of jointly Gaussian variables (due to Slepian [1972]) can be regarded as a generalization of the result given in problem 5 of this chapter. This result is usually called Mehler's formula.

3.5 The discussion of section 5 was originally presented in Rosenblatt [1961] to give an example of a process that is not strongly mixing. It was later shown by Taqqu [1975], [1979] and Dobrushin and Major [1979] that such exotic limit distributions are obtained for a large class of processes with long-range dependence. We mention some results of this type in a formulation given by

Dobrushin and Major. Let $\{X_n\}$, $E\,X_n \equiv 0$, $E\,X_n^2 \equiv 1$, be a Gaussian stationary sequence with correlation function $r(n) = E\,X_0\,X_n = n^{-\alpha}\,L(n)$, $0 < \alpha < 1$, with $L(n)$ a slowly varying function. Let $H(x)$ be a real-valued function such that $E\,H(X_0) = 0$, $0 < E\,H(X_0)^2 < \infty$. Expand $H(x)$ in terms of the Hermite polynomials $H_j(x)$ (taken with highest coefficient 1)

$$H(x) = \sum_{j=1}^{\infty} c_j\,H_j(x)\,, \quad \sum_{j=1}^{\infty} c_j^2\,j! < \infty\,.$$

Let $Y_j = H(X_j)$. Suppose $\alpha < 1/k$ where k is the smallest index of the Hermite expansion of $H(x)$ for which $c_k \neq 0$. Consider the partial sums $S_N = \Sigma_{j=1}^N Y_j$. Let $A_N = N^{1-k\alpha/2}\,L(N)^{k/2}$. One can show that $A_N^{-1}\,S_N$ has a limiting distribution as $N \to \infty$, the distribution of the multiple Wiener-Ito integral

$$Y^* = D^{-k/2}\,c_k \int \frac{e^{i(x_1+\cdots+x_k)}-1}{i(x_1+\cdots+x_k)}\,|\,x_1\,|^{\frac{\alpha-1}{2}} \cdots |\,x_k\,|^{\frac{\alpha-1}{2}}\,dW(x_1)\,\ldots\,dW(x_k)$$

with W the random spectral measure of the Gaussian white-noise process and

$$D = \int_{-\infty}^{\infty} \exp(i\,x)\,|\,x\,|^{\alpha-1}\,dx = 2\,\Gamma(\alpha)\,\cos\frac{\alpha\,\pi}{2}\,.$$

Notice that the principal term in the normalization is $N^{1-k\alpha/2}$ so that one normalizes by N^{β}, $\beta = 1 - k\,\alpha/2 > 1/2$. The fact that $\beta > 1/2$ could be taken as a sign of the long-range dependence. However in Rosenblatt [1979] processes are constructed for which there are exotic limiting distributions with a normalization N^{β} with $\beta < 1/2$. These results are extended in a paper of P. Major [1981]. The lecture notes of Major [1980] contain a good exposition of useful background for the development of these results.

3.6 The proof of Theorem 7 was essentially given in Rosenblatt [1972]. The derivation follows the proof given in Kolmogorov and Rozanov [1960] for the 1-dimensional case.

Chapter IV

Estimation of Parameters of Finite Parameter Models

1. Maximum Likelihood Estimates

For convenience let us assume that $\{X_m, m = \ldots, -1, 0, 1, \ldots\}$ is a strictly stationary process whose probability distribution is parameterized by a k-dimensional parameter θ with real components $\theta_1, \ldots, \theta_k$. Assume that the finite dimensional distributions of the process $\{X_m\}$ are either all absolutely continuous with respect to the corresponding Lebesgue measure or are all discrete. Suppose that one can observe $X_1, \ldots, X_n$. On the basis of these observations one should like to obtain an effective estimate $\theta_n(X_1, \ldots, X_n)$ of the unknown parameter θ^0. The maximum likelihood estimate $\hat{\theta}_n(X_1, \ldots, X_n)$ is obtained by considering the likelihood function

$$L_n(\theta) = L_n(X_1, \ldots, X_n; \theta) ,$$

the joint probability density of the potential observations $X_1, \ldots, X_n$. The maximum likelihood estimate $\hat{\theta}_n(X_1, \ldots, X_n)$ is that value of θ maximizing (assuming that such a maximum exists)

$$L_n(X_1, \ldots, X_n; \theta) .$$

In the case of independent, identically distributed random variables, under appropriate regularity conditions, the sequence of maximum likelihood estimates $\hat{\theta}_n(X_1, \ldots, X_n)$ have been shown to have certain asymptotic optimality properties as $n \to \infty$. It should be emphasized that these are asymptotic rather than finite sample optimality properties. The maximum likelihood estimator is shown to be the best consistent (in the sense of convergence in probability to the true parameter θ^0) continuously asymptotically normal estimator in the sense of having minimal asymptotic covariance matrix (see C. R. Rao [1973]). There are limited generalizations of the results in the independent, identically distributed case to dependent processes. Discussions of these generalizations can be found in Roussas [1972] (for stationary ergodic Markov chains) and Hall and Heyde [1980]. Many of the estimators in finite parameter time series analysis, though not literally maximum likelihood estimates, are suggested by consideration of maximum likelihood estimators in the case of normal processes.

Let us for the moment assume that $\{X_t\}$ is a stationary autoregressive normal process with mean zero, that is,

$$X_t = \sum_{j=1}^{k} \beta_j X_{t-j} + V_t \tag{1}$$

where the $\{V_t\}$ are independent identically distributed normal random variables with mean zero and variance $\sigma^2 > 0$. The object is to obtain estimators of the parameters β_s, $s = 0, 1, \ldots, k$, and σ^2. It has already been noted that the β_s's in the normal case are not uniquely determined by the joint probability structure of the X_t's. However, they are uniquely determined if we assume that the polynomial

$$\beta(z) = 1 - \sum_{j=1}^{k} \beta_j Z^j \tag{2}$$

has all its zeros outside the unit disc in the complex plane. Assume that this is the case. Then, the random variable V_t is orthogonal (and because of the normal context independent) of the past X_τ, $\tau < t$, of the $\{X_t\}$ process. Let $f(x_{1-k}, \ldots, x_0)$ be the joint density of the random variables $X_{1-k}, \ldots, X_0$. Of course, f will depend on the parameters β_s and σ^2 even though this is not explicitly indicated. Because of the assumptions on the roots of $\beta(z)$, the joint probability density of $X_{1-k}, \ldots, X_0, X_1, \ldots, X_n$ is

$$f(x_{1-k}, \ldots, x_0)\, (2\,\pi)^{-n/2}\, \sigma^{-n} \exp\left\{ -\frac{1}{2\,\sigma^2} \sum_{t=1}^{n} (x_t - \beta_1\, x_{t-1} - \cdots - \beta_k\, x_{t-k})^2 \right\}.$$

Even in this simple situation, it is complicated to write out the exact form of the maximum likelihood estimate. However, it is easy to maximize the conditional likelihood function of $X_1, \ldots, X_n$ given $X_{1-k}, \ldots, X_0$, that is, to maximize

$$(2\,\pi)^{-n/2}\, \sigma^{-n} \exp\left\{ -\frac{1}{2\,\sigma^2} \sum_{t=1}^{n} (X_t - \beta_1\, X_{t-1} - \cdots - \beta_k\, X_{t-k})^2 \right\}.$$

Differentiate the logarithm of this conditional density with respect to the parameters. The following system of equations in the estimates $b_1, \ldots, b_k, s^2$ of the unknown parameters $\beta_1, \ldots, \beta_k, \sigma^2$ is obtained

$$\sum_{t=1}^{n} X_t\, X_{t-j} - \sum_{i=1}^{k} b_i \sum_{t=1}^{n} X_{t-i}\, X_{t-j} = 0\,, \quad j = 1, \ldots, k\,,$$

$$s^2 = \frac{1}{n} \sum_{t=1}^{n} (X_t - b_1\, X_{t-1} - \cdots - b_k\, X_{t-k})^2\,.$$

Let

$$Y_j = n^{-1/2} \sum_{t=1}^{n} (X_t - \beta_1\, X_{t-1} - \cdots - \beta_k\, X_{t-k})\, X_{t-j}$$

$$= n^{-1/2} \sum_{t=1}^{n} V_t\, X_{t-j}\,.$$

Then Y_j can be rewritten as

$$Y_j = \sum_{l=1}^{k} n^{1/2}\, (b_l - \beta_l) \cdot \frac{1}{n} \sum_{t=1}^{n} X_{t-j}\, X_{t-l}\,, \quad j = 1, \ldots, k\,. \tag{3}$$

The equations (3), occasionally called the Yule-Walker equations, were obtained by a modification of a maximum likelihood argument under the assumption that the V_t's and hence the X_t's are normal random variables. Suppose that $\{X_t\}$ now is a stationary autoregressive process with the V_t's independent identically distributed nonnormal with mean zero and variance $\sigma^2 > 0$, and that all zeros of (2) were outside the unit disc in the complex plane. We shall still use the Yule-Walker equations to estimate the unknown parameters of the process in this nonnormal context even though they are not the equations one would be led to by an application of the conditional maximum likelihood procedure (assuming that one knew the density g of the V_t variables). At this point it will be assumed that the first four moments of the V_t variables exist. Let

$$D_{ijn} = \frac{1}{n} \sum_{t=1}^{n} X_{t-i} X_{t-j} .$$

The system of equations (3) can then be written

$$Y_i = \sum_{j=1}^{k} n^{1/2}(b_j - \beta_j) D_{ijn} . \tag{4}$$

Now

$$D_{ijn} \to r_{i-j} = \operatorname{cov}(X_i, X_j)$$

in probability as $n \to \infty$ and so the limiting matrix in probability as $n \to \infty$ of the system (4) is nonsingular since the determinant $|r_{i-j}; i, j = 1, \ldots, k| \neq 0$. The determinant is nonzero since the process $\{X_t\}$ has a strictly positive spectral density. The random variables Y_j, $j = 1, \ldots, k$ are asymptotically normally distributed with mean zero and covariance matrix

$$\sigma^2 \{r_{i-j}; \quad i, j = 1, \ldots, k\}$$

since

$$\operatorname{cov}(V_t X_{t-j}, V_\tau X_{\tau-l}) = \delta_{t,\tau} r_{t-\tau-j+l} \sigma^2 .$$

Solve the equations (4) for

$$n^{1/2}(b_i - \beta_i), \quad i = 1, \ldots, k , \tag{5}$$

in terms of the Y_j's. It is then clear that the random variables $n^{1/2}(b_i - \beta_i)$ asymptotically have mean zero and covariance matrix

$$\sigma^2 \{r_{i-j}; \quad i, j = 1, \ldots, k\}^{-1} \tag{6}$$

as $n \to \infty$. Also notice that

$$(7) \qquad s^2 = \frac{1}{n} \sum_{t=1}^{n} (X_t - \beta_1 X_{t-1} - \cdots - \beta_k X_{t-k} - (b_1 - \beta_1) X_{t-1} - \cdots - (b_k - \beta_k) X_{t-k})^2$$

$$= \frac{1}{n} \sum_{t=1}^{n} V_t^2 - \frac{2}{n} \sum_{t=1}^{n} V_t \{(b_1 - \beta_1) X_{t-1} + \cdots + (b_k - \beta_k) X_{t-k}\} + \frac{1}{n} \sum_{t=1}^{n} \{(b_1 - \beta_1) X_{t-1} + \cdots + (b_k - \beta_k) X_{t-k}\}^2$$

$$= \frac{1}{n} \sum_{t=1}^{n} V_t^2 + o\left(\frac{1}{n}\right).$$

Since

$$(8) \qquad E(\{V_t^2 - \sigma^2\} V_t) X_{t-i} = E(\{V_t^2 - \sigma^2\} V_t) E X_{t-i} = 0,$$

it follows from (7) and (8) that $n^{1/2}(s^2 - \sigma^2)$ is asymptotically uncorrelated with the Y_i's and hence with the random variables $n^{1/2}(b_j - \beta_j)$. Also $n^{1/2}(s^2 - \sigma^2)$ and the random variables $n^{1/2}(b_j - \beta_j)$ are jointly asymptotically normal with $n^{1/2}(s^2 - \sigma^2)$ having limiting mean o and variance $\mu_4 = E V_t^4$ as $n \to \infty$. We therefore have the following result.

Theorem 1. *Let $\{X_t\}$ be a stationary autoregressive process, that is, it satisfies* (1) *with the $\{V_t\}$ sequence independent, identically distributed random variables with mean $E V_t \equiv 0$, variance $\sigma^2 = E V_t^2 > 0$ and finite fourth moment $\mu_4 = E V_t^4 < \infty$. Assume that the polynomial* (2) *has all its roots outside the unit disc in the complex plane. Then the solutions b_i of the Yule-Walker equations* (3) *and s^2* (7) *provide estimates of the β_i and σ^2 and are such that*

$$n^{1/2}(b_i - \beta_i), \quad i = 1, \ldots, k,$$

and

$$n^{1/2}(s^2 - \sigma^2)$$

are jointly asymptotically normal as $n \to \infty$ with the random variables (5) *having limiting covariance matrix* (6), (7) *having limiting variance μ_4. Further the random variables* (5) *and* (7) *are asymptotically independent.*

Mann and Wald [1943] derived this result under broader conditions. First of all, they considered an autoregressive process with nonzero mean. Taking care of this requires a simple modification of the argument. They also allowed a nonstationary solution of the system (1) under the assumption that the roots of (2) are outside the unit circle in the complex plane. It can be shown that such a solution tends to a stationary solution as $t \to \infty$.

If some of the roots of the equation (2) are inside the unit disc in the complex plane, it is clear from the discussion in Section II. 3 that there is still a strictly stationary solution. However, if we are dealing with a non-stationary solution of (1), because of the instability of the recursive autoregressive system (1) when some of the roots of the equation (2) are inside the unit disc, the asymptotic behavior of the solution will be explosive.

The Yule-Walker equations are a linear system of equations. Later on we shall see that the equations one is led to, for example, in the case of moving average schemas on the basis of approximate maximum likelihood procedures (in the normal case) are nonlinear.

It is of some interest to look at the case of an autoregressive scheme

$$\sum_{k=0}^{p} \beta_k Y_{t-k} = V_t, \quad \beta_0 = 1,$$

where the polynomial $\sum_{j=0}^{p} \beta_j z^j$ has all its zeros outside $|z| \leq 1$ and the V_s's are independent, identically distributed random variables with common known non Gaussian density function f and mean zero and variance one. Admittedly the assumption that the density function f is known is an idealization. We shall consider conditional maximum likelihood estimation of the parameters β_k assuming f known and see how this compares with the procedure used earlier in which one carried out a computation as if the variables were Gaussian. Notice that here to ease notation the variance σ^2 is assumed to be one. The joint density function of $Y_0, \ldots, Y_n$ can then be written

$$g(y_{1-p}, \ldots, y_0) \prod_{s=1}^{n} f\left(\sum_{j=0}^{p} \beta_j y_{s-j}\right)$$

where it is understood that $g(y_0, \ldots, y_{p-1})$ is the joint density function of $y_0, \ldots, y_{p-1}$. Suppose that one can neglect $g(Y_0, \ldots, Y_{p-1})$ and just maximize

$$\sum_{t=1}^{n} \log f\left(\sum_{j=0}^{p} \beta_j Y_{t-j}\right). \tag{9}$$

So as to be able to carry out formal aspects of an argument, let us assume f positive, twice continuously differentiable, and $-\log f$ strictly convex. The argument surely holds under weaker conditions. If the derivatives of the logarithm of the conditional density (9) are taken with respect to the parameters $\beta_1, \ldots, \beta_p$ and set equal to zero, the following system of equations in estimates $b_1, \ldots, b_p$ of the parameters is obtained

$$\sum_{t=1}^{n} Y_{t-k} f'\left(\sum_{j=0}^{p} b_j Y_{t-j}\right) \Big/ f\left(\sum_{j=0}^{p} b_j Y_{t-j}\right) = 0, \quad k = 1, \ldots, p.$$

The solution $\mathbf{b}$ of this system of equations converges to $\boldsymbol{\beta}$ in probability as $n \to \infty$ by an argument like that given in the Lemmas of section 3 of this chapter. For convenience, the derivative will at times be indicated by D. Notice that second order partials of (9) are given by

$$\sum_{t=1}^{p} Y_{t-k}\, Y_{t-k'}\, D^2 \log f\left(\sum_{t=1}^{n} b_j\, Y_{t-j}\right).$$

The standard argument using a Taylor expansion about $\boldsymbol{\beta} = (\beta_j)$ with appropriate approximations suggests that

$$(10) \qquad -\left\{n^{-1} \sum_{k=1}^{p} \sum_{t=1}^{n} Y_{t-k}\, Y_{t-j}\, D^2 \log f(V_t)\right\} n^{1/2}(b_k - \beta_k)$$
$$\cong n^{-1/2} \sum_{t=1}^{n} Y_{t-j}\, D \log f(V_t)\,, \quad j = 1, \ldots, p\,.$$

Notice that

$$E\, D \log f(V_t) = E\{f'(V_t) \,/\, f(V_t)\} = \int f'(v)\, dv = 0\,.$$

Now if $j, k = 1, \ldots, p$,

$$E\,\{Y_{t-k}\, Y_{t-j}\, D^2 \log f(V_t)\}$$
$$= E\,\{Y_{t-k}\, Y_{t-j}\}\, E\,\{D^2 \log f(V_t)\}$$
$$= r_{j-k}\, E\,\{D^2 \log f(V_t)\}$$

because of the independence of V_t and Y_τ for $\tau < t$. Also

$$E\,\{D^2 \log f(V_t)\} = -\, E\,\{D \log f(V_t)\}^2\,.$$

The ergodicity of the Y_t sequence then implies that

$$n^{-1} \sum_{t=1}^{n} Y_{t-k}\, Y_{t-j}\, D^2 \log f(V_t)$$
$$\to -\, r_{j-k}\, E\,\{D \log f(V_t)\}^2\,,$$

$j, k = 1, \ldots, p$, as $n \to \infty$ with probability one. Further

$$E\,\{Y_{t-j}\, Y_{\tau-j'}\, D \log f(V_t)\, D \log f(V_\tau)\}\,,$$

$j, j' = 1, \ldots, p$, is zero if $\tau < t$ and is equal to

$$E\,\{Y_{t-j}\, Y_{t-j'}\}\, E\,\{D \log f(V_t)\}^2$$
$$= r_{j-j'}\, E\,\{D \log f(V_t)\}^2$$

if $t = \tau$. This indicates that the covariance matrix of the right hand side of relation (10) is

$$R\, E\,\{D \log f(V_t)\}^2\,.$$

The usual argument then implies that $n^{1/2}(b_k - \beta_k)$, $k = 1, \ldots, p$, are asymptotically jointly normal with mean zero and covariance matrix

$$R^{-1}\,(E\,\{D \log f(V_t)\}^2)^{-1}\,. \tag{11}$$

In the case of the estimates obtained by the argument of Mann and Wald the covariance matrix derived by the asymptotic derivation is

$$R^{-1}\,. \tag{12}$$

Now

$$1 = \int f(v)\,dv = -\int v f'(v)\,dv$$

and

$$\left|\int v f'(v)\,dv\right| = \left|\int v\sqrt{f(v)}\,\frac{f'(v)}{\sqrt{f(v)}}\,dv\right|$$

$$\leq \left\{\int v^2 f(v)\,dv \int \frac{\{f'(v)\}^2}{f(v)}\,dv\right\}^{1/2}.$$

This implies that

$$E(D \log f(V))^2 = \int \{f'(v)\}^2 / f(v)\,dv$$

$$\geq \int v^2 f(v)\,dv = 1\,.$$

Equality of (11) and (12) as expected is attained only in the Gaussian case.

In problem 5 one is asked to interpret the results of sections 1 and 3 for the stationary autoregressive sequence $\{X_t\}$

$$X_t = a\,X_{t-1} + V_t,\quad |\,a\,| < 1\,,$$

when the random variables V_t are independent and identically distributed with $E\,V_t \equiv 0$, $E\,V_t^2 \equiv 1$, $E\,V_t^4 < \infty$. The statistic

$$\hat{a}_N = \sum_{t=1}^{N} X_t\,X_{t-1} \Big/ \sum_{t=1}^{N} X_{t-1}^2$$

is considered as an estimate of a and one readily concludes that $N^{1/2}(\hat{a}_N - a)$ is asymptotically normally distributed as $N \to \infty$. It is of some interest to consider the nonstationary situation in which $|\,a\,| \geq 1$ and $X_0 = 0$. First notice that

$$\hat{a}_N - a = \sum_{t=1}^{N} V_t\,X_{t-1} \Big/ \sum_{t=1}^{N} X_{t-1}^2\,.$$

The case in which we have $|\,a\,| > 1$ will first be dealt with. Now

$$X_t = V_t + a\,V_{t-1} + \cdots + a^{t-1}\,V_1\,.$$

This implies that

$$\lim_{t \to \infty} \alpha^{-(t-1)} X_t = V_1 + \alpha^{-1} V_2 + \alpha^{-2} V_3 + \cdots \tag{13}$$
$$= U$$

Therefore

$$\lim_{N \to \infty} \alpha^{-2(N-2)} \sum_{t=1}^{N} X_{t-1}^2$$
$$= U^2 + \alpha^{-2} U^2 + \cdots = U^2 / (1 - \alpha^{-2}) .$$

Similarly the limiting distribution of

$$\alpha^{-(N-1)} \sum_{t=1}^{N} V_t X_{t-1}$$

as $N \to \infty$ is the same as that of

$$\{ V_N U + \alpha^{-1} V_{N-1} U + \cdots \}$$

and this indicates that the limiting distribution must be that of

$$U' U$$

where U and U' are independent and identically distributed with U given by (13). Thus the limiting distribution of

$$\alpha^{N-1} \{ \hat{\alpha}_N - \alpha \}$$

is the same as that of

$$\{ U' / U \} (1 - \alpha^{-2})^{-1} .$$

The case we have just considered is an exponentially explosive case because $| \alpha | > 1$. Let us now examine the boundary case $| \alpha | = 1$ in which one still has nonstationarity. Since the case $\alpha = -1$ is quite similar to $\alpha = 1$ we shall give a detailed discussion only for $\alpha = 1$. Now

$$\sum_{t=1}^{N} V_t X_{t-1} = \frac{1}{2} (V_1 + \cdots V_N)^2 - \frac{1}{2} (V_1^2 + \cdots + V_N^2)$$
$$= \frac{N}{2} \left(\frac{1}{\sqrt{N}} \{ V_1 + \cdots + V_N \} \right)^2 - \frac{N}{2}$$
$$- \sqrt{N} \frac{1}{\sqrt{N}} [V_1^2 - 1) + \cdots + (V_n^2 - 1)] .$$

and

$$\sum_{t=1}^{N} X_{t-1}^2 = N^2 \frac{1}{N} \sum_{t=1}^{N} \left(\frac{X_{t-1}}{\sqrt{N}} \right)^2 .$$

If $k/N \to \tau$, $0 \leq \tau \leq 1$, the distribution of

$$N^{-(1/2)} X_k = N^{-(1/2)} \{V_1 + \cdots + V_k\} \tag{14}$$

tends to that of $B(\tau)$ where $B(\cdot)$ is the Brownian motion process, that is, $B(\cdot)$ is the Gaussian process with mean zero and covariance

$$E\, B(\tau)\, B(\tau') = \min(\tau, \tau') .$$

The joint distribution of (14) for any finite number of values k tends to the joint distribution of the corresponding $B(\cdot)$ variables. This suggests that the joint distribution of

$$\left(\frac{N}{2}\right)^{-1} \sum_{t=1}^{N} V_t X_{t-1}$$

and

$$N^{-2} \sum_{t=1}^{N} X_{t-1}^2$$

tends to that of

$$B(1)^2 - 1$$

and

$$\int_0^1 B^2(u)\, du .$$

This can be verified by making use of a so-called "invariance" principle (see Billingsley [1968]). Thus, the asymptotic distribution of

$$(2\, N)\, \{\hat{a}_N - a\}$$

when $a = 1$ should be the same as that of

$$\{B^2(1) - 1\} \Big/ \int_0^1 B^2(u)\, du .$$

2. The Newton-Rafson Procedure and Gaussian ARMA Schemes

We have seen that a modification of the maximum likelihood procedure in the case of a stationary autoregressive Gaussian process leads to a linear system of equations for the parameter estimates. However, in the more general case of Gaussian ARMA schemes the maximum likelihood method will generally yield a system of nonlinear equations for the parameter estimates. For this reason it seemed appropriate to given a discussion of a method for solution of such a system of equations, specifically the Newton-Rafson method.

A Newton-Rafson procedure of first order is obtained by linearization of f in setting up an iteration to solve the system of equations

$$\mathbf{f}(\mathbf{x}) = \begin{bmatrix} f_1(x^1, \ldots, x^n) \\ \vdots \\ f_n(x^1, \ldots, x^n) \end{bmatrix} = 0 .$$

Let $\boldsymbol{\xi}$ be a zero of $\mathbf{f}$, $\mathbf{x}_0$ a neighboring point of $\boldsymbol{\xi}$ and $\mathbf{f}$ differentiable at $\mathbf{x} = \mathbf{x}_0$. We consider the approximation

$$0 = \mathbf{f}(\boldsymbol{\xi}) \approx \mathbf{f}(\mathbf{x}_0) + D\mathbf{f}(\mathbf{x}_0)\,(\boldsymbol{\xi} - \mathbf{x}_0)$$

with

$$D\mathbf{f}(\mathbf{x}_0) = \begin{bmatrix} \frac{\partial f_1}{\partial x^1} & \cdots & \frac{\partial f_1}{\partial x^n} \\ & \vdots & \\ \frac{\partial f_n}{\partial x^1} & \cdots & \frac{\partial f_n}{\partial x^n} \end{bmatrix}, \quad \boldsymbol{\xi} - \mathbf{x}_0 = \begin{bmatrix} \xi^1 - x_0^1 \\ \vdots \\ \xi^n - x_0^n \end{bmatrix} .$$

In searching for the minimum of a function g, one will look for a zero of $\mathbf{f}$, the gradient of g.

If $D\mathbf{f}(\mathbf{x}_0)$ is nonsingular we can solve

$$\mathbf{f}(\mathbf{x}_0) + D\mathbf{f}(\mathbf{x}_0)\,(\mathbf{x}_1 - \mathbf{x}_0) = 0$$

for $\mathbf{x}_1$ to get

$$\mathbf{x}_1 = \mathbf{x}_0 - \big(D\mathbf{f}(\mathbf{x}_0)\big)^{-1}\,\mathbf{f}(\mathbf{x}_0)$$

and recursively

$$\mathbf{x}_{i+1} = \mathbf{x}_i - \big(D\mathbf{f}(\mathbf{x}_i)\big)^{-1}\,\mathbf{f}(\mathbf{y}_i)\,, \quad i = 0, 1, \ldots .$$

Convergence of the procedure can be demonstrated by the following argument under appropriate conditions (see Stoer and Bulirsch [1980]). Use is made of the following lemma. Let $\|\cdot\|$ denote Euclidean distance in R^n.

Lemma 1. Assume that $D\mathbf{f}(\mathbf{x})$ *exists for all* $\mathbf{x} \in D_0$ *with* D_0 *a convex region in* R^n *and that there is a constant* γ *such that for all* $\mathbf{x}, \mathbf{y} \in D_0$

$$\| D\mathbf{f}(\mathbf{x}) - D\mathbf{f}(\mathbf{y}) \| \le \gamma \| \mathbf{x} - \mathbf{y} \| .$$

Then for all $\mathbf{x}, \mathbf{y} \in D_0$ *one has*

$$\| \mathbf{f}(\mathbf{x}) - \mathbf{f}(\mathbf{y}) - D\mathbf{f}(\mathbf{y})\,(\mathbf{x} - \mathbf{y}) \| \le \frac{\gamma}{2} \| \mathbf{x} - \mathbf{y} \|^2 .$$

Proof of the Lemma. Consider $\boldsymbol{\varphi}\colon [0, 1] \to R^n$

$$\boldsymbol{\varphi}(t) = \mathbf{f}\big(\mathbf{y} + t(\mathbf{x} - \mathbf{y})\big) .$$

This is differentiable for $0 \leq t \leq 1$ given any $\mathbf{x}, \mathbf{y} \in D_0$ and

$$\boldsymbol{\varphi}'(t) = D\mathbf{f}(\mathbf{y} + t(\mathbf{x} - \mathbf{y}))\,(\mathbf{x} - \mathbf{y})\,.$$

Then

$$\begin{aligned} \| \boldsymbol{\varphi}'(t) - \boldsymbol{\varphi}'(0) \| &= \| (D\mathbf{f}(\mathbf{y} + t(\mathbf{x} - \mathbf{y})) - D\mathbf{f}(\mathbf{y}))\,(\mathbf{x} - \mathbf{y}) \| \\ &\leq \| D\mathbf{f}(\mathbf{y} + t(\mathbf{x} - \mathbf{y})) - D\mathbf{f}(\mathbf{y}) \| \; \| \mathbf{x} - \mathbf{y} \| \\ &\leq \gamma\, t \| \mathbf{x} - \mathbf{y} \|^2 . \end{aligned}$$

Consequently

$$\begin{aligned} \boldsymbol{\Delta} &= \mathbf{f}(\mathbf{x}) - \mathbf{f}(\mathbf{y}) - D\mathbf{f}(\mathbf{y})\,(\mathbf{x} - \mathbf{y}) = \boldsymbol{\varphi}(1) - \boldsymbol{\varphi}(0) - \boldsymbol{\varphi}'(0) \\ &= \int_0^1 (\boldsymbol{\varphi}'(t) - \boldsymbol{\varphi}'(0))\, dt \end{aligned}$$

and so we have

$$\begin{aligned} \| \boldsymbol{\Delta} \| &\leq \int_0^1 \| \boldsymbol{\varphi}'(t) - \boldsymbol{\varphi}'(0) \|\, dt \leq \gamma \| \mathbf{x} - \mathbf{y} \|^2 \int_0^1 t\, dt \\ &= \frac{\gamma}{2} \| \mathbf{x} - \mathbf{y} \|^2 . \end{aligned}$$

From this one can obtain the following result.

Theorem 2. *Let D be an open set in R^n with a convex subset $D_0 \subset D$ such that f: $D \to \mathbb{R}_n$ is for all $\mathbf{x} \in D_0$ a differentiable and for all $\mathbf{x} \in D$ continuous function.*

For a point $\mathbf{x}_0 \in D_0$ let there be positive constants $r, \alpha, \beta, \gamma, h$ with the following properties:

$$\begin{aligned} S_r(\mathbf{x}_0) &= \{\mathbf{x} \colon \| \mathbf{x} - \mathbf{x}_0 \| < r\} \subset D_0 \\ h &= \alpha \beta \gamma / 2 < 1 \\ r &= \alpha / (1 - h)\,. \end{aligned}$$

Let $\mathbf{f}$ satisfy

(a) $\| D\mathbf{f}(\mathbf{x}) - D\mathbf{f}(\mathbf{y}) \| \leq \gamma \| \mathbf{x} - \mathbf{y} \|$ *for all* $\mathbf{x}, \mathbf{y} \in D_0$.

(b) $D\mathbf{f}(\mathbf{x})^{-1}$ *exists and* $\| D\mathbf{f}(\mathbf{x})^{-1} \| \leq \beta$ *for all* $\mathbf{x} \in D_0$.

(c) $\| D\mathbf{f}(\mathbf{x}_0)^{-1} \mathbf{f}(\mathbf{x}_0) \| \leq \alpha$.

Then

(A) *starting with* $\mathbf{x}_0$

$$\mathbf{x}_{k+1} = \mathbf{x}_k - D\mathbf{f}(\mathbf{x}_k)^{-1}\,\mathbf{f}(\mathbf{x}_k)\,, \quad k = 0, 1, \ldots$$

is well-defined and $\mathbf{x}_k \in S_r(\mathbf{x}_0)$ *for all* $k > 0$.

(B) $\lim_{k\to\infty} \mathbf{x}_k = \boldsymbol{\xi}$ *exists and*

$$\boldsymbol{\xi} \in S_r(\mathbf{x}_0) \quad \text{and} \quad \mathbf{f}(\boldsymbol{\xi}) = 0\,.$$

(C) *For all* $k \geq 0$

$$\| \mathbf{x}_k - \boldsymbol{\xi} \| \leq \alpha \frac{h^{2^k-1}}{1 - h^{2^k}}\,.$$

Proof. $D\mathbf{f}(\mathbf{x})^{-1}$ exists for $\mathbf{x} = \mathbf{x}_0,\ \mathbf{x}_1$ by the assumptions. Then if $\mathbf{x}_j \in S_r(\mathbf{x}_0)$, $j = 0, 1, \ldots, k$

$$\begin{aligned}\| \mathbf{x}_{k+1} - \mathbf{x}_k \| &= \| -D\mathbf{f}(\mathbf{x}_k)^{-1}\,\mathbf{f}(\mathbf{x}_k) \| \leq \beta\,\| \mathbf{f}(\mathbf{x}_k) \| \\ &= \beta\,\| \mathbf{f}(\mathbf{x}_k) - \mathbf{f}(\mathbf{x}_{k-1}) - D\mathbf{f}(\mathbf{x}_{k-1})\,(\mathbf{x}_k - \mathbf{x}_{k-1}) \|\end{aligned}$$

and by the lemma

$$\| \mathbf{x}_{k+1} - \mathbf{x}_k \| \leq \frac{\beta\,\gamma}{2}\,\| \mathbf{x}_k - \mathbf{x}_{k-1} \|^2\,.$$

We wish to show

$$\| \mathbf{x}_{k+1} - \mathbf{x}_k \| \leq \alpha\,h^{2^k-1}\,.$$

It holds for $k = 0$ by (c). If it is valid for $k \geq 0$ then

$$\| \mathbf{x}_{k+1} - \mathbf{x}_k \| \leq \frac{\beta\,\gamma}{2}\,\| \mathbf{x}_k - \mathbf{x}_{k-1} \|^2 \leq \frac{\beta\,\gamma}{2}\,\alpha^2\,h^{2^k-2} = \alpha\,h^{2^k-1}\,.$$

Also

$$\begin{aligned}\| \mathbf{x}_{k+1} - \mathbf{x}_0 \| &\leq \| \mathbf{x}_{k+1} - \mathbf{x}_k \| + \cdots + \| \mathbf{x}_1 - \mathbf{x}_0 \| \\ &\leq \alpha(1 + h + h^3 + h^7 + \cdots + h^{2^k-1}) \\ &\leq \alpha\,/\,(1 - h) = r\end{aligned}$$

so $\mathbf{x}_{k+1} \in S_r(\mathbf{x}_0)$. $\{\mathbf{x}_k\}$ is a Cauchy sequence since

$$\begin{aligned}\| \mathbf{x}_{m+1} - \mathbf{x}_n \| &\leq \| \mathbf{x}_{m+1} - \mathbf{x}_m \| + \| \mathbf{x}_m - \mathbf{x}_{m-1} \| + \cdots + \| \mathbf{x}_{n+1} - \mathbf{x}_n \| \\ &\leq \alpha\,h^{2^n-1}\big(1 + h^{2^n} + (h^{2^n})^2 + \cdots\big) \leq \alpha\,\frac{h^{2^n-1}}{1 - h^{2^n}} < \varepsilon\end{aligned}$$

if n is large enough. Also $\boldsymbol{\xi} = \lim_{k\to\infty} \mathbf{x}_k \in \overline{S_r(\mathbf{x}_0)}$ the closure of $S_r(\mathbf{x}_0)$. Further

$$\lim_{m\to\infty} || \mathbf{x}_m - \mathbf{x}_n || = || \boldsymbol{\xi} - \mathbf{x}_n || \leq \frac{\alpha\, h^{2^n-1}}{1 - h^{2^n}} .$$

Because of (a)

$$|| D\mathbf{f}(\mathbf{x}_k) - D\mathbf{f}(\mathbf{x}_0) || \leq \gamma \, || \mathbf{x}_k - \mathbf{x}_0 || < \gamma\, r$$

and hence

$$|| D\mathbf{f}(\mathbf{x}_k) || \leq \gamma\, r + || D\mathbf{f}(\mathbf{x}_0) || = k .$$

From

$$\mathbf{f}(\mathbf{x}_k) = - D\mathbf{f}(\mathbf{x}_k)\, (\mathbf{x}_{k+1} - \mathbf{x}_k)$$

follows $|| \mathbf{f}(\mathbf{x}_k) || \leq k \, || \mathbf{x}_{k+1} - \mathbf{x}_k ||$ so that $\lim_{k\to\infty} || \mathbf{f}(\mathbf{x}_k) || = 0$.

The use of the Newton-Rafson method in a minimization problem requires the computation of the matrix of second partial derivatives, the Hessian, at each step of the computation. In many problems this is too complicated and quasi-Newton methods replacing computation of the Hessian by easier matrices have been proposed (see Kennedy and Gentle [1980] and Stoer and Bulirsch [1980]).

Assume that one has an autoregressive moving average process $\{Y_t\}$ satisfying

$$\sum_{k=0}^{p} \beta_k\, Y_{t-k} = \sum_{g=0}^{q} a_g\, V_{t-g}$$

that is stationary Gaussian, and that the V_t's are independent Gaussian. Let

$$A(z) = \sum_{g=0}^{q} a_g\, z^g$$

$$B(z) = \sum_{k=0}^{p} \beta_k\, z^k .$$

Also assume that the roots of $A(z) = 0$, $B(z) = 0$ have modulus greater than one and that there are no roots in common. Because of the Gaussian assumption one can't distinguish between roots inside the unit circle and roots outside. If the V_t's were independent and nonGaussian one could distinguish between roots inside the unit disc and roots outside.

We shall now give a discussion modeled on that of T. W. Anderson [1977]. Basically simple variations (or modifications) of maximum likelihood estimates will be used. The modifications are required because attempting to solve for the exact maximum likelihood estimate is too difficult. The first modification runs as follows. Let L be $T \times T$, I of order $T - 1$

$$L = \begin{bmatrix} 0 & 0 \\ I & 0 \end{bmatrix}.$$

Then the T vectors $\mathbf{Y} = \begin{pmatrix} Y_1 \\ \vdots \\ Y_T \end{pmatrix}$ and $\mathbf{V} = \begin{pmatrix} V_1 \\ \vdots \\ V_T \end{pmatrix}$ are to satisfy

$$\sum_{k=0}^{p} \beta_k L^k \mathbf{Y} = \sum_{g=0}^{q} \alpha_g L^g \mathbf{V}$$

where $\mathbf{V}$ has the distribution $N(0, \sigma^2 I)$. This corresponds to setting $Y_0 = Y_{-1} = \cdots = Y_{1-p} = 0$ and $V_0 = V_{-1} = \cdots = V_{1-q} = 0$.

A second modified model is circular. Let M be the circulant matrix

$$M = \begin{pmatrix} 0 & 1 \\ I & 0 \end{pmatrix}$$

with I of order $T - 1$. The circular model for $\mathbf{Y}$ is then

$$\sum_{k=0}^{p} \beta_k M^k \mathbf{Y} = \sum_{g=0}^{q} \alpha_g M^g \mathbf{V} .$$

Here $Y_{-k} = Y_{T-k}$, $k = 0, 1, \ldots, p - 1$ and $V_{-g} = V_{T-g}$, $g = 0, \ldots, q$. The covariance matrix of $\mathbf{Y}$ is

$$E\,\mathbf{Y}\,\mathbf{Y}' = \sigma^2 B(M)^{-1} A(M)\, A(M')\, B(M')^{-1} .$$

Notice that $M' = M^{-1} = M^{T-1}$. The unitary matrix

$$U = (T^{-1/2} \exp \{i\, 2\, \pi\, t\, s\, /\, T\})$$

diagonalizes M since

$$MU = U D$$

where D is the diagonal matrix with sth diagonal entry $\exp\{- i\, 2\, \pi\, s/T\}$. The quadratic form in the density of $\mathbf{Y}$ is $-1/2$ times

$$\frac{1}{\sigma^2}\, \mathbf{Y}'\, U\, B(\overline{D})\, A^{-1}(\overline{D})\, A^{-1}(D)\, B(D)\, \overline{U}'\, \mathbf{Y} .$$

If we set $z = U'\,\mathbf{Y}$ then the periodogram

$$I(\lambda_t) = \frac{1}{2\,\pi} \mid z_t \mid^2, \quad \lambda_t = \frac{2\,\pi\, t}{T},$$

and the exponent in the density is

$$\text{(15)} \qquad \frac{1}{\sigma^2} \sum_{t=1}^{T} \mid z_t \mid^2 \frac{\mid B(e^{i\lambda_t}) \mid^2}{\mid A(e^{i\lambda_t}) \mid^2} = \sum_{t=1}^{T} \frac{I(\lambda_t)}{f(\lambda_t)} .$$

The logarithm of the modified likelihood function in the case of both models is

$$\log L = -\frac{1}{2} T \log 2\pi - \frac{1}{2} T \log \sigma^2 + \log |B| - \log |A| - \frac{1}{2\sigma^2} \mathbf{Y}' B' A'^{-1} A^{-1} B \mathbf{Y} \tag{16}$$

where $A = \sum_{g=0}^{q} \alpha_g J_g$, $B = \sum_{k=0}^{p} \beta_k J_k$ have positive determinants and J_g is L^g or M^g. We note that $\partial \log |A| / \partial a_h = \operatorname{tr} \{A^{-1} \partial A / \partial a_h\}$ because

$$\frac{\partial \log |A|}{\partial \alpha_h} = \frac{1}{|A|} \frac{\partial |A|}{\partial \alpha_h},$$

$$\frac{\partial |A|}{\partial \alpha_h} = \sum_{i,j}^{n} A_{i,j} \frac{\partial a_{i,j}}{\partial \alpha_h}$$

and $A^{-1} = (A_{j,i} / |A|; i, j = 1, \ldots, n)$. Also

$$\begin{aligned}
\frac{\partial A^{-1}}{\partial \alpha} &= \lim_{h \downarrow 0} \frac{A^{-1}(\alpha + h) - A^{-1}(\alpha)}{h} \\
&= \lim_{h \downarrow 0} \frac{A^{-1}(\alpha)\, A(\alpha)\, A^{-1}(\alpha + h) - A^{-1}(\alpha)}{h} = \lim_{h \downarrow 0} \frac{A^{-1}(\alpha)\, [A(\alpha)\, A^{-1}(\alpha + h) - I]}{h} \\
&= \lim_{h \downarrow 0} \frac{A^{-1}(\alpha)\, [A(\alpha) - A(\alpha + h)]\, A^{-1}(\alpha + h)}{h} \\
&= - A^{-1} \frac{\partial A}{\partial \alpha} A^{-1}.
\end{aligned}$$

By making use of these expressions for the derivatives, on differentiating the expression (16) one obtains

$$\frac{\partial}{\partial \alpha_g} \log L = - \operatorname{tr} A^{-1} J_g + \frac{1}{\sigma^2} \mathbf{Y}' B' A'^{-1} A^{-1} J_g A^{-1} B \mathbf{Y}, \quad g = 1, \ldots, q$$

$$\frac{\partial}{\partial \beta_k} \log L = \operatorname{tr} B^{-1} J_k - \frac{1}{\sigma^2} \mathbf{Y}' B' A'^{-1} A^{-1} J_k \mathbf{Y}, \quad k = 1, \ldots, p$$

$$\frac{\partial}{\partial \sigma^2} \log L = - \frac{T}{2\sigma^2} + \frac{1}{2\sigma^4} \mathbf{Y}' B' A'^{-1} A^{-1} B \mathbf{Y}.$$

Generally (except under appropriate modification in the autoregressive case) these equations will be nonlinear.

Several iterative methods of solving the likelihood equations numerically are based on a Taylor's expansion

$$\frac{\partial}{\partial \theta} \log L(y \mid \theta) = \frac{\partial}{\partial \theta} \log L(y \mid \theta)_{|\theta = \theta_0} + \frac{\partial^2}{\partial \theta\, \partial \theta'} \log L(y \mid \theta)_{|\theta = \theta_0} (\theta - \theta_0) + R(y \mid \theta, \theta_0).$$

Notice that $\partial / \partial \theta$ and $\partial^2 / \partial \theta\, \partial \theta'$ represent a vector and matrix of derivatives.

In the Newton-Rafson method the right side of the equation and $R(y \mid \theta, \theta_0)$ are set equal to zero. At the ith step of the iteration one solves for θ in terms of θ_0. In the method of scoring the matrix of second derivatives is replaced by

$$\left[E_\theta \frac{\partial^2}{\partial\theta\,\partial\theta'} \log L\,(y \mid \theta) \right]_{\theta=\theta_0}$$

and the iteration solves the equations

$$-\left[E_\theta \frac{\partial^2}{\partial\theta\,\partial\theta'} \log L(y \mid \theta) \right]_{\theta=\theta_0} (\theta - \theta_0) = \frac{\partial}{\partial\theta} \log L(y \mid \theta)_{|\theta=\theta_0}$$

for θ.

Let $\hat{\boldsymbol{\theta}}_i' = (\hat{\alpha}_i, \hat{\beta}_i, \hat{\sigma}_i^2)$ be the vector of estimates determined at the ith iteration. The equations for $\hat{\alpha}_i, \hat{\beta}_i$ at the kth stage are

$$\begin{bmatrix} \hat{\Phi}_{i-1} & \hat{\Omega}_{i-1} \\ \hat{\Omega}_{i-1} & \hat{\Psi}_{i-1} \end{bmatrix} \begin{bmatrix} \hat{\alpha}_i - \hat{\alpha}_{i-1} \\ \hat{\beta}_i - \hat{\beta}_{i-1} \end{bmatrix} = \begin{bmatrix} \hat{q}_{i-1} \\ \hat{p}_{i-1} \end{bmatrix} \tag{17}$$

where the matrix on the left is an estimate of the information matrix for α, β based on the $(j-1)$st estimates and the vector on the right consists of estimates of $\partial \log L/\partial\alpha_g$, $\partial \log L/\partial\beta_k$. The estimate $\hat{\sigma}_i^2$ is a function of y and $\hat{\theta}_{i-1}$.

One can rewrite the Newton-Rafson equations as

$$-\frac{1}{T}\frac{\partial^2 \log L}{\partial\theta\,\partial\theta'}\bigg|_{\hat{\theta}_{i-1}} T^{1/2}(\hat{\boldsymbol{\theta}}_i - \boldsymbol{\theta})$$

$$= T^{-1/2}\frac{\log L}{\partial\theta}\bigg|_{\hat{\theta}_{i-1}} - \frac{1}{T}\frac{\partial^2 \log L}{\partial\theta\,\partial\theta'}\bigg|_{\hat{\theta}_{i-1}} T^{1/2}(\hat{\boldsymbol{\theta}}_{i-1} - \boldsymbol{\theta})\,.$$

In computing various expressions one will often drop terms which are negligible in probability asymptotically.

Let

$$\hat{v}_{i-1} = \hat{A}_{i-1}^{-1}\hat{B}_{i-1}y = \hat{B}_{i-1}\hat{A}_{i-1}^{-1}y\,.$$

In the Newton-Rafson and scoring methods the components on the right side of (17) are taken as

$$[\hat{q}_{i-1}]_g = \frac{\hat{v}_{i-1}' L^g \hat{A}_{i-1}^{-1}\hat{v}_{i-1}}{\hat{\sigma}_{i-1}^2}$$

$$[\hat{p}_{i-1}]_k = -\frac{\hat{v}_{i-1}' L^k \hat{B}_{i-1}^{-1}\hat{v}_{i-1}}{\hat{\sigma}_{i-1}^2} = -\frac{\hat{v}_{i-1}' L^k \hat{A}_{i-1}^{-1} y}{\hat{\sigma}_{i-1}^2}$$

in the case of the first model.

In the Newton-Rafson method

$$[\hat{\Phi}_{i-1}]_{gf} = \frac{\hat{v}'_{i-1}\,\hat{A}'^{-1}_{i-1}\,L'^{g}\,L^{f}\,\hat{A}^{-1}_{i-1}\,\hat{v}_{i-1}}{\hat{\sigma}^2_{i-1}}$$

$$[\hat{\Omega}_{i-1}]_{gl} = -\,\frac{\hat{v}'_{i-1}\,\hat{A}'^{-1}_{i-1}\,L'^{g}\,L^{l}\,\hat{B}^{-1}_{i-1}\,\hat{v}_{i-1}}{\hat{\sigma}^2_{i-1}} = -\,\frac{\hat{v}'_{i-1}\,\hat{A}'^{-1}_{i-1}\,L'^{g}\,L^{l}\,\hat{A}_{i-1}\,y}{\hat{\sigma}^2_{i-1}}$$

$$[\hat{\Psi}_{i-1}]_{kl} = \frac{\hat{v}'_{i-1}\,\hat{B}'^{-1}_{i-1}\,L'^{k}\,L^{l}\,\hat{B}^{-1}_{i-1}\,\hat{v}_{i-1}}{\hat{\sigma}^2_{i-1}} = \frac{y'\,\hat{A}'^{-1}_{i-1}\,L'^{k}\,L^{l}\,\hat{A}^{-1}_{i-1}\,y}{\hat{\sigma}^2_{i-1}}$$

while the equation for $\hat{\sigma}^2_i$ is

$$\left[2\,\frac{\hat{v}'_{i-1}\,\hat{v}_{i-1}}{\hat{\sigma}^2_{i-1}} - T\right]\hat{\sigma}^2_i = 2\,\hat{v}'_{i-1}\,\hat{v}_{i-1} - 2\,T\,\hat{\sigma}^2_{i-1}\,.$$

In the equations given above

$$\hat{v}_{i-1} = \hat{A}^{-1}_{i-1}\,B_{i-1}\,y = B_{i-1}\,A^{-1}_{i-1}\,y\,.$$

In the method of scoring

$$[\hat{\Phi}_{i-1}]_{gf} = \operatorname{tr}\hat{A}'^{-1}_{i-1}\,L'^{g}\,L^{f}\,\hat{A}^{-1}_{i-1}$$

$$[\hat{\Omega}_{i-1}]_{gl} = -\operatorname{tr}\hat{A}'^{-1}_{i-1}\,L'^{g}\,L^{l}\,\hat{B}^{-1}_{i-1}$$

$$[\hat{\Psi}_{i-1}]_{kl} = \operatorname{tr}\hat{B}'^{-1}\,L'^{k}\,L^{l}\,\hat{B}^{-1}_{i-1}\,.$$

We shall now briefly indicate the corresponding expression for the Newton-Rafson procedure in the frequency domain for the second model. The right hand sides of (17) are

$$[\hat{g}_{i-1}]_g = \sum_{t=1}^{T}\frac{I(\lambda_t)\,e^{i\lambda_t g}}{\hat{f}_{i-1}(\lambda_t)\,\hat{A}_{i-1}(e^{i\lambda_t})}$$

$$[\hat{p}_{i-1}]_k = -\sum_{t=1}^{T}\frac{I(\lambda_t)\,e^{i\lambda_t k}}{\hat{f}_{i-1}(\lambda_t)\,\hat{B}_{i-1}(e^{i\lambda_t})}$$

while the components of the matrix on the left hand side are

$$[\hat{\Phi}_{i-1}]_{gf} = \sum_{t=1}^{T}\frac{I(\lambda_t)\,e^{i\lambda_t g}\,e^{-i\lambda_t f}}{\hat{f}_{i-1}(\lambda_t)\,|\,\hat{A}_{i-1}(e^{i\lambda_t})\,|^2}$$

$$[\hat{\Omega}_{i-1}]_{gl} = -\sum_{t=1}^{T}\frac{I(\lambda_t)\,e^{i\lambda_t g}\,e^{-i\lambda_t l}}{\hat{f}_{i-1}(\lambda_t)\,\hat{A}_{i-1}(e^{i\lambda_t})\,\hat{B}_{i-1}(e^{-i\lambda_t})}$$

$$[\hat{\Psi}_{i-1}]_{kl} = \sum_{t=1}^{T}\frac{I(\lambda_t)\,e^{i\lambda_t k}\,e^{-i\lambda_t l}}{\hat{f}_{i-1}(\lambda_t)\,|\,\hat{B}_{i-1}(e^{i\lambda_t})\,|^2}$$

The equation for $\hat{\sigma}_i^2$ is

$$\left[2 \sum_{t=1}^{T} \frac{I(\lambda_t)}{\hat{f}_{i-1}(\lambda_t)} - T\right] \sigma_i^2 = \left[3 \sum_{t=1}^{T} \frac{I(\lambda_t)}{\hat{f}_{i-1}(\lambda_t)} - 2\,T\right] \hat{\sigma}_{i-1}^2 .$$

3. Asymptotic Properties of Some Finite Parameter Estimates

In this section we shall derive some results of P. Whittle [1954] under somewhat more general assumptions by using an appropriate modification of arguments of A.M. Walker [1964].

If X_t, $t = \ldots, -1, 0, 1, \ldots$ is a weakly stationary process, $E\,X_t \equiv 0$, with absolutely continuous spectrum and spectral density $f(\lambda)$, the one-step prediction error variance for the process is given by

$$\sigma^2 = 2\,\pi \,\exp\left\{\frac{1}{2\,\pi} \int_{-\pi}^{\pi} \log f(\lambda)\, d\lambda\right\}.$$

Assume that $\log f(\lambda) \in L$ so that $\sigma^2 > 0$ and let

$$g(\lambda) = \left(\frac{\sigma^2}{2\,\pi}\right)^{-1} f(\lambda)\,. \tag{18}$$

We now suppose that the spectral density f and hence that g is a function of a finite number of unknown parameters $\theta_1, \ldots, \theta_p$. The object is to obtain estimates of the parameters from n consecutive observations on the process $X_1, \ldots, X_n$. At times to emphasize the dependence of g on the parameters we shall write $g = g(\lambda, \theta)$.

First consider the case of $\{X_t\}$ a normal process. The likelihood function $L_n(\theta, \sigma^2)$ of the process is given by

$$\frac{1}{n} \log L_n(\theta, \sigma^2) = -\frac{1}{2}\frac{1}{n} \log |\,B_n(\theta)\,| - \frac{1}{2} \log 2\,\pi\,\sigma^2 - (2\,n\,\sigma^2)^{-1}\, Q_n(x, \theta) \tag{19}$$

where

$$B_n(\theta) = \left(b_{r-s}(\theta)\,;\ \ r, s = 1, \ldots, n\right)$$

$$b_{r-s}(\theta) = E(X_r\, X_s)\,/\,\sigma^2$$

$$= \frac{1}{2\,\pi} \int_{-\pi}^{\pi} e^{i\lambda(r-s)}\, g(\lambda, \theta)\, d\lambda$$

and

$$Q_n(x, \theta) = x'\,\{B_n(\theta)\}^{-1}\, x$$

where $x' = (x_1, \ldots, x_n)$. As noted earlier, it is difficult to obtain the maximum likelihood estimates in this general normal context. Certain approximations have been suggested by Whittle. First of all if f is bounded

$$\frac{1}{n} \log | B_n(\theta) | \to \frac{1}{2\pi} \int_{-\pi}^{\pi} \log \left\{ \left(\frac{\sigma^2}{2\pi} \right)^{-1} f(\lambda) \right\} d\lambda = 0$$

as $n \to \infty$ (see problem 12 of Chapter II). This suggests that if the sample size n is sufficiently large, it would make only a small difference if one maximized $-\frac{1}{2} \log 2\pi\sigma^2 - Q_n(X, \theta)(2n\sigma^2)^{-1}$ instead of (19). Further, if g is continuous and bounded away from zero, for large n, $\frac{1}{n} Q_n$ is well approximated by

$$U_n(X, \theta) = \frac{1}{2\pi n} \int_{-\pi}^{\pi} \left| \sum_{t=1}^{n} X_t e^{i\lambda t} \right|^2 g(\lambda, \theta)^{-1} d\lambda$$

$$= \sum_{j=-n+1}^{n-1} \gamma_j(\theta) r_j(n)$$

with

$$r_j(n) = \frac{1}{n} \sum_{t=1}^{n-|j|} X_t X_{t+j}$$

the sample covariance with lag j and

$$\gamma_j(\theta) = \frac{1}{2\pi} \int_{-\pi}^{\pi} e^{i\lambda j} \{g(\lambda, \theta)\}^{-1} d\lambda$$

(see problems 3 and 4). This suggests obtaining estimates of θ_j, $j = 1, \ldots, p$, and σ^2 by maximizing

(20) $$-\frac{1}{2} \log 2\pi\sigma^2 - (2\sigma^2)^{-1} U_n(X, \theta) .$$

The following set of assumptions will be made throughout this section:

Assumptions:

(1) *The process X_t should satisfy the assumptions of Corollary 2 of Chapter III. The assumptions of this Theorem could be replaced by corresponding ones involving strong mixing.*

(2) *The estimates $\hat{\theta}_n$, $\hat{\sigma}_n^2$ are such that* (20) *is an absolute maximum for $\theta = \hat{\theta}_n$, $\sigma^2 = \hat{\sigma}_n^2$. It is possible that there may be more than one set of* estimates maximizing (20).

(3) *The actual values θ, σ^2 of the parameters lie in a region $0 < \sigma^2 < \infty$, $\theta \in \Theta$ where Θ is a bounded closed set in an open set S of p dimensional Euclidean space.*

(4) *For* $\theta \in S$, $g(\lambda, \theta)$ *and* $\{g(\lambda, \theta)\}^{-1}$ *are continuous functions of* λ.

(5) *Given any two distinct points* $\theta_1, \theta_2 \in \Theta$, $g(\lambda, \theta_1)$ *and* $g(\lambda, \theta_2)$ *are not identically equal.*

One should note that in the case of ARMA normal stationary processes the assumption that the roots of $A(z) = \sum_{j=0}^{p} \alpha_j z^j$, $B(z) = \sum_{j=0}^{q} \beta_j z^j$ are outside the unit circle is made so that condition (5) will be satisfied. In the normal case one cannot, in estimating the coefficients α_j and β_k of an ARMA scheme distinguish between coefficients arising from roots outside the unit disc and corresponding sets of coefficients of appropriate polynomials with some roots inside the unit disc (see sections 3 and 4 of Chapter II). In the normal case the complete probability structure is determined by the spectral density of the process. *If the random variables* V_t *are independent and identically distributed but not normal, one can distinguish between roots outside the unit disc and nonzero roots inside the unit disc.* But for this one requires infoımation on moments of the process that are of higher order than the second.

Consider the moving average

$$X_t = \frac{1}{6} V_t - \frac{5}{6} V_{t-1} + V_{t-2}$$

with the random variables V_t independent exponential random variables. The corresponding polynomial $z^2 - \frac{5}{6} z^2 + \frac{1}{6}$ has zeros $\frac{1}{2}$ and $\frac{1}{3}$. This process has the same spectıal density as the moving average

$$Y_t = V_t - \frac{5}{6} V_{t-1} + \frac{1}{6} V_{t-2}$$

but a different probability structure. The polynomial corresponding to this second process has zeros 2 and 3. The techniques discussed in this section cannot distinguish between these two processes. In fact, if the methods were applied to the initial process $\{X_t\}$ one would typically estimate the zeros as outside the unit disc in the complex plane or the coefficients corresponding to these zeros. In Chapter VIII section 1 methods that can distinguish between $\{X_t\}$ and $\{Y_t\}$ will be implemented. The condition that the polynomial of such a scheme have all zeros outside the unit disc is called a *minimum phase condition.* In a variety of geophysical problems non-Gaussian processes that do not satisfy a minimum phase condition are encountered (see Donoho [1981] and Wiggins [1978]).

Under appropriate further conditions we shall show that

(i) $\hat{\theta}_n$ is a consistent estimate of θ, that is, $\lim_{n \to \infty} \hat{\theta}_n = \theta$ in probability.

(ii) Let

$$W = (w_{j,k}; \quad j, k = 1, \ldots, p)$$

and

$$Q = (q_{j,k}; \quad j, k = 1, \ldots, p)$$

be the matrices with elements

$$(21) \qquad w_{j,k} = \frac{1}{4\pi} \int_{-\pi}^{\pi} \frac{\partial \log g(\lambda, \theta)}{\partial \theta_j} \frac{\partial \log g(\lambda, \theta)}{\partial \theta_k} \, d\lambda$$

and

$$(22) \quad q_{j,k} = \left\{ 4\sigma^4 w_{j,k} + 2\pi \int_{-\pi}^{\pi} \int_{-\pi}^{\pi} q(\lambda, \mu, -\mu) \frac{\partial}{\partial \theta_j} g^{-1}(\lambda, \theta) \frac{\partial}{\partial \theta_k} g^{-1}(\mu, \theta) \, d\lambda \, d\mu \right\}$$

respectively where $q(\lambda, \mu, \eta)$ is the fourth order cumulant spectral density of the process $\{X_t\}$. We shall show that $n^{1/2}(\hat{\theta}_n - \theta)$ is asymptotically normal with mean zero and covariance matrix

$$(23) \qquad (4\sigma^4)^{-1} W^{-1} Q W^{-1} .$$

If the process $\{X_t\}$ is a linear process, the matrix (21) is equal to W^{-1}.

(iii) $\hat{\sigma}_n^2$ is a consistent estimate of σ^2.

(iv) $n^{1/2}(\hat{\sigma}^2 - \sigma^2)$ is asymptotically normal with mean zero and variance

$$(24) \qquad 2\sigma^4 + 2\pi \int_{-\pi}^{\pi} \int_{-\pi}^{\pi} q(\lambda, \mu, -\mu) g^{-1}(\lambda, \theta) g^{-1}(\mu, \theta) \, d\lambda \, d\mu .$$

If $\{X_t\}$ is a linear process,

$$(25) \qquad q(\lambda, \mu, -\mu) = (2\pi)^{-1} \gamma_4 f(\lambda) f(\mu)$$

(with γ_4 the fourth cumulant of the independent variables $\{V_t\}$) and this implies that the second term of the variance is then $\sigma^4 \gamma_4$. $n^{1/2}(\hat{\sigma}_n^2 - \sigma^2)$ and $n^{1/2}(\hat{\theta}_n - \theta)$ are jointly asymptotically normal with the limiting covariance of $n^{1/2}(\hat{\sigma}_n^2 - \sigma^2)$ and $n^{1/2}(\hat{\theta}_{n,j} - \theta_j)$ given by the row vector with jth component

$$(26) \qquad 2\pi \int_{-\pi}^{\pi} \int_{-\pi}^{\pi} q(\lambda, \mu, -\mu) g^{-1}(\lambda, \theta) \frac{\partial}{\partial \theta_j} \left(g^{-1}(\mu, \theta)\right) d\lambda \, d\mu$$

postmultiplied by the matrix $-\{2\sigma^2 W\}^{-1}$. These covariances are zero in the case of a linear process and so in that case $n^{1/2}(\hat{\sigma}_n^2 - \sigma^2)$ and $n^{1/2}(\hat{\theta}_n - \theta)$ are asymptotically independent.

In the case of an ARMA scheme

$$g(\lambda, \theta) = | \alpha(e^{-i\lambda}) |^2 / | \beta(e^{-i\lambda}) |^2$$

with $\theta = (\alpha_1, \ldots, \alpha_q, \beta_1, \ldots, \beta_p)$. Then

$$\log g(\lambda, \theta) = \log | \alpha(e^{-i\lambda}) |^2 - \log | \beta(e^{-i\lambda}) |^2$$

and this implies that

$$\int_{-\pi}^{\pi} \frac{\partial \log g(\lambda, \theta)}{\partial \alpha_u} \frac{\partial \log g(\lambda, \theta)}{\partial \alpha_v} d\lambda = 2 \int_{-\pi}^{\pi} e^{i(u-v)\lambda} \, | \alpha(e^{-i\lambda}) |^{-2} \, d\lambda, \; u, v = 1, \ldots, q \,,$$

and that

$$\int_{-\pi}^{\pi} \frac{\partial \log g(\lambda, \theta)}{\partial \alpha_u} \frac{\partial \log g(\lambda, \theta)}{\partial \beta_v} d\lambda = \operatorname{Re} \int_{-\pi}^{\pi} e^{i(u-v)\lambda} \{\alpha(e^{-i\lambda}) \, \overline{\beta(e^{-i\lambda})}\}^{-1} \, d\lambda \,,$$

$u = 1, \ldots, q, v = 1, \ldots, p$ with a corresponding expression for

$$\int_{-\pi}^{\pi} \frac{\partial \log g(\lambda, \theta)}{\partial \beta_u} \frac{\partial \log g(\lambda, \theta)}{\partial \beta_v} \,.$$

A number of preliminary lemmas are first derived.

Lemma 2. Consider θ_0 the actual value of θ with $\theta^* \neq \theta_0$ any other point in Θ. There is then a positive constant $k(\theta_0, \theta^*)$ such that

$$\lim_{n \to \infty} P[U_n(\theta_0) - U_n(\theta^*) < - k(\theta_0, \theta^*)] = 1$$

[From this point on, we write $U_n(\theta)$ instead of $U_n(X, \theta)$].

Proof. Let

$$Y_n = U_n(\theta_0) - U_n(\theta^*) = \int_{-\pi}^{\pi} I_n(\lambda) \left(g_0^{-1} - (g^*)^{-1}\right) d\lambda \tag{27}$$

$$= \sum_{|s| \leq n-1} r_s^{(n)} \{\gamma_s(\theta_0) - \gamma_s(\theta^*)\}$$

where it is understood that f_0, g_0, and g^* stand for $f(\lambda, \theta_0)$, $g(\lambda, \theta_0)$, and $g(\lambda, \theta^*)$. Notice that

$$E(Y_n) = \sum_{|s| \leq n-1} {}_0r_s \left(1 - \frac{|s|}{n}\right) \{\gamma_s(\theta_0) - \gamma_s(\theta^*)\}$$

with ${}_0r_s = \operatorname{cov}(X_t, X_{t+s})$ when θ_0 and σ_0^2 are the actual values of θ and σ^2. The Parseval relation implies that

$$\sum_{s=-\infty}^{\infty} {}_0r_s \, \gamma_s(\theta_0) = \frac{1}{2\pi} \int 2\pi f_0 / g_0 \, d\lambda = \sigma_0^2$$

and

$$\sum_s {}_0r_s \, \gamma_s(\theta^*) = \frac{1}{2\pi} \int_{-\pi}^{\pi} 2\pi f_0 / g^* \, d\lambda = \frac{\sigma_0^2}{2\pi} \int_{-\pi}^{\pi} g_0 / g^* \, d\lambda \,.$$

Therefore

$$\lim_{n\to\infty} E(Y_n) = \sigma_0^2 \left\{1 - \frac{1}{2\pi}\int_{-\pi}^{\pi} g_0 / g^* \, d\lambda\right\}.$$

From (18) it follows that

$$\int_{-\pi}^{\pi} \log g(\lambda)\, d\lambda = 0 .$$

Thus

$$0 = \frac{1}{2\pi}\int_{-\pi}^{\pi} \log(g_0 / g^*) < \log\left\{\frac{1}{2\pi}\int_{-\pi}^{\pi} (g_0 / g^*)\, d\lambda\right\}$$

by Jensen's inequality and the fact that g_0 and g^* differ on a set of positive measure. From this we conclude that

$$\lim_{n\to\infty} E(Y_n) = -\mu(\theta_0, \theta^*) < 0 .$$

However

$$\operatorname{var}(Y_n) = O\left(\frac{1}{n}\right)$$

as $n \to \infty$ (see Theorem 3 of Chapter III) since it is a weighted integral of the periodogram $I_n(\lambda)$ by (27). The lemma follows by applying the Chebyshev inequality. This lemma implies that $\hat{\theta}_n$ is consistent when Θ is a finite set.

Lemma 3 is helpful in obtaining a condition for consistency of $\hat{\theta}_n$ when Θ is an arbitrary bounded closed set.

Lemma 3. Let

$$| U_n(\theta_2) - U_n(\theta_1) | < H_{\delta, n}(x, \theta_1)$$

for each $\theta_1 \in \Theta$ and all $\theta_2 \in S$ such that $| \theta_2 - \theta_1 | < \delta(\theta_1)$ with the random bound $H_{\delta, n}$ satisfying

(28) $$\lim_{\delta\to 0} E(H_{\delta, n}) = 0$$

uniformly in n and

(29) $$\lim_{n\to\infty} \sigma^2(H_{\delta, n}) = 0$$

for each δ. It then follows that $\hat{\theta}_n \to \theta_0$ in probability as $n \to \infty$.

If $\theta_1 \neq \theta_0$ Lemma 2 implies that

(30) $$\lim_{n\to\infty} P[U_n(\theta_0) - U_n(\theta_1) < -k(\theta_0, \theta_1)] = 1 .$$

Choose δ so that $E\{H_{\delta, n}(x, \theta_1)\} \leq \frac{1}{2} k(\theta_0, \theta_1)$. Then (28), (29) and an application of the Chebyshev inequality implies that

(31) $$\lim_{n \to \infty} P[H_{n, \delta}(x, \theta_1) < k(\theta_0, \theta_1)] = 1 .$$

Let $N(\theta_1) = \{\theta : | \theta - \theta_1 | < \delta_1\}$. If the events in (30) and (31) occur simultaneously we have for some $\delta_1 > 0$

$$[U_n(\theta_0) - U_n(\theta_2)] < 0 \quad \text{for} \quad | \theta_2 - \theta_1 | < \delta_1 .$$

But this implies that

$$\lim_{n \to \infty} P\left\{ \sup_{\theta_2 \in N(\theta_1)} [U_n(\theta_0) - U_n(\theta_2)] < 0 \right\} = 1 .$$

An open covering of the compact set Θ is obtained by considering the collection of sets $\{N(\theta_1) : \theta_1 \neq \theta_0, \theta_1 \in \Theta\}$ and the set $N(\theta_0) = \{\theta : | \theta - \theta_0 | < \delta_0\}$ with δ_0 arbitrary. A finite open subcovering $\{N(\theta_j) ; j = 0, 1, \ldots, m\}$ of Θ exists by the compactness of Θ and therefore

$$\lim_{n \to \infty} P\left\{ \sup_{\theta_2 \in \bigcup_{j=1}^{m} N(\theta_n)} [U_n(\theta_0) - U_n(\theta_2)] < 0 \right\} = 1$$

and

$$\lim_{n \to \infty} P\left\{ \inf_{\theta \in \Theta} U_n(\theta) = \inf_{\theta \in N(\theta_0)} U_n(\theta) \right\} = 1$$

hold. It then follows that

$$\lim_{n \to \infty} P\{ | \hat{\theta}_n - \theta_0 | < \delta_0\} = 1 .$$

Since δ_0 can be made arbitrarily small, Lemma 3 follows.

Let $h(\lambda, \theta) = \{g(\lambda, \theta)\}^{-1}$. In Lemma 4 it will be shown that the condition of Lemma 3 is satisfied if h is continuously differentiable with respect to the components of θ. This, of course, implies consistency of $\hat{\theta}_n$.

Lemma 4. Let $h^{(i)}(\lambda, \theta) = \partial h(\lambda, \theta) / \partial \theta_i$, $i = 1, \ldots, p$, *be continuous functions of* (λ, θ), $| \lambda | \leq \pi$, $\theta \in S$. *Then* $\hat{\theta}_n \to \theta_0$ *in probability as* $n \to \infty$.

Now

(32) $$U_n(\theta_2) - U_n(\theta_1) = \int_{-\pi}^{\pi} I_n(\lambda) \{h(\lambda, \theta_2) - h(\lambda, \theta_1)\} \, d\lambda$$

and the mean value theorem implies that

$$h(\lambda, \theta_2) - h(\lambda, \theta_1) = \sum_{i=1}^{p} (\theta_{2,i} - \theta_{1,i})\, h^{(i)}\,(\lambda, \eta\,\theta_1 + (1-\eta)\,\theta_2)$$

for some η, $0 < \eta < 1$. Thus, if $|\theta_2 - \theta_1| < \delta$

$$h(\lambda, \theta_2) - h(\lambda, \theta_1) \leq \delta\, M(\theta_1)$$

where $M(\theta_1) < \infty$ is the supremum of $\Sigma_i\, |\, h^{(i)}(\theta)\,|$ over $|\,\lambda\,| \leq \pi$, $|\,\theta - \theta_1\,| < \delta(\theta_1)$ where $\delta(\theta_1) > \delta$ is chosen so that $\{\theta: |\,\theta - \theta_1\,| \leq \delta(\theta_1)\}$ is contained in S. This implies that

$$|\,U_n(\theta_2) - U_n(\theta_1)\,| \leq \delta\, M(\theta_1) \int_{-\pi}^{\pi} I_n(\lambda)\, d\lambda \tag{33}$$

for $|\,\theta_2 - \theta_1\,| < \delta$. Notice that

$$\int_{-\pi}^{\pi} I_n(\lambda)\, d\lambda = C_0\,,$$

$E(C_0) = {}_0r_0$ and

$$\sigma^2(C_0) = O\left(\frac{1}{n}\right)$$

as $n \to \infty$. This implies that if $H_{\delta,n}(x, \theta_1)$ is set equal to the right hand side of (33) the conditions (28) and (29) will be satisfied and the result will follow from Lemma 3.

The consistency of $\hat{\theta}_n$ has been established under appropriate conditions. The asymptotic normality of $n^{1/2}(\hat{\theta}_n - \theta_0)$ as $n \to \infty$ will now be established by assuming additional smoothness conditions and making use of the mean value theorem.

Theorem 3. *Let $h^{(i)}(\lambda, \theta)$, $i = 1, \ldots, p$, be continuous in (λ, θ) for $|\,\lambda\,| \leq \pi$, $\theta \in S$ and assume that $h^{(ij)}(\lambda, \theta) = \partial^2 h \,/\, \partial\theta_i\, \partial\theta_j$, $h^{(ijk)}(\lambda, \theta) = \partial^3 h \,/\, \partial\theta_i\, \partial\theta_j\, \partial\theta_k$, $i, j, k = 1, \ldots, p$, are continuous in (λ, θ) for $|\,\lambda\,| \leq \pi$, $\theta \in N_\delta(\theta_0) = \{\theta: |\,\theta - \theta_0\,| < \delta\}$. Further let $f(\lambda, \theta_0)$ have a bounded derivative in λ for $|\,\lambda\,| \leq \pi$. Assume that the matrix $W_0 = (w_{ij}^{(0)})$, $i, j = 1, \ldots, p$, with*

$$w_{ij}^{(0)} = \frac{1}{4\pi} \int_{-\pi}^{\pi} \{h^{(i)}(\lambda, \theta_0)\, h^{(j)}(\lambda, \theta_0) \,/\, h^2(\lambda, \theta_0)\}\, d\lambda \tag{22}$$

$$= \frac{1}{4\pi} \int_{-\pi}^{\pi} \left(\frac{\partial \log g}{\partial \theta_i}\right)_0 \left(\frac{\partial \log g}{\partial \theta_j}\right)_0 d\lambda$$

is nonsingular. Then, the limiting distribution of $n^{1/2}(\hat{\theta}_n - \theta_0)$ *as* $n \to \infty$ *is normal with mean o and covariance matrix* $(4\,\sigma^4)^{-1}\, W^{-1}\, Q\, W^{-1}$ with Q given by formula (22).

The assumptions on differentiability of h with respect to the θ_j imply that one can interchange integration and differentiation and conclude that

$$\partial \gamma_s \,/\, \partial \theta_i = \gamma_s^{(i)}(\theta) = (1/2\,\pi) \int_{-\pi}^{\pi} e^{i\lambda s}\, h^{(i)}(\lambda, \theta)\, d\lambda\,,$$

$$\partial^2 \gamma_s \,/\, \partial \theta_i\, \partial \theta_j = \gamma_s^{(ij)}(\theta) = (1/2\,\pi) \int_{-\pi}^{\pi} e^{i\lambda s}\, h^{(ij)}(\lambda, \theta)\, d\lambda\,,$$

$$\partial^3 \gamma_s \,/\, \partial \theta_i\, \partial \theta_j\, \partial \theta_k = \gamma^{(ijk)}(\theta) = (1/2\,\pi) \int_{-\pi}^{\pi} e^{i\lambda s}\, h^{(ijk)}(\lambda, \theta)\, d\lambda\,,$$

exist when $\theta \in N_\delta(\theta_0)$. This in turn implies that the corresponding partial derivatives $U_n^{(i)}(\theta)$, $U_n^{(ij)}(\theta)$, and $U_n^{(ijk)}(\theta)$ of $U_n(\theta)$ exist when $\theta \in N_\delta(\theta_0)$.

The consistency of $\hat{\theta}_n$ follows by Lemma 4. With probability tending to 1 as $n \to \infty$, $\hat{\theta}_n \in N_\delta(\theta_0)$. Suppose $\hat{\theta}_n \in N_\delta(\theta_0)$. It is clear that $U_n^{(j)}(\hat{\theta}_n) = 0$, $j = 1, \ldots, p$, and by the mean value theorem

$$0 = U_n^{(j)}(\theta_0) + \sum_{i=1}^{p} (\hat{\theta}_{n,i} - \theta_{0,i})\, U_n^{(ij)}(\theta_n^*)$$

with

$$\theta_n^* = \lambda\, \hat{\theta}_n + (1 - \lambda)\, \theta_0 \in N_\delta(\theta_0)\,.$$

The dependence of θ_n^* on j has been deleted to avoid cumbersome notation. This can be rewritten as

$$\sum_{i=1}^{p} \{ - U_n^{(ij)}(\theta_n^*)\}\, n^{1/2}(\hat{\theta}_{n,i} - \theta_{0,i}) = n^{1/2}\, U_n^{(j)}(\theta_0)\,. \tag{34}$$

We now wish to show that

(i) $U_n^{(ij)}(\theta_n^*) \to \lim_{n \to \infty} E\{U_n^{(ij)}(\theta_0\} = 2\,\sigma_0^2\, w_{ij}^{(0)}$

in probability as $n \to \infty$ and

(ii) that $n^{1/2}\, U_n^{(j)}(\theta_0)$, $j = 1, \ldots, p$ is asymptotically normal with mean 0 and covariance matrix given by (22).

The conclusion of the Theorem would then follow from (34). We first prove (i). Just as in the proof of Lemma 4 one has

$$|\, U_n^{(ij)}(\theta_n^*) - U_n^{(ij)}(\theta_0)\,| \le |\, \theta_n^* - \theta_0\,|\, M(\theta_0)\, C_0 \tag{35}$$

where now

$$M(\theta_0) = \sum_{k=1}^{p} \sup_{|\lambda| \leq \pi,\, \theta \in N_\delta(\theta_0),\, i,\, j,\, k} | h^{(ijk)}(\theta_0) | .$$

and (35) tends to zero in probability as $n \to \infty$. Further

$$U_n^{(ij)}(\theta_0) - E\{U_n^{(ij)}(\theta_0)\}$$

tends to zero in probability as $n \to \infty$ also since $\sigma^2\{U_n^{(ij)}(\theta_0)\} \to 0$ by (32). Also

$$\lim_{n\to\infty} E\{U_n^{(ij)}(\theta_0)\} = \lim_{n\to\infty} \sum_{|s| \leq n-1} \left(1 - \frac{|s|}{n}\right) {}_0r_s\, \gamma_s^{(ij)}(\theta_0) \tag{36}$$

$$= \sum_{s=-\infty}^{\infty} {}_0r_s\, \gamma_s^{(ij)}(\theta_0)$$

converges (by the assumptions made on f and h) and the Parseval relation implies that (36) equals

$$\frac{1}{2\pi} \int_{-\pi}^{\pi} h^{(ij)}(w, \theta_0)\, 2\pi f(\lambda, \theta_0)\, d\lambda = \frac{\sigma_0^2}{2\pi} \int_{-\pi}^{\pi} (h_0^{(ij)} / h_0)\, d\lambda .$$

Notice that

$$\int_{-\pi}^{\pi} \log h(\lambda, \theta)\, d\lambda = 0 \tag{37}$$

for $\theta \in N_\delta(\theta_0)$. If we differentiate (37) with respect to θ_j and θ_i

$$\int_{-\pi}^{\pi} \{h^{(ij)}(\lambda, \theta)\, g(\lambda, \theta) + h^{(j)}(\lambda, \theta)\, g^{(i)}(\lambda, \theta)\}\, d\lambda = 0$$

or

$$\int_{-\pi}^{\pi} \{(h^{(ij)} / h) - (h^{(i)} h^{(j)} / h^2)\}\, d\lambda = 0 \tag{38}$$

is obtained. Equation (38) can be rewritten as

$$\frac{\sigma_0^2}{2\pi} \int_{-\pi}^{\pi} (h_0^{(i)} h_0^{(j)} / h_0^2)\, d\lambda = 2\,\sigma_0^2\, w_{ij}^{(0)}$$

and the remark (i) follows. Remark (ii) is a direct consequence of Corollary 2 of Chapter III.

Theorem 4. *Under the conditions of Theorem 3 $n^{1/2}(\hat{\sigma}_n^2 - \sigma^2)$ is asymptotically normal with mean zero and variance* (24) *as $n \to \infty$. $n^{1/2}(\hat{\sigma}_n^2 - \sigma^2)$ and $n^{1/2}(\hat{\theta}_n - \theta)$ are jointly asymptotically normal with covariances as specified in statement* (iv).

$\hat{\theta}_n$ converges to θ_0 in probability as $n \to \infty$. This together with an estimate based on Taylor's formula implies that $\hat{\sigma}_n^2$ converges to σ_0^2 in probability as $n \to \infty$. A second application of Taylor's formula indicates that

$$U_n(\theta_0) = U_n(\hat{\theta}_n) + \sum_{i,j=1}^{p} (\theta_{0,i} - \hat{\theta}_{n,i}) (\theta_{0,j} - \hat{\theta}_{n,j}) U_n^{(ij)}(\theta_n^*) \tag{39}$$

with $\theta_n^* = \lambda \theta_0 + (1 - \lambda) \hat{\theta}_n$. The second term on the right of (39) is of order $n^{-1/2}$ and so we have

$$n^{1/2}(\hat{\sigma}_n^2 - \sigma^2) = U_n(\theta_0) - n^{1/2} \sigma_0^2 + O(n^{1/2}) . \tag{40}$$

It follows from problem 1 of Chapter III that

$$n^{1/2} E\{U_n(\theta_0)\} = n^{1/2} \sigma_0^2 (2\pi)^{-1} \int_{-\pi}^{\pi} g_0 h_0 \, d\lambda + O(\log n / n^{1/2}) . \tag{41}$$

Relations (40) and (41) imply that $n^{1/2}(\hat{\sigma}_n^2 - \sigma^2)$ is asymptotically normal with mean zero and variance

$$2\pi \left\{ 2 \int_{-\pi}^{\pi} g_0^{-2} f_0^2 \, d\lambda + \iint_{-\pi}^{\pi} q(\lambda, \mu, -\mu) g_0^{-1}(\mu) g_0^{-1}(\mu) \, d\lambda \, d\mu \right\} . \tag{42}$$

The first term of (42) can be seen to be $2\sigma^4$. $n^{1/2}(\hat{\sigma}_n^2 - \sigma^2)$ and $n^{1/2}(\hat{\theta}_n^2 - \theta)$ are asymptotically jointly asymptotically normal because the components of $\hat{\theta}_n$ can be well approximated by smoothed periodograms. The form of the limiting covariances follows from Corollary 2 of Chapter III which describes the asymptotic behaviour of smoothed periodograms.

4. Sample Computations Using Monte Carlo Simulation

A number of simulations of autoregressive and moving average models have been made using Monte Carlo. A normal and exponential routine in IMSL was used.

The autoregressive models we represent in the form

$$f(B) X_t = V_t$$

with B the backshift operator

$$B X_t = X_{t-1} , \quad B V_t = V_{t-1}$$

and $f(B)$ a polynomial in B. The moving average models have the form

$$X_t = f(B)\, V_t\,.$$

There are two choices for the polynomials

$$f_1(B) = 1 - 1.333\, B + 0.8333\, B^2 - 0.1666\, B^3$$

$$f_2(B) = 1 - 0.90909\, B - 0.25\, B^2 + 0.22727\, B^3\,.$$

The roots of $f_1(z)$ are

$$1+i,\ 1-i,\ 3$$

and those of $f^2(z)$

$$1.1,\ 2 \text{ and } -2\,.$$

There are also two choices for the independent, identically distributed V_t's, the first normal with mean zero and variance one while the second is the symmetrized exponential with density $2^{-1} \exp(- \mid v \mid)$. Notice that in the cases considered all zeros of the corresponding polynomials lie outside the unit disc so that the commonly assumed minimum phase condition is satisfied.

First we consider autoregressive schemes and contrast estimation of the coefficients using the Yule-Walker equations with a procedure suggested by Box and Jenkins [1970]. In this procedure one first makes an initial estimate of the coefficients using the Yule-Walker equations. One then estimates $V_{-l+1}, \ldots, V_0$ by their conditional expectations given the X's and the estimated coefficients and then runs the procedure forward again to get improved estimates of the coefficients. This is sometimes referred to as "back forecasting". In the following tables the Yule-Walker procedure is contrasted with the Box-Jenkins estimates for sample sizes of 64, 128, and 256.

Autoregressive Scheme:

Model 1. $f_1(B)\, X_t = V_t \quad N(0, 1)$

Sample Size	Yule-Walker Solution			Box-Jenkins Solution		
64	–0.9413	0.6295	–0.1465	–1.225	1.005	–0.3233
128	–1.1653	0.6880	–0.0592	–1.188	0.7293	–0.0832
256	–1.3266	0.8495	–0.1745	–1.414	0.9580	–0.2294

Model 2. $f_1(B)\, X_t = V_t$ Symmetric exponential

$N = 64$	–1.3264	0.6385	0.0444	–1.355	0.6402	0.0746
128	–1.3574	0.8662	–0.2224	–1.396	0.9316	–0.0563
256	–1.2346	0.7474	–0.1062	–1.247	0.7561	–0.1067

Model 3. $f_2(B)\, X_t = V_t \quad N(0, 1)$

$N = 64$	-0.88007	0.065473	-0.060975	-0.8943	0.1901	-0.1553
128	-0.78862	-0.108385	0.037191	-0.7943	-0.1572	0.0715
256	-0.96080	-0.080440	0.105723	-0.9911	-0.1233	0.1686

Model 4. $f_2(B)\, X_t = V_t$ Symmetric exponential

$N = 64$	-0.96079	-0.236547	0.389607	-0.9369	-0.3292	00.4523
128	-1.01002	-0.201022	0.278869	-1.009	-0.1241	00.1976
256	-0.83110	-0.246766	0.164607	-0.8337	-0.2455	00.1651

In the following tables the results of computations carried out for moving average processes are given. Estimates of the coefficients are obtained. The technique is one developed by Box and Jenkins [1970]. Given the observations $X_1, \ldots, X_n$ one sets $V_0, \ldots, V_{-(p-1)} = 0$ (assuming the scheme is of order p) and solves for $V_1, \ldots, V_n$

$$V_1 = a_0^{-1} X_1$$

$$V_2 = a_0^{-1}(X_2 - a_1 V_1)$$

$$\ldots$$

$$V_p = a_0^{-1}(X_p - a_1 V_{p-1} - \cdots - a_{p-1} V_1)$$

$$\ldots$$

$$V_t = a_0^{-1}(X_t - a_1 V_{t-1} - \cdots - a_{p-1} V_{t-p+1}), \quad 1 \leq t \leq n .$$

The conditional log likelihood assuming a Gaussian distribution and ignoring initial conditions ($V_0 = \cdots = V_{-(p-1)} = 0$) is a constant minus

$$\frac{1}{2\sigma^2} Q_n(X, a)$$

with

$$Q_n(X, a) = \sum_{t=1}^{n} V_t^2 .$$

The conditional maximum likelihood estimates of $a_0, \ldots, a_{p-1}$ are obtained by minimizing $Q_n(X, a)$ as a function of the parameters a_i. Given values of $a_0, \ldots, a_{p-1}$ one can determine the V_t's. By plotting values of Q numerically one can determine values $\hat{a}_0, \ldots, \hat{a}_{p-1}$ minimizing Q. Estimates of the coefficients for the moving average schemes are carried out using this procedure.

Moving Average Scheme: ε $N(0, 1)$ and η symmetric exponential

Model 1. $X_t = f_1(B)\ \varepsilon_t$

Sample Size	Box-Jenkins Solution		
$N = 64$	–1.322	0.4232	0.0246
128	–1.331	0.7153	–0.1264
256	–1.237	0.6953	–0.1061

Model 2. $X_t = f_1(B)\ \eta_t$

$N = 64$	–1.272	0.8632	–0.3077
128	–1.204	0.638	–0.0691
256	–1.350	0.8596	–0.2091

Model 3. $X_t = f_2(B)\ \varepsilon_t$

$N = 64$	–0.8568	–0.4637	0.4850
128	–0.938	–0.3143	0.3649
256	–0.7556	–0.3531	0.2416

Model 4. $X_t = f_2(B)\ \eta_t$

$N = 64$	–0.8096	–0.1053	0.0090
128	–0.6327	–0.3341	0.2823
256	–0.8307	–0.2664	0.2373

5. *Estimating the Order of a Model*

A variety of procedures have been suggested for the estimation of the order of a finite parameter model. We shall briefly mention only a few. Of the several introduced by Akaike [1971, 1974], consider the one usually referred to as AIC (Akaike's information criterion). Suppose the model has q independent parameters that are estimated. Then the AIC is defined as

$$\mathrm{AIC}(q) = (-2) \log[\text{maximized likelihood}] \\ + 2q .$$

In the case of a *Gaussian process* the logarithm of the likelihood is

$$-\frac{n}{2} \log \sigma^2 - \frac{1}{2\sigma^2} Q(\theta)$$

where θ denotes the set of parameters other than the variance. Let $\hat{\theta}$ be the maximum likelihood estimate of θ. The estimate of the variance is

$$\hat{\sigma}^2 = \frac{1}{n} Q(\hat{\theta})$$

and so the maximum of the logarithm of the likelihood is

$$-\frac{n}{2}\log\hat{\sigma}^2 - \frac{n}{2}.$$

Neglecting the second term of this expression (because it is a constant) the AIC measure can be seen to be equivalent in the case of a Gaussian process with

$$n\log\hat{\sigma}^2 + 2q.$$

If this last measure is plotted as a function of q, there will be a minimum value. The order of the model is taken to be that value of q at which the minimum value is assumed.

There are a number of ways in which one can interpret the specification of the order of a model. One could assume that there is in fact a finite parameter model with q independent parameters and we wish to estimate the "order" q in a precise manner. In another formulation there is no a priori assumption about a model of finite order. One simply wishes to "approximate" the process by a model of finite order in an appropriate way. For example, one may wish to approximate a Gaussian stationary process by an autoregressive process of order q. Such estimates were suggested as a possibility in Grenander and Rosenblatt [1957] (p. 270) if the primary object was prediction. A specific procedure of this type has been suggested by Parzen [1974]. It's suggested that one compute

$$h(k) = \begin{cases} \left(\dfrac{1}{n}\displaystyle\sum_{j=1}^{k}\dfrac{1}{\hat{\sigma}_j^2}\right) - \dfrac{1}{\hat{\sigma}_k^2}, & k = 1, 2, \ldots, \\ -1 + n^{-1} \quad \text{if} \quad k = 0, \end{cases}$$

with $\hat{\sigma}_j^2$ an estimate of residual variance with a fitted autoregressive model of order j. One chooses the order k as the one for which this sequence of values assumes a minimal value.

6. Finite Parameter Stationary Random Fields

Let $\{Y_{\mathbf{n}}\}$, $\mathbf{n} = (n_1, \ldots, n_k)$, $n_i = \ldots, -1, 0, 1, \ldots$, $i = 1, \ldots, k$, be a stationary random field with $E\,Y_{\mathbf{n}} \equiv 0$, $E\,Y_{\mathbf{n}}^2 < \infty$. Assume that

$$(43) \qquad E\left|\sum_{\substack{n_i=a_i\\ i=1,\ldots,k}}^{b_i} Y_{\mathbf{n}}\right|^2 = h(b_1 - a_1, \ldots, b_k - a_k) = h(\mathbf{b} - \mathbf{a}) \to \infty$$

as $b_i - a_i \to \infty$, $i = 1, \ldots, k$. Also let

$$h(\alpha_1, \ldots, \alpha_k) = o\big(h(\beta_1, \ldots, \beta_k)\big) \tag{44}$$

if $\alpha_1, \ldots, \alpha_k \to \infty$, $\alpha_1 = O(\beta_i)$, $i = 1, \ldots, k$ but for some j, $\alpha_j = o(\beta_j)$. We give a central limit theorem for random fields that is a parallel to the central limit theorem for random sequences indicated in problem 4 of Chapter III. The derivation is not given here because it follows from an appropriate adaptation of the proof of Theorem 4 of section III. 3. Our object is to apply this result to get a generalization of the result obtained in section 3 on the asymptotic distribution of estimates of parameters of finite parameter stationary sequences for finite parameter random fields.

Theorem 5. *Let $\{Y_n\}$ be a strictly stationary strongly mixing random field with $E\,Y_n \equiv 0$ and $E\,|\,Y_n\,|^{2+\delta}$ for some $\delta > 0$. Assume conditions (43) and (44) are satisfied. Then if*

$$E\left|\sum_{\substack{n_i = a_i\\ i = 1, \ldots, k}}^{b_i} Y_{\mathbf{n}}\right|^{2+\delta} = O\big(h(\mathbf{b} - \mathbf{a})\big)^{1+\delta/2} \tag{45}$$

as $b_i - a_i \to \infty$, $i = 1, \ldots, k$

$$\sum_{\substack{n_i = 1\\ i = 1, \ldots, k}}^{\alpha_i m} Y_{\mathbf{n}}$$

when appropriately normalized is asymptotically $N(0,\,1)$ as $m \to \infty$ for each $\alpha_1, \ldots, \alpha_k > 0$. Specifically, there are sequences $q_1(m), \ldots, q_k(m), p(m) \to \infty$ with $m/q_j(m) - p(m) = o\big(p(m)\big)$, $q_j(m)\,p(m) \cong \alpha_j\,m$, $j = 1, \ldots, k$ with the proper normalization

$$\left\{\left(\prod_{j=1}^{k} q_j(m)\right) h\big(p(m), \ldots, p(m)\big)\right\}^{1/2}.$$

If $q_1(m) \ldots q_k(m)\,h\big(p(m), \ldots, p(m)\big) \cong h(\alpha_1 m, \ldots, \alpha_k m)$ one can normalize by $\{h(\alpha_1 m, \ldots, \alpha_k m)\}^{1/2}$.

Suppose the stationary random field $X_{\mathbf{n}}$, $E\,X_{\mathbf{n}} \equiv 0$ is observed over the cube of lattice points

$$n_i = 1, \ldots, m, \quad i = 1, \ldots, k\,.$$

Covariance estimates are given by

$$r_{\mathbf{u}}(m) = \frac{1}{m^k} \sum_{\substack{j_i = 1\\ i = 1, \ldots, k}}^{m}{}' X_{\mathbf{j}}\,X_{\mathbf{j}+\mathbf{u}}$$

where the primed sum represents a sum over $j_1, \ldots, j_k$ such that $j_1 + u_1, \ldots, j_k + u_k$ also fall in the range $1, \ldots, m$. Set

$$r_{\mathbf{u}} = \operatorname{cov}(X_{\mathbf{j}}, X_{\mathbf{j}+\mathbf{u}}) .$$

Assume that moments (or equivalently cumulants) up to eighth order exist. Let $c_{\mathbf{a},\mathbf{b},\mathbf{d}}$ be the fourth cumulant

$$c_{\mathbf{a},\mathbf{b},\mathbf{d}} = \operatorname{cum}(X_{\mathbf{n}}, X_{\mathbf{n}+\mathbf{a}}, X_{\mathbf{n}+\mathbf{b}}, X_{\mathbf{n}+\mathbf{d}}) .$$

Theorem 6. *Assume $\{X_{\mathbf{n}}\}$ a strongly mixing strictly stationary random field with $E X_{\mathbf{n}} \equiv 0$, $E X_{\mathbf{n}}^8 < \infty$ and cumulants up to eighth order absolutely summable. Then the random variables $m^{k/2}(r_{\mathbf{u}}(m) - r_{\mathbf{u}})$ for a fixed* (but finite) *number of lags u are jointly asymptotically normal as $m \to \infty$ with mean zero and covariances*

$$R_{\mathbf{u},\mathbf{v}} = \sum_{\mathbf{a}} \{r_{\mathbf{a}} r_{\mathbf{a}+\mathbf{v}-\mathbf{u}} + r_{\mathbf{a}+\mathbf{v}} r_{\mathbf{a}-\mathbf{u}} + c_{\mathbf{u},\mathbf{a},\mathbf{a}+\mathbf{v}}\} .$$

The result can be derived by applying Theorem 5 to every linear combination of the terms $r_{\mathbf{u}}(m) - r_{\mathbf{u}}$. Then up to an error that can be disregarded, one is dealing with a partial sum

$$\sum_{j_1, \ldots, j_k = 1}^{m} a_s X_{\mathbf{j}} X_{\mathbf{j}+\mathbf{u}(s)}$$

with weights a_s. The random variables

$$Y_{\mathbf{j}} = \sum_{s} a_s X_{\mathbf{j}} X_{\mathbf{j}+\mathbf{u}(s)} .$$

One can estimate $h(\boldsymbol{\alpha})$ as well as the moment (45) by making use of the summability conditions on the cumulants. The cumulant conditions imply that one normalizes $r_{\mathbf{u}}(m) - r_{\mathbf{u}}$ by $m^{k/2}$.

We will now see that conditions of this type imply asymptotic normality of a class of smoothed periodograms. Let

$$I_m(\boldsymbol{\lambda}) = \frac{1}{(2\pi m^k)} \left| \sum_{j_1, \ldots, j_k = 1}^{m} X_{\mathbf{j}} e^{-i\mathbf{j}\cdot\boldsymbol{\lambda}} \right|^2 = \frac{1}{(2\pi)^k} \sum_{\substack{|s_j| \le m-1 \\ j=1,\ldots,k}} r_{\mathbf{s}}(m) e^{-i\mathbf{s}\cdot\boldsymbol{\lambda}}$$

be a k-dimensional periodogram. Consider smoothed periodograms

$$\underbrace{\int_{-\pi}^{\pi} \cdots \int_{-\pi}^{\pi}}_{k} I_m(\boldsymbol{\lambda}) A(\boldsymbol{\lambda}) \, d\boldsymbol{\lambda}$$

with the smoothing function $A(\boldsymbol{\lambda})$ symmetric about zero, that is,

$$A(\boldsymbol{\lambda}) = A(-\boldsymbol{\lambda}) .$$

Let $f_4(\boldsymbol{\lambda}, \boldsymbol{\mu}, \boldsymbol{\eta})$ be the fourth order cumulant spectral density

$$f_4(\boldsymbol{\lambda}, \boldsymbol{\mu}, \boldsymbol{\eta}) = (2\pi)^{-3k} \sum_{\mathbf{a}, \mathbf{b}, \mathbf{d}} c_{\mathbf{a}, \mathbf{b}, \mathbf{d}}\, e^{-i(\mathbf{a} \cdot \boldsymbol{\lambda} + \mathbf{b} \cdot \boldsymbol{\mu} + \mathbf{d} \cdot \boldsymbol{\eta})}$$

Theorem 7. *Let $X = \{X_{\mathbf{n}}\}$ be a strictly stationary random field that satisfies the assumptions of Theorem 6. Consider smoothed periodograms*

$$\int I_m(\boldsymbol{\lambda})\, A_j(\boldsymbol{\lambda})\, d\boldsymbol{\lambda}, \quad j = 1, \ldots, s \tag{46}$$

with weight functions $A_j(\boldsymbol{\lambda})$ symmetric about zero and square integrable. The quadratic forms (46) *are then asymptotically normal with means*

$$\int f(\boldsymbol{\lambda})\, A_j(\boldsymbol{\lambda})\, d\boldsymbol{\lambda}, \quad j = 1, \ldots, s\,,$$

and covariances

$$\cong \left(\frac{2\pi}{m}\right)^k \left\{2 \int A_j(\boldsymbol{\lambda})\, A_k(\boldsymbol{\lambda})\, f^2(\boldsymbol{\lambda})\, d\boldsymbol{\lambda} + \iint f_4(\boldsymbol{\lambda}, -\boldsymbol{\mu}, \boldsymbol{\mu})\, A_j(\boldsymbol{\lambda})\, A_k(\boldsymbol{\mu})\, d\boldsymbol{\lambda}\, d\boldsymbol{\mu}\right\},$$

$j, k = 1, \ldots, s$.

The theorem can be derived by approximating the integrals (46) by expansions in terms of a finite number of Fourier coefficients of the weight functions $A_j(\boldsymbol{\lambda})$. One can then apply Theorem 6 to these approximations.

Let us now assume that $X = (X_{\mathbf{n}})$, $E\, X_{\mathbf{n}} \equiv 0$, is a finite parameter random field with spectral density $f(\boldsymbol{\lambda}; \theta, \sigma^2)$,

$$\sigma^2 = (2\pi)^k \exp\left\{\frac{1}{(2\pi)^k} \int \log f(\boldsymbol{\lambda}; \theta, \sigma^2)\, d\boldsymbol{\lambda}\right\}$$

and

$$g(\boldsymbol{\lambda}, \theta) = \{\sigma^2 / (2\pi)^k\}^{-1} f(\boldsymbol{\lambda}; \theta, \sigma^2)\,.$$

Notice that σ^2 is the prediction error in the half-plane prediction problem considered by Helson and Lowdenslager [1958] (also see section VIII. 2). Here $\theta = (\theta_1, \ldots, \theta_p)$ as p unknown parameters other than the variance. Assume that $X_{\mathbf{n}}$, $n_i = 1, \ldots, m$, $i = 1, \ldots, k$ are observed and that $I_m(\boldsymbol{\lambda})$ is the periodogram computed in terms of this data set. If the random field X is Gaussian it is plausible that m^{-k} times the log likelihood of the observations could be approximated by

$$-\frac{1}{2} \log 2\pi\sigma^2 - (2\sigma^2)^{-1} \int I_m(\boldsymbol{\lambda})\, g(\boldsymbol{\lambda}, \theta)^{-1}\, d\boldsymbol{\lambda} \tag{47}$$

under appropriate conditions. Estimates of the unknown parameters are obtained by computing the values $\hat{\theta}_m$ and $\hat{\sigma}^2_m$ maximizing (47). One can show that these estimates are asymptotically consistent and Gaussian even for a large class of nonGaussian and nonlinear models. The assumptions made parallel those given in section 3.

Assumptions:

(1) *An absolute maximum of* (47) *is assumed at* $\theta = \hat{\theta}_m$, $\sigma^2 = \hat{\sigma}^2_m$.

(2) *The possible values of* (θ, σ^2) *lie in a region* $0 < \sigma^2 < \infty$, $\theta \in \Theta$ *with* Θ *a closed bounded subset of an open set* S *of* p*-dimensional Euclidean space.*

(3) $g(\boldsymbol{\lambda}, \theta)$ *and* $g(\boldsymbol{\lambda}, \theta)^{-1}$ *are continuous functions of* $\boldsymbol{\lambda}$.

(4) *Given any two distinct points* θ_1 *and* θ_2, *the functions* $g(\boldsymbol{\lambda}, \theta_1)$ *and* $g(\boldsymbol{\lambda}, \theta_2)$ *are not the same functions of* $\boldsymbol{\lambda}$. It is to satisfy this requirement that a condition like minimum phase is required. One should note that in the case of random fields ($k \geq 2$) there is no uniquely persuasive generalization of the one dimensional minimum phase concept There are many possible concepts.

(5) *The random field satisfies the conditions of Theorem 7.*

(6) *If* $h(\boldsymbol{\lambda}, \theta) = \{g(\boldsymbol{\lambda}, \theta)\}^{-1}$ *then* h *is continuously differentiable in* θ *up to third order as a function of* $(\boldsymbol{\lambda}, \theta)$. *Also* $f(\boldsymbol{\lambda}, \theta)$ *is assumed to have bounded first order derivatives in* $\boldsymbol{\lambda}$.

These assumptions are not intended as minimal assumptions.

Theorem 8. *Let the assumptions given above hold. Also assume that the matrix* $W = \{w_{ij}\}$ *with elements*

$$w_{ij} = \frac{\sigma_0^2}{(2\pi)^k} \int \frac{\partial}{\partial \theta_i} \log g(\boldsymbol{\lambda}, \theta) \frac{\partial}{\partial \theta_j} \log g(\boldsymbol{\lambda}, \theta)\, d\boldsymbol{\lambda} \Big|_{\theta = \theta_0}$$

is nonsingular where θ_0 *and* σ_0^2 *are the true values of* θ *and* σ^2. *Then* $m^{k/2}\{\hat{\theta}_m - \theta_0\}$ *is asymptotically normal* (as $m \to \infty$) *with mean zero and covariance matrix*

$$W^{-1}\, Q\, W^{-1}$$

where $Q = (q_{ij})$ *is given by*

$$q_{ij} = \left\{ 2\sigma_0^2 w_{ij} + \iint f_4(\boldsymbol{\lambda}, \boldsymbol{\mu}, -\boldsymbol{\mu}) \frac{\partial}{\partial \theta_i} g^{-1} \frac{\partial}{\partial \theta_j} g^{-1}\, d\boldsymbol{\lambda}\, d\boldsymbol{\mu} \right\}_{\theta = \theta_0}.$$

The proof of this theorem follows the line of the derivation of Theorem 3 and makes use of Theorem 7. It is clear that $m^{k/2}(\hat{\theta}_m - \theta_0)$ and $m^{k/2}(\hat{\sigma}^2_m - \sigma_0^2)$ are jointly asymptotically normal as $m \to \infty$. Since

$$\hat{\sigma}^2_m = \int I_m(\boldsymbol{\lambda})\, g(\boldsymbol{\lambda}, \theta)^{-1}\, d\boldsymbol{\lambda},$$

the limit of the variance of $m^{k/2}(\hat{\sigma}^2_m - \sigma_0^2)$ as $m \to \infty$ is seen to be

$$(2\pi)^k \left\{ 2 \int g_0(\boldsymbol{\lambda})^{-2} f_0^2(\boldsymbol{\lambda})\, d\boldsymbol{\lambda} + \iint q(\boldsymbol{\lambda}, \boldsymbol{\mu}, -\boldsymbol{\mu})\, g_0^{-1}(\boldsymbol{\lambda})\, g_0^{-1}(\boldsymbol{\mu})\, d\boldsymbol{\lambda}\, d\boldsymbol{\mu} \right\}$$

with g_0, f_0 the true g and f functions.

The results discussed in this section have been for a cubic layout for the indices of the data points. The same arguments can be used for a rectangular parallelepiped as the layout of indices of data points.

Problems

1. Let $\{X_j\}$ be an autoregressive stationary sequence with spectral density

$$f(\lambda) = \frac{1}{2\pi} \mid a(e^{-i\lambda}) \mid^{-2}$$

where $a(z)$ is a polynomial of degree p with all its zeros outside the unit disc in the complex plane. Generate from $X_1, X_2, \ldots, X_n$ an orthonormal sequence of random variables by the Gramm-Schmidt procedure in a recursive manner so that the jth is generated from $X_1, \ldots, X_j$. If the orthonoımal random variables are V_j, $j = 1, 2, \ldots$, show that $\Sigma_{s=0}^{p} a_s X_{t-s} = V_t$ for $t \geq p$. If R_n is the covariance matrix of $X_1, \ldots, X_n$ show that this implies that $R_n^{-1} = B' B$ where B is a lower triangular matrix with $b_{i,j} = a_{j-i}$ if $j \geq p$, $0 \leq j - i \leq p$ and $b_{i,j} = 0$ if $j \geq p$, $j - i > p$.

2. Let R_n be defined as in the previous problem. Assume that $\{Y_j\}$ is a stationary ergodic sequence with finite second moments. Set

$${}_n\mathbf{Y} = \begin{pmatrix} Y_1 \\ \vdots \\ Y_n \end{pmatrix}, \quad C = \{c_{j-k}; j, k = 1, \ldots, n\}$$

with

$$c_s = \int_{-\pi}^{\pi} e^{is\lambda} (2\pi)^{-2} f(\lambda)^{-1} \, d\lambda .$$

Show that

$$E \left| \frac{1}{n} {}_n\mathbf{Y}' R_n^{-1} {}_n\mathbf{Y} - \frac{1}{n} {}_n\mathbf{Y}' C {}_n\mathbf{Y} \right| \leq \frac{K p^2}{n}$$

as $n \to \infty$. This tells us that a quadratic form with matrix R_n^{-1} can be approximated by one with a matrix C that is a Toeplitz matrix when f is the reciprocal of a trionometric polynomial.

3. Show that

$$E \left| \frac{1}{n} {}_n\mathbf{Y} R_n^{-1} {}_n\mathbf{Y} - \frac{1}{n} {}_n\mathbf{Y}' C {}_n\mathbf{Y} \right| \to 0$$

as $n \to \infty$ under the conditions on the process $\{Y_j\}$ specified in problem 2 when $f(\lambda)$ is a positive continuous function.

4. Show that under the conditions of Corollary 2 of Chapter III

$$n \sigma^2 \left(\frac{1}{n} {}_n\mathbf{Y}' \{R_n^{-1} - C\} {}_n\mathbf{Y} \right) \to 0$$

as $n \to \infty$.

Notes

4.1 Related recent results of Lai and Wei [1983] are of some interest. They consider the system of equations

$$y_n = \beta_1 x_{n1} + \cdots + \beta_p x_{np} + \varepsilon_n , \quad n = 1, 2, \ldots$$

where it is assumed that the residuals ε_n form a martingale difference sequence relative to an increasing sequence of σ-fields $\mathscr{F}_n$ and the design levels $x_{n1}, \ldots, x_{np}$ at stage n are $\mathscr{F}_{n-1}$ measurable random variables. Let $X_n = \{x_{ij}; 1 \leq i \leq n, 1 \leq j \leq p\}$ and $\beta = (\beta_1, \ldots, \beta_p)'$. If $X_n' X_n$ is nonsingular, the least square estimate of β is given by

$$b_n = (X_n' X_n)^{-1} X_n' {}_n y$$

where ${}_n y = (y_1, \ldots, y_n)'$. Let

$$\sup_n E(|\varepsilon_n|^\alpha \mid \mathscr{F}_{n-1}) < \infty$$

with probability one for some $\alpha > 2$. Let $\lambda_{\min}(A)$, $\lambda_{\max}(A)$ denote the smallest and largest eigenvalues of a symmetric matrix A. They show that then if $\lambda_{\min}(X_n' X_n) \to \infty$, $\{\log \lambda_{\max}(X_n' X_n) / \lambda_{\min}(X_n' X_n)\} \to 0$ with probability one, it follows that $b_n \to \beta$ as $n \to \infty$ with probability one.

A number of people have considered stationary sequences X_n satisfying the autoregressive scheme

$$\sum_{j=0}^{p} \beta_j X_{n-j} = V_n$$

where the V_n are independent, identically distributed random variables with distribution function in the domain of attraction of a stable law of index $\alpha \in (0, 2)$. As usual, the β_j are constants with $\beta_0 = 1$ and such that the equation $\Sigma \beta_j z^j = 0$ has no roots z_j with $|z_j| \leq 1$. If $\alpha > 1$ it's assumed that $E(z_n) = 0$ for all n. Hannan and Kanter [1977] consider the least squares estimates b_j of the parameters β_j obtained from the equations

$$\sum_{j=0}^{p} b_j \left\{ \sum_{n=1}^{N} X_{n-j} X_{n-k} \right\} = 0, \quad k = 1, \ldots, p$$

with $b_0 = 1$. They show that for any $\delta > \alpha$, $N^{1/\delta}(b_j - \beta_j) \to 0$, $j = 1, \ldots, p$, as $N \to \infty$ with probability one.

4.2 The simplest example of a point process with stationary increments is the Poisson process. In many applications one is interested in point processes with stationary increments having a more complicated structure. Such models have been suggested in the analysis of earthquake shocks, of nerve impulses, etc. We briefly mention some relevant notions and refer to Cox and Lewis [1966], Lewis [1970], Ogata [1978], and Vere-Jones and Ozaki [1982] for a more detailed discussion. Let $P(\cdot)$ be a point process with stationary increments on the real line. Assume that there are no fixed atoms on the real line and that the singleton random points are at $\ldots < t_{-1} < 0 \le t_0 < t_1 < \ldots$ with no limit points. The counting function $N(A) = N(A, w)$ counts the number of points in the set A. The point process is said to be orderly if $\lim_{\delta \downarrow 0} (1/\delta)\, P[N\{[0, \delta)\} \ge 2] = 0$. Let $\mathscr{B}_{s,t}$ denote the σ-field generated by $\{N\{(u, t]\}; s < u \le t\}$. The following two intensity functions (assuming they exist) can then be introduced

$$\lambda(t, w) = \lim_{\delta \downarrow 0} \frac{1}{\delta} P[N\{[t, t+\delta)\} > 0 \mid \mathscr{B}_{-\infty, t}]$$

$$\begin{aligned} \lambda^*(t, w) &= \lim_{\delta \downarrow 0} \frac{1}{\delta} P[N\{[t, t+\delta)\} > 0 \mid \mathscr{B}_{0, t}] \\ &= E\{\lambda(t, w) \mid \mathscr{B}_{0, t}\}\,. \end{aligned}$$

In the estimation problem one considers a parametrized family of intensity functions $\{\lambda_\theta(t, w);\, \theta \in \Theta \subset R^d\}$ which are assumed to correspond to point processes $\{P_\theta;\, \theta \in \Theta\}$ with stationary increments. The exact log-likelihood on the interval $[0, T]$ has the form

$$L_T^*(\theta) = -\int_0^T \lambda_\theta^*(t, w)\, dt + \int_0^T \log \lambda_\theta^*(t, w)\, dN(t)\,.$$

The maximum likelihood estimator $\hat{\theta}_T = \hat{\theta}(t_i;\, 0 \le t_i \le T)$ is obtained by maximizing this likelihood function as a function of θ when one has observations from the point process P_{θ_0} with θ_0 the true parameter value. Much of the discussion given above still has substance even if the point process doesn't have stationary increments. This is the case if one has, for example, a Poisson process with nonconstant intensity function. A Poisson process with cyclically varying intensity function

$$\lambda(t) = A \exp\{\varrho \cos(w_0 t + \Phi)\}$$

has been considered at some length by Cox and Lewis [1966] and Lewis [1970].

4.3 The term minimum phase can be partly motivated in a continuous version of the problem. The discrete case condition that a polynomial have all its zeros outside the unit disc was discussed in section II. 3 in the context of a linear prediction problem. In the continuous analogue of this prediction problem one can map the interior of the unit circle into the upper half plane. The condition is then that the function $f(w)$ obtained be analytic in the upper half plane and have no zeros there. If f has zeros at w_i in the upper half plane they can be removed so as to obtain a function $f_0(w)$ without them

$$f(w) = f_0(w) \prod \frac{w - w_i}{w - \overline{w}_i}\,.$$

The relation between the phase $\Phi(w)$ of f and $\Phi_0(w)$ of f_0 (for w real) can be given by

$$\Phi(w) = \Phi_0(w) - \sum 2 \tan^{-1} \left(\frac{\operatorname{Im} w_i}{w - \operatorname{Re} w_i} \right)$$

and

$$\frac{d}{dw} \{\Phi(w) - \Phi_0(w)\} = 2 \sum \frac{\operatorname{Im} w_i}{\{(\operatorname{Im} w_i)^2 + (w - \operatorname{Re} w_i)^2} \geq 0\,.$$

The phase of f and f_0 at $w = -\infty$ can be taken to be the same.

We briefly discuss likelihood estimates for point processes. Ogata [1978] under appropriate conditions shows that the log-likelihood $L_T^*(\theta)$ (introduced in note 4.2) can be replaced by

$$L_T(\theta) = -\int_0^T \lambda_\theta(t, w)\, dt + \int_0^T \log \lambda_\theta(t, w)\, dN(t)$$

for large T when the point processes $\{P_\theta;\, \theta \in \Theta\}$ have stationary increments. Further conditions for consistency and asymptotic normality of likelihood estimates for these models have been given in Ogata's paper. These conditions require boundedness and differentiability with respect to the parameters of the conditional intensity as well as a law of large numbers and central limit theorem for the first derivatives of the log-likelihood with respect to the parameters. In one of the models discussed in the paper of Vere-Jones and Ozaki [1982]

$$\lambda(t, w) = \mu + \int_{-\infty}^{t} g(t - u)\, dN(u)$$

with

$$g(t) = \sum_{m=0}^{M} a_m\, t^m\, e^{-\beta t}\,.$$

The parameters a_m and β are restrained so that $g(t) \geq 0$ for $t \geq 0$ and $\int_0^\infty g(t)\, dt < 1$

4.4 It is clear that in the case of moving average models and ARMA models (which are not autoregressive) the maximum likelihood procedure (approximate) based on a Gaussian assumption leads to a nonlinear extremal problem. The asymptotic theory assures convergence and asymptotic normality even in a broad range of nonnormal cases if the minimum phase condition is satisfied. A number of popular computing programs have been based on essentially this asymptotic theory. However, it is curious that no extensive Monte Carlo simulation has been carried out even in the moderate sample Gaussian context to see how effective these programs are from a finite sample perspective. Their performance in the non-Gaussian case would also be of considerable interest to assess. Even in the Gaussian case the effectiveness of such computational methods would depend on the character of the likelihood surface and how close to the extremal value the initial guess is.

Chapter V

Spectral Density Estimates

1. The Periodogram

The periodogram was introduced in Chapter III. Here we shall consider it in greater detail. It was originally introduced by Schuster [1898] as a tool to detect hidden periodicities. Suppose that

$$Y_k = X_k + m_k$$

where X_k *is a weakly stationary process with mean zero and a continuous spectral density* f (with $f(\pi) = f(-\pi)$), *and* m_k *is a trigonometric series with a finite number of frequencies* μ_j

$$m_k = \sum_j a_j e^{ik\mu_j} . \tag{1}$$

If this is a real trigonometric regression, the pairs μ_j, $\mu_{-j} = -\mu_j$ will both occur with $a_{-j} = \bar{a}_j$. Assume that Y_k, $k = 0, 1, \ldots, N-1$, has been observed. Consider the periodogram computed for this stretch of data. Then if

$$d_Y^{(N)}(\lambda) = \sum_{k=0}^{N-1} Y_k e^{-ik\lambda}$$

is the *finite Fourier transform* of this stretch of Y data, the periodogram

$$I_N(\lambda) = \frac{1}{2\pi N} \left| d_Y^{(N)}(\lambda) \right|^2 .$$

The mean value of the periodogram

$$E\, I_N(\lambda) = \frac{1}{2\pi N} E\, | d_X^{(N)}(\lambda) |^2 + \frac{1}{2\pi N} | d_m^{(N)}(\lambda) |^2 . \tag{2}$$

The spectral representation of the weakly stationary process $X = (X_k)$ implies that

$$\frac{1}{2\pi N} E\, | d_X^{(N)}(\lambda) |^2 = \frac{1}{2\pi N} E \left| \int_{-\pi}^{\pi} \frac{e^{iN(\alpha-\lambda)} - 1}{e^{i(\alpha-\lambda)} - 1} dZ_x(\alpha) \right|^2 \tag{3}$$

$$= \int_{-\pi}^{\pi} \frac{1}{2\pi N} \frac{\sin^2 \frac{N}{2}(\alpha - \lambda)}{\sin^2 \frac{1}{2}(\alpha - \lambda)} f(\alpha)\, d\alpha .$$

The kernel

$$F_N(\alpha) = \frac{1}{2\pi N} \frac{\sin^2 \frac{N}{2}\alpha}{\sin^2 \frac{1}{2}\alpha}$$

is the Fejer kernel. It is a nonnegative kernel of total mass one (over the interval $[-\pi, \pi]$) with the property that

$$\sup_{0 < \varepsilon < |\alpha| \leq \pi} F_N(\alpha) \to 0$$

as $N \to \infty$ for each fixed $\varepsilon > 0$. All the mass concentrates in the immediate vicinity of zero as $N \to \infty$. Therefore (3) tends to $f(\lambda)$ as $N \to \infty$. However

(4) $$d_m^{(N)}(\lambda) = \sum_j a_j \frac{e^{iN(\mu_j - \lambda)} - 1}{e^{i(\mu_j - \lambda)} - 1} .$$

The main contribution to (2) arising from $d_m^{(N)}(\lambda)$ in the neighborhood of $\lambda = \mu_j$ is

(5) $$| a_j |^2 F_N(\mu_j - \lambda) .$$

The magnitude of this last expression at $\lambda = \mu_j$ is $| a_j |^2 N/2\pi$ and its bandwidth is of the order of magnitude of N^{-1}. If the spectral density f of X is close to constant and $| a_j |^2$ is not small, one expects the contribution from (5) to stand out. This is basically the idea behind Schuster's procedure for detecting discrete spectral lines. Of course, if some of the frequencies μ_j are too close to each other or if f is too variable, there will be difficulties with such a procedure.

Theorem 1. *Consider $Y_k = X_k + m_k$ with $X = (X_k)$, $E X_k \equiv 0$, weakly stationary with a continuous spectral density f and m_k a discrete sum of harmonics* (1). *The mean of the periodogram of Y_k, $k = 0, 1, \ldots, N - 1$, equals* (2) *with $d_m^{(N)}(\lambda)$ given by* (4). *As $N \to \infty$*

$$\frac{1}{2\pi N} E | d_X^{(N)}(\lambda) |^2 \to f(\lambda) .$$

Thus if $m_k \equiv 0$, $I_N(\lambda)$ is asymptotically an unbiased estimate of the spectral density.

However, as we shall see the periodogram is a bad estimate of the spectral density. This will follow on examining the variance and the asymptotic distribution of the periodogram. Thus, even though the periodogram may be useful in trying to isolate the discrete spectrum, it is not effective as an estimate of the continuous spectrum of the process. Nonetheless, certain results on the behavior of the covariance function of the periodogram will suggest effective estimates of the spectral density function f.

Assume that fourth order moments of the Y_k's (or equivalently the X_k's) exist. The covariance function of the periodogram

(6) $$\operatorname{cov}[I_N(\lambda), I_N(\lambda')] = (2\pi N)^{-2} \sum_{t,\tau=0}^{N-1} \sum_{t',\tau'=0}^{N-1} e^{i(t-\tau)\lambda} e^{-i(t'-\tau')\lambda} \operatorname{cov}[X_t X_\tau, X_{t'} X_{\tau'}] .$$

Consider the array of random variables

$$X_t X_\tau$$
$$X_{t'} X_{\tau'} .$$

Theorem 2 of Chapter II implies that

$$
\begin{aligned}
&\operatorname{cov}[X_t X_\tau, X_{t'} X_{\tau'}] \\
(7)\qquad &= r_{t-t'} r_{\tau-\tau'} + r_{t-\tau'} r_{\tau-t'} \\
&+ r_{t-t'} m_\tau m_{\tau'} + r_{t-\tau'} m_{t'} m_\tau \\
&+ r_{\tau-t'} m_t m_{\tau'} + r_{\tau-\tau'} m_t m_{t'} \\
&+ \operatorname{cum}(X_t, X_{t'}, X_{\tau'})\, m_\tau \\
&+ \operatorname{cum}(X_t, X_\tau, X_{\tau'})\, m_{t'} \\
&+ \operatorname{cum}(X_\tau, X_{t'}, X_{\tau'})\, m_t \\
&+ \operatorname{cum}(X_t, X_\tau, X_{t'}, X_{\tau'})\,.
\end{aligned}
$$

Let us first consider the contribution to (7) due to the products of covariances $r_{t-t'} r_{\tau-\tau'}$ and $r_{t-\tau'} r_{\tau-t'}$. The contribution from the first term can be written as

$$
\begin{aligned}
(8)\qquad &(2\pi N)^{-2} \sum_{t,t'=0}^{N-1} r_{t-t'} e^{it\lambda} e^{-it'\lambda'} \sum_{\tau,\tau'=0}^{N-1} r_{\tau-\tau'} e^{-i\tau\lambda} e^{i\tau'\lambda'} \\
&= (2\pi N)^{-2} \int_{-\pi}^{\pi} g_N(\alpha+\lambda)\, g_N(-\alpha-\lambda') f(\alpha)\, d\alpha \\
&\qquad \int_{-\pi}^{\pi} g_N(\alpha-\lambda)\, g_N(-\alpha+\lambda') f(\alpha)\, d\alpha
\end{aligned}
$$

where

$$
g_N(\alpha) = \frac{e^{iN\alpha}-1}{e^{i\alpha}-1}\,.
$$

If $0 \le \lambda, \lambda' \le \pi$ but $\lambda \ne \lambda'$

$$
\begin{aligned}
(9)\qquad &\left| (2\pi N)^{-1} \int_{-\pi}^{\pi} g_N(\alpha+\lambda)\, g_N(-\alpha-\lambda') f(\alpha)\, d\alpha \right| \\
&\le \left\{ \int_{-\pi}^{\pi} F_N(\alpha+\lambda)\, F_N(-\alpha-\lambda')\, f(\alpha)\, d\alpha \int_{-\pi}^{\pi} f(\alpha)\, d\alpha \right\}^{1/2}.
\end{aligned}
$$

The properties of the Fejer kernel imply that (9) approaches zero as $N \to \infty$. However, if $\lambda = \lambda'$, the expression equals

$$
\int_{-\pi}^{\pi} F_N(\alpha+\lambda) f(\alpha)\, d\alpha \int_{-\pi}^{\pi} F_N(\alpha-\lambda) f(\alpha)\, d\alpha
$$

and this approaches $f^2(\lambda)$ as $N \to \infty$. The contribution to (7) from the term with the factor $r_{t-\tau'}\, r_{\tau-t'}$ is

$$(2\pi N)^{-2} \int_{-\pi}^{\pi} g_N(a+\lambda)\, g_N(-a-\lambda')\, f(a)\, da$$

$$\int_{-\pi}^{\pi} g_N(a-\lambda)\, g_N(-a-\lambda')\, f(a)\, da\,.$$

An inequality like (9) indicates that this expression tends to zero if $0 < \lambda$, $\lambda' < \pi$ as $N \to \infty$. However, if $\lambda = \lambda' = 0$ or $\lambda = \lambda' = \pi$ it tends to $f^2(\lambda)$ as $N \to \infty$. Thus the contribution of the first two terms on the right of (7) to the covariance $\operatorname{cov}[I_N(\lambda), I_N(\lambda')]$, $0 \le \lambda, \lambda' \le \pi$, as $N \to \infty$ is

$$[\delta(\lambda-\lambda') + \delta(\lambda+\lambda')]\, f^2(\lambda)\,. \tag{10}$$

Notice that this is the limiting covariance if $m_t \equiv 0$. This already indicates that the periodogram is useless as an estimate of the spectral density even when $m_t \equiv 0$ since it is not consistent. This follows from the fact that the variance of the periodogram does not tend to zero as $N \to \infty$.

Let us now consider the contribution of the other terms in (7) to (6) when $m_t \not\equiv 0$. From here on it will be assumed that second, third and fourth cumulant functions are all summable. This implies that second, third and fourth order (cumulant) spectral densities are continuous. The contribution of $r_{t-t'}\, m_\tau\, m_{\tau'}$ to (6) is

$$\begin{aligned}(2\pi N)^{-2} \sum_{t,\tau,t',\tau'=0}^{N-1} & r_{t-t'}\, e^{i(t-t')\lambda}\, e^{it'(\lambda-\lambda')}\, m_\tau\, m_{\tau'}\, e^{-i\tau\lambda} e^{i\tau'\lambda'} \\ &= \left[\frac{1}{N}\, \frac{e^{iN(\lambda-\lambda')}-1}{e^{i(\lambda-\lambda')}-1}\, f(\lambda) + o(1)\right] \frac{1}{2\pi N}\, d_m^{(N)}(\lambda)\, d_m^{(N)}(-\lambda')\,.\end{aligned} \tag{11}$$

Since the other three terms of the same type can be assessed in a similar manner, one finds that the net contribution from the four terms is

$$\begin{aligned}&\left\{\frac{1}{N}\, d_1^{(N)}(\lambda'-\lambda)\, f(\lambda) + o(1)\right\} \frac{1}{2\pi N}\, d_m^{(N)}(\lambda)\, d_m^{(N)}(-\lambda') \\ &+ \left\{\frac{1}{N}\, d_1^{(N)}(-\lambda-\lambda')\, f(\lambda) + o(1)\right\} \frac{1}{2\pi N}\, d_m^{(N)}(\lambda')\, d_m^{(N)}(\lambda) \\ &+ \left\{\frac{1}{N}\, d_1^{(N)}(\lambda+\lambda')\, f(\lambda) + o(1)\right\} \frac{1}{2\pi N}\, d_m^{(N)}(-\lambda)\, d_m^{(N)}(-\lambda') \\ &+ \left\{\frac{1}{N}\, d_1^{(N)}(-\lambda'+\lambda)\, f(\lambda) + o(1)\right\} \frac{1}{2\pi N}\, d_m^{(N)}(-\lambda)\, d_m^{(N)}(\lambda')\,.\end{aligned} \tag{12}$$

The contribution from the seventh term on the righthand side of (5) is

$$(13) \qquad (2\pi N)^{-2} \sum_{t,t',\tau'=0}^{N-1} c_3(t'-t, \tau'-t)\, e^{it\lambda}\, e^{-i(t'-\tau')\lambda'} \sum_{\tau=0}^{N-1} m_\tau\, e^{-i\tau\lambda}$$

$$= \left\{ \frac{1}{N} d_1^{(N)}(-\lambda)\, f_3(\lambda', -\lambda') + o(1) \right\} \frac{1}{N} d_m^{(N)}(\lambda)\,.$$

The other two terms of the same type yield contributions of the same character as (13). The very last term on the right of (7) gives us

$$\sum_{t,t',\tau,\tau'=0}^{N-1} c_4(\tau-t, t'-t, \tau'-t)\, e^{i(t-\tau)\lambda}\, e^{-i(t'-\tau')\lambda'} = \frac{1}{N} f_4(\lambda, \lambda', -\lambda')\,(1+o(1))\,.$$

Theorem 2. *Let $Y_k = X_k + m_k$ with $X = (X_k)$, $E\,X_k \equiv 0$, $E\,X_k^4 < \infty$, stationary up to fourth order and with cumulants up to fourth order summable. m_k is a discrete sum of harmonics* (1). *If $m_k \equiv 0$ the limit of* $\operatorname{cov}[I_N(\lambda), I_N(\lambda')]$, $0 \le \lambda$, $\lambda' \le \pi$ *as $N \to \infty$ is given by* (10) *where f is the spectral density of X. Even if $m_k \not\equiv 0$, this limit behavior still holds as long as λ, λ' are not equal to any of the frequencies μ_j. The primary contribution to the variance of the periodogram in the immediate neighborhood of a frequency μ_j is given by* (11) *and* (12).

Notice that the maximal contribution to the variance of the periodogram in the neighborhood of a frequency μ_j is of the order of magnitude of N. The magnitude of the random fluctuation corresponding to this is thus $O(N^{1/2})$, of smaller magnitude than a multiple of N which was the contribution of (5) arising from the mean of the periodogram.

We shall now consider the asymptotic distribution of the periodogram values when $E\,Y_k \equiv 0$, the process $Y = \{Y_k\}$ is strongly mixing and the second and fourth cumulant functions are summable. First notice that

$$I_N(\lambda) = (2\pi N)^{-1} \left\{ \left(\sum_{k=0}^{N-1} Y_k \cos k\lambda \right)^2 + \left(\sum_{k=0}^{N-1} Y_k \sin k\lambda \right)^2 \right\}.$$

In determining the asymptotic distribution (joint) of $I_N(\lambda)$ at $\lambda = \lambda_0 = 0 < \lambda_1 < \cdots < \lambda_s \le \pi$ it is convenient to consider

$$(14) \qquad \frac{1}{\sqrt{N}} \sum_{k=0}^{N-1} Y_k h_k$$

with

$$h_k = \sum_{j=0}^{s} (\alpha_j \cos k\lambda_j + \beta_j \sin k\lambda_j)\,.$$

Now

$$(15) \qquad E\left|\frac{1}{\sqrt{N}}\sum_{k=0}^{N-1} Y_k h_k\right|^2 = \frac{1}{N}\sum_{k,k'=0}^{N-1} r_{k-k'}\, h_k\, h_{k'}$$

$$\to |\alpha_0|^2 f(0) + |\alpha_s|^2 f(\pi) + \sum_{j=1}^{s-1} \frac{1}{2}\left(|\alpha_j|^2 + |\beta_j|^2\right) f(\lambda_j)$$

as N $\to \infty$. Further

$$E\left|\frac{1}{\sqrt{N}}\sum_{k=0}^{N-1} Y_k h_k\right|^4 = \sum_{k_1,\ldots,k_4=0}^{N-1} c_4(k_2-k_1, k_3-k_1, k_4-k_1)\, h_{k_1} h_{k_2} h_{k_3} h_{k_4} N^{-2}$$

$$\le A \sum_{u_1,u_2,u_3} |c_4(u_1,u_2,u_3)|\, N^{-1}.$$

The conditions for asymptotic normality in the central limit theorem of problem 4 of Chapter III are satisfied. Thus (14) is asymptotically normal with mean zero and variance (15). This in turn implies that

$$N^{-1/2}\sum_{k=0}^{N-1} Y_k, \quad N^{-1/2}\sum_{k=0}^{N-1}(-1)^k Y_k,$$

$$N^{-1/2}\sum_{k=0}^{N-1} Y_k \cos k\,\lambda_j, \quad N^{-1/2}\sum_{k=0}^{N-1} Y_k \sin k\,\lambda_j, \quad j = 1, \ldots, s-1$$

are asymptotically normal and independent with means zero and variances

$$2\pi f(0), \quad 2\pi f(\pi)$$

$$\pi f(\lambda_j), \quad \pi f(\lambda_j), \quad j = 1, \ldots, s-1$$

respectively. Therefore $I_N(0)$, $I_N(\pi)$, and $I_N(\lambda_j)$, $j = 1, \ldots, s-1$, are asymptotically independent with chi-square distributions having one degree of freedom with means $f(0)$, $f(\pi)$, and then chi-square distributions having two degrees of freedom with means $f(\lambda_j)$, $j = 1, \ldots, s-1$.

Theorem 3. *Let $Y = (Y_k)$ be strictly stationary with $E\, Y_k \equiv 0$ and summable second and fourth order cumulant functions. Assume that Y is strongly mixing. The periodogram values $I_N(0)$, $I_N(\pi)$, $I_N(\lambda_j)$, $0 < \lambda_1 < \cdots < \lambda_{s-1}$, are asymptotically independent as $N \to \infty$. The distributions of $I_N(0)$, $I_N(\pi)$ are asymptotically chi-square with one degree of freedom and means $f(0)$ and $f(\pi)$. The distributions of $I_N(\lambda_j)$, $j = 1, \ldots, s-1$, are asymptotically chi-square with two degrees of freedom and means $f(\lambda_j)$, $j = 1, \ldots, s-1$.*

2. Bias and Variance of Spectral Density Estimates

The results on the asymptotic properties of the periodogram suggest plausible ways in which to construct reasonably good estimates of the spectral density function. Assume for convenience that $Y = (Y_k)$ is strictly stationary with mean zero and summable second and fourth order cumulant functions. Theorem 1 implies that the periodogram is asymptotically unbiased as an estimate of the spectral density. In fact, if

$$\sum_k |k|\,|r_k| < \infty \tag{16}$$

(this implies continuous differentiability of the spectral density function) then

$$E\,I_N(\lambda) = f(\lambda) + O(N^{-1}) \tag{17}$$

since the mean of the periodogram (see (3)) can be rewritten as

$$E\,I_N(\lambda) = \frac{1}{2\pi} \sum_{|k| \le N} r_k \left(1 - \frac{|k|}{N}\right) e^{-ik\lambda}.$$

This means that the bias of the periodogram (as an estimate of the spectral density) tends to zero rapidly as $N \to \infty$ under reasonable smoothness conditions on f. However, Theorem 2 tells us that the variance of the periodogram does not tend to zero as $N \to \infty$. However, the covariance $\operatorname{cov}[I_N(\lambda_1), I_N(\lambda_2)]$, $0 \le \lambda_1, \lambda_2 \le \pi$, $\lambda_1 \ne \lambda_2$ tends to zero as $N \to \infty$. This suggests that an appropriate smoothing of the periodogram with a weight function $W_N(\mu - \lambda)$ might lead to a reasonable estimate

$$f_N(\mu) = \int_{-\pi}^{\pi} W_N(\mu - \lambda)\, I_N(\lambda)\, d\lambda \tag{18}$$

of $f(\mu)$. Smoothed periodograms with a fixed weight function have already been considered in Chapter IV. Here the object is to consider a sequence of weight functions $W_N(\mu)$ that concentrate all mass in the neighborhood of zero as $N \to \infty$. However, the mass should not accumulate too rapidly in the neighborhood of zero since one would then be led to an estimate with the inconsistency of the periodogram.

Notice that if the smoothing weight function

$$W_N(\lambda) = \frac{1}{2\pi} \sum w_k^{(N)} e^{-ik\lambda}$$

has Fourier coefficients

$$w_k^{(N)} = \int_{-\pi}^{\pi} e^{ik\lambda}\, W_N(\lambda)\, d\lambda$$

the spectral density estimate (18) can be rewritten in the form

$$f_N(\mu) = \frac{1}{2\pi} \sum_{|k| \leq N-1} w_k^{(N)} r_k^{(N)} e^{-ik\mu} . \tag{19}$$

In representation (18) $f_N(\mu)$ is given as a smoothed periodogram with weight function $W_N(\mu - \lambda)$. In the alternate representation (19), it is given in terms of covariance estimates $r_k^{(N)}$ with weights $w_k^{(N)}$. As we shall later see a discretized version of the first representation is useful when the periodogram is computed by making use of a fast Fourier transform (see section 2 of Chapter VI). The following conditions are often assumed to hold for the sequence of weight functions $W_N(\lambda)$ for spectral density estimates:

(i) symmetry about zero, $W_N(-\lambda) = W_N(\lambda)$,

(ii) $\int_{-\pi}^{\pi} W_N(\lambda)\, d\lambda \equiv 1$ for all N,

(iii) $\int_{-\pi}^{\pi} W_N^2(\lambda)\, d\lambda < \infty$ for all N,

(iv) Given any $\varepsilon > 0$, $W_N(\lambda) \to 0$ uniformly as $N \to \infty$ for $|\lambda| \geq \varepsilon$,

(v) For any positive A

$$\left| \frac{\int W_N(\lambda - \mu)\, W_N(\mu)\, d\mu}{\int W_N^2(\mu)\, d\mu} - 1 \right| \to 0 \quad \text{for } |\lambda| \leq A/N$$

as $N \to \infty$.

Condition (ii) simply says that the total signed mass of the weight function is one. Conditions (ii) and (iv) imply that the sequence of weight functions behave asymptotically like a δ-function as $N \to \infty$. In fact they also indicate that

$$\int_{-\pi}^{\pi} W_N^2(\lambda)\, d\lambda \to \infty \tag{20}$$

as $N \to \infty$. Let $A_N(\varepsilon) = \{\lambda\colon |\lambda| \leq \varepsilon,\ W_N(\lambda) \geq \varepsilon^{-1}\}$. Then

$$\int_{A_N(\varepsilon)} |W_N(\lambda)|\, d\lambda \geq 1 - \varepsilon$$

by (ii) for N sufficiently large. However,

$$\int_{-\pi}^{\pi} W_N^2(\lambda)\, d\lambda \geq \varepsilon^{-1} \int_{A_N(\varepsilon)} |W_N(\lambda)|\, d\lambda \geq \varepsilon^{-1}/2$$

as $N \to \infty$. Since this holds for any $\varepsilon > 0$, (20) follows. Condition (v) is made to insure that the mass of the weight functions does not concentrate about zero in a bandwidth of magnitude $O(N^{-1})$ but rather somewhat more slowly. Notice that condition (16) implies that

$$E f_N(\mu) = \int_{-\pi}^{\pi} W_N(\mu - \lambda) f(\lambda)\, d\lambda + O\left(\frac{1}{N}\right).$$

In Corollary 2 of Chapter III the asymptotic behavior of the covariance of smoothed periodogram estimates was obtained under appropriate assumptions when the weight functions are independent of N and symmetric about zero. The result stated in that Theorem suggest an asymptotic approximation for the covariance of spectral density estimates. First notice that $f_N(\mu)$ can be specified in terms of a symmetric weight function

$$f_N(\mu) = \int_{-\pi}^{\pi} \frac{1}{2} \{W_N(\mu - \lambda) + W_N(\mu + \lambda)\}\, I_N(\lambda)\, d\lambda .$$

The suggested approximation for the covariance then is

$$\text{(21)} \qquad \operatorname{cov}[f_N(\mu), f_N(\mu')] = \frac{\pi}{N} \int_{-\pi}^{\pi} \{W_N(\mu - \lambda) + W_N(\mu + \lambda)\}^2 f^2(\lambda)\, d\lambda + o\left(\int_{-\pi}^{\pi} W_N^2(\lambda)\, d\lambda\right)$$

as $N \to \infty$. Actually one can show that this approximation is valid if conditions (i) to (v) for the weight functions are satisfied together with summability of second and fourth order cumulants. Notice that the term corresponding to the expression involving the fourth order cumulant spectrum on the right of (6) does not appear in (21) because that term is

$$o\left(\int_{-\pi}^{\pi} W_N^2(\lambda)\, d\lambda\right).$$

The approximation (21) implies that

$$\text{(22)} \qquad \operatorname{var}[f_N(\mu)] = \frac{2\pi}{N} (1 + \eta(2\mu)) f^2(\mu) (1 + o(1)) \int_{-\pi}^{\pi} W_N^2(\lambda)\, d\lambda$$

where

$$\eta(\lambda) = \begin{cases} 1 & \text{if } \lambda = 2k\pi,\ k \text{ integer} \\ 0 & \text{otherwise}. \end{cases}$$

Actually this approximation for the variance will be derived for a class of weight functions in the next section rigorously in the course of deriving a central limit theorem for spectral density estimates. The approximation (21) also indicates that the estimates $f_N(\mu), f_N(\mu'), 0 \leq \mu, \mu' \leq \pi$, are uncorrelated asymptotically as $N \to \infty$ if $\mu \neq \mu'$.

We shall now briefly discuss a number of weight functions that have been suggested. A simple estimate is obtained by truncating the periodogram, that is, by setting

$$w_k^{(N)} = \begin{cases} 1 & \text{if } |k| \leq M(N) \\ 0 & \text{otherwise} \end{cases}$$

where $M(N) = o(N)$, $M(N) \to \infty$ as $N \to \infty$. The corresponding weight function

$$W_N(\lambda) = \frac{1}{2\pi} \sum_{|k| \leq M} e^{ik\lambda} = \frac{1}{2\pi} \frac{\sin\left(M + \frac{1}{2}\right)\lambda}{\sin \frac{\lambda}{2}}$$

is the Dirichlet kernel. Notice that the weight function is not nonnegative.

Another weight function is basically a rescaled Fejer kernel. Let

$$w_k^{(N)} = \begin{cases} (1 - |k|/M) & \text{if } |k| \leq M(N) \\ 0 & \text{otherwise} \end{cases}$$

with $M(N) = o(N)$, $M(N) \to \infty$ as $N \to \infty$. The weight function

$$W_N(\lambda) = \frac{1}{2\pi} \sum_{|k| \leq M} (1 - |k|/M)\, e^{ik\lambda} = (2\pi M)^{-1} \left| \sum_{k=0}^{M-1} e^{ik\lambda} \right|^2$$

$$= (2\pi M)^{-1} \frac{\sin^2 \frac{M}{2}\lambda}{\sin^2 \frac{\lambda}{2}} .$$

This weight function is nonnegative.

In the case of both the Dirichlet and Fejer kernel weight functions discussed above, one would naturally say that their bandwidth is of magnitude $O(M^{-1})$. However, they both have sidelobes that extend far beyond this range. They can permit leakage of spectral mass from spectral peaks that may be far away from the frequency μ at which one is interested in getting an estimate. Partly for this reason a rectangular weight function of the form

$$W_N(\lambda) = \begin{cases} \frac{M}{2\pi} & \text{if } |\lambda| \leq \pi M^{-1} \\ 0 & \text{otherwise} \end{cases}$$

with $M = M(N) = o(N)$, $M(N) \to \infty$ as $N \to \infty$ has been proposed. The corresponding weights

$$w_k^{(N)} = \int_{-\pi/M}^{\pi/M} e^{ik\lambda} \frac{M}{2\pi} d\lambda$$

$$= \frac{\sin(\pi k/M)}{(\pi k/M)} .$$

With such a weight function, the weights $w_k^{(N)}$ are nonzero over the full range $|k| \le N$ and this contrasts with the first two weight functions mentioned.

A class of weight functions that can be seen to be linear combinations of Dirichlet kernels have been suggested by Tukey [1949]. The weights

$$w_k^{(N)} = \begin{cases} 1 - 2a + 2a\cos(\pi k/M) & \text{if } |k| \le M \\ 0 & \text{otherwise} \end{cases}$$

with $M(N) = o(N)$, $M(N) \to \infty$ as $N \to \infty$. The weight function

$$W_N(\lambda) = \frac{1}{2\pi} \sum_{|k| \le M} \{1 - 2a + 2a\cos(\pi k/M)\} e^{ik\lambda}$$

$$= \frac{1}{2\pi} \left\{ a \frac{\sin\left(M + \frac{1}{2}\right)\left(\lambda - \frac{\pi}{M}\right)}{\sin \frac{1}{2}\left(\lambda - \frac{\pi}{M}\right)} + (1 - 2a) \frac{\sin\left(M + \frac{1}{2}\right)\lambda}{\sin \frac{1}{2}\lambda} \right.$$

$$\left. + a \frac{\sin\left(M + \frac{1}{2}\right)\left(\lambda + \frac{\pi}{M}\right)}{\sin \frac{1}{2}\left(\lambda + \frac{\pi}{M}\right)} \right\}.$$

This weight function is still not nonnegative. However, some of the effect of sidelobes can be mitigated, for example, if $a = .23$ or $a = .25$.

In the case of many estimates, *the weights* $w_k^{(N)}$ *are of the form*

$$w_k^{(N)} = a(k\, b_N)$$

with $a(x)$ *a given function continuous at zero, piecewise continuous and such that* $x\, a(x)$ *is bounded.* In fact, a condition of this type will be assumed in obtaining asymptotic results on the variance and distribution of spectral estimates in the next section. *Here* $b_N \to 0$ *with* $N^{-1} = o(b_N)$. b_N can be considered a bandwidth parameter of the estimate. All of the weight functions mentioned in the previous paragraphs are of this type. In the case of the truncated periodogram

$$a(x) = \begin{cases} 1 & \text{if } |x| \le 1 \\ 0 & \text{otherwise} \end{cases}$$

and the bandwidth $b_N = M(N)^{-1}$. For the second example we have

$$a(x) = \begin{cases} 1 - |x| & \text{if } |x| \le 1 \\ 0 & \text{otherwise} \end{cases}$$

with $b_N = M(N)^{-1}$ again. The function $a(x)$ corresponding to the rectangular weight function is

$$a(x) = \frac{\sin x}{x}$$

with $b_N = \pi M(N)^{-1}$. The following result can give reasonable estimates for the bias of spectral estimates computed in terms of such a function $a(x)$ under appropriate assumptions. This result of this type was initially given in Parzen [1957]. *Assume that there also is a* $q \geq 1$ *such that*

$$\lim_{x \to 0} \frac{1 - a(x)}{|x|^q} = c_q$$

with $c_q > 0$ *and that* $b_N \to 0 +$ *with* $N^{-1} = o(b_N^q)$. *Further let*

$$\sum_s |s|^q |r_s| < \infty . \tag{23}$$

It then follows that the bias of a spectral density estimate with weights satisfying these assumptions is given by

$$f(\lambda) - E f_N(\lambda) = c_q b_N^q k^{(q)}(\lambda) + o(b_N^q) \tag{24}$$

with

$$k^{(q)}(\lambda) = \frac{1}{2\pi} \sum_s |s|^q r_s e^{-ik\lambda}$$

if the covariances of the process satisfy (23). The derivation of this result will now be given. Notice that

$$\begin{aligned} f(\lambda) - E f_N(\lambda) &= \frac{1}{2\pi} \sum_{|k| \leq N} \big(1 - a(k b_N)\big) r_k e^{-ik\lambda} \\ &\quad - \frac{1}{2\pi N} \sum_{|k| \leq N} a(k b_N) |k| r_k e^{-ik\lambda} + \frac{1}{2\pi} \sum_{|k| > N} r_k e^{-ik\lambda} \\ &= (1) + (2) + (3) . \end{aligned}$$

The second term $(2) = O(N^{-1})$ and the third term $(3) = o(N^{-q})$. The estimate for the bias (24) is obtained directly from the first term (1).

One can obviously suggest many measures of deviation for $f_N(\lambda)$ as an estimate of $f(\lambda)$. A convenient measure suggested by Grenander and Rosenblatt [1957] is the mean square deviation

$$E |f_N(\lambda) - f(\lambda)|^2 = \sigma^2\big(f_N(\lambda)\big) + |f(\lambda) - E f_N(\lambda)|^2 .$$

If we make use of the estimates of variance and bias given by (22) and (24), the mean square error to the first order will be given by

$$\frac{2\pi}{N b_N} f^2(\lambda) \int W^2(\alpha)\, d\alpha + b_N^{2q} \big(c_q k^{(q)}(\lambda)\big)^2$$

where

$$W(\alpha) = \frac{1}{2\pi} \int a(u)\, e^{-iu\alpha}\, du$$

(see section 3). It is of some interest to consider the case $q = 2$. Then

$$k^{(q)}(\lambda) = f''(\lambda)$$

and the condition (23) implies that the spectral density $f(\lambda)$ is continuously differentiable up to second order.

The method of getting bias estimates given above is effective in many cases. Notice, however, that it does not work in the case of the truncated periodogram.

3. *Asymptotic Distribution of Spectral Density Estimates*

The bias and covariance properties of a class of spectral density estimates have been discussed to some extent in the preceding section of this Chapter. In this section, the asymptotic distribution of such estimates will be considered. Here the spectral density estimates $f_N(\lambda)$ are assumed to be of the form

$$f_N(\lambda) = \frac{1}{2\pi} \sum_{k=-N+1}^{N-1} r_k^{(N)}\, w_k^{(N)} \cos k\lambda$$

with

$$w_k^{(N)} = a(k\, b_N),\ a(0) = 1\,.$$

The function $a(x)$ is assumed to be an even function of finite support, piecewise continuous and continuous at zero. Further, we shall have $b_N \to 0$ with $N^{-1} = o(b_N)$ as $N \to \infty$. Some summability conditions on cumulants

(25) $$\sum_{s_1, \ldots, s_{k-1}} |\, c(s_1, \ldots, s_{k-1})\, | < \infty$$

will be utilized. The following theorem will be derived.

Theorem 4. *Let $X = \{X_n\}$ be a strictly stationary strongly mixing process with $E\, X_j \equiv 0$. Let the cumulant functions* (25) *up to order eight be summable. Further assume that the spectral density estimate $f_N(\lambda)$ has weights $w_k^{(N)}$ defined in terms of a function $a(\cdot)$ satisfying the conditions specified above. It then follows that $(N\, b_N)^{1/2}[f_N(\lambda) - E\, f_N(\lambda)]$ is asymptotically normally distributed with mean zero and variance*

$$2\pi(1 + \eta(2\lambda))\, f^2(\lambda) \int W^2(\alpha)\, d\alpha$$

where

$$W(\alpha) = \frac{1}{2\pi} \int a(u)\, e^{-iu\alpha}\, du\,.$$

In the proof we first consider replacing $f_N(\lambda)$ by

$$\tilde{f}_N(\lambda) = \frac{1}{2\pi} \sum_{k=-b_N^{-1}}^{b_N^{-1}} \frac{1}{N} \sum_{j=1}^{N} X_j X_{j+k} w_k^{(N)} \cos k\lambda .$$

It is assumed here for ease in notation that the support of the function $a(\cdot)$ is $[-1, 1]$ but the argument goes through in exactly the same manner for any other finite interval. Notice that

$$2\pi N \tilde{f}_N(\lambda) = \sum_{u=1}^{N} Y_u^{(N)}$$

with

$$Y_u^{(N)} = \sum_{k=-b_N^{-1}}^{b_N^{-1}} X_u X_{u+k} w_k^{(N)} \cos k\lambda . \tag{26}$$

Expression (26) is a partial sum of the type considered in Theorem 4 of Chapter III.

Our first object is to get an estimate for the variance of $f_N(\lambda) - \tilde{f}_N(\lambda)$. Notice that

$$\sigma^2\left(f_N(\lambda) - \tilde{f}_N(\lambda)\right) \leq (\pi N)^{-2} \sigma^2 \left\{ \sum_{k=1}^{b_N^{-1}} \sum_{j=N-k}^{N} X_j X_{j+k} w_k^{(N)} \cos k\lambda \right\} \tag{27}$$

and the expression on the right hand side of (27) is bounded by

$$(\pi N)^{-2} \sum_{k,k'=1}^{b_N^{-1}} \sum_{j=N-k}^{N} \sum_{j'=N-k'}^{N} \{ | r_{j-j'} r_{j-j'+k-k'} | + | r_{j-j'-k'} r_{j-j'+k} |$$

$$+ | c(k, j'-j, j'+k'-j) | \} | w_k^{(N)} w_{k'}^{(N)} |$$

$$\leq (\pi N)^{-2} \sum_{k,k'=1}^{b_N^{-1}} \sum_{s} \min(k', k) \{ | r_s | \, | r_{s+k'-k} |$$

$$+ | r_{s-k'} r_{s+k} | + | c(k, -s, -s+k') | \} | w_k^{(N)} | \, | w_{k'}^{(N)} | .$$

The conditions on the weights $w_k^{(N)}$ and the function $a(\cdot)$ imply that the weights are bounded and

$$| k | \, | w_k^{(N)} | \leq L \, b_N^{-1}$$

for all k with L an absolute constant. This bound and the summability of the cumulants for $k = 2, 4$ imply that

$$\sigma^2\big(f_N(\lambda) - \tilde{f}_N(\lambda)\big) = O\big((N\, b_N)^{-2}\big) = o\big((N\, b_N)^{-1}\big)\,.$$

This will later be shown to be of a smaller order of magnitude than $\sigma^2\big(\tilde{f}_N(\lambda)\big)$. Consider now

$$(28)\quad \sigma^2\left(\sum_{u=1}^{m} Y_u^{(N)}\right) = h_N\big(m(N)\big) = \sum_{u,\,u'=1}^{m}\ \sum_{k,\,k'=-b_N^{-1}}^{b_N^{-1}} \{r_{u-u'}\, r_{u-u'+k-k'}$$

$$+ r_{u'-u+k'}\, r_{u'-u-k} + c(k, u'-u, u'-u+k)\} \cos k\,\lambda \cos k'\,\lambda\, w_k^{(N)}\, w_{k'}^{(N)}$$

$$= (1) + (2) + (3)$$

where $m = m(N) \le N$ with $b_N^{-1} = o\big(m(N)\big)$ as $N \to \infty$. The initial term on the right side of equation (28) is

$$(1) = \iint_{-\pi}^{\pi} \frac{\sin^2 \frac{m}{2}(\alpha - \beta)}{\sin^2 \frac{1}{2}(\alpha - \beta)} \left| \sum_{v=-b_N^{-1}}^{b_N^{-1}} w_v^{(N)} \cos v\,\lambda\, e^{iv\beta}\right|^2 f(\alpha)\, f(\beta)\, d\alpha\, d\beta\,.$$

The sequence of weight functions $W_N(\lambda)$ satisfy the condition (v) of section 2 (on page 133). Because $b_N^{-1} = o\big(m(N)\big)$ and the continuity of the spectral density f, it follows that (using an argument like that on page 176 of Rosenblatt [1974])

$$(29)\quad (1) = (2\,\pi)^3\, m\big(1 + o(1)\big) \int_{-\pi}^{\pi} f^2(\beta) \left| \frac{1}{2} W_N(\beta + \lambda) + \frac{1}{2} W_N(\beta - \lambda)\right|^2 d\beta$$

where

$$W_N(u) = \frac{1}{2\,\pi} \sum_k w_k^{(N)}\, e^{iku}\,.$$

These properties imply that

$$(30)\qquad (1) = 4\,\pi^3\, m\big(1 + \eta(2\,\lambda) + o(1)\big)\, f^2(\lambda) \int_{-\pi}^{\pi} W_N^2(v)\, dv\,.$$

The second term on the right hand side of equation (28) is

$$(2) = \iint_{-\pi}^{\pi} \frac{\sin^2 \frac{m}{2}(\alpha - \beta)}{\sin^2 \frac{1}{2}(\alpha - \beta)}\, f(\alpha)\, f(\beta) \left(\frac{1}{2} W_N(\alpha + \lambda) + \frac{1}{2} W_N(\alpha - \lambda)\right)$$

$$\times \left(\frac{1}{2} W_N(\beta + \lambda) + \frac{1}{2} W_N(\beta - \lambda)\right) d\alpha\, d\beta\,.$$

Property (v) of the weight functions and the continuity of the spectral density f imply that relation (29) is valid for expression (2) as well as expression (1). We therefore find that the estimate (30) holds for (2) as well as for (1). A simple estimate shows that

$$|(3)| \leq m \sum_{k,s,k'} |c(k,s,k')|$$

and this bound is of a smaller order of magnitude than (30) under the assumptions made. Also notice that

$$\int_{-\pi}^{\pi} W_N(u)^2\,du = (1+o(1))\, b_N^{-1} \int W(u)^2\,du\,.$$

The estimates of the expressions (1), (2), and (3) obtained above imply that $\sigma^2(\tilde{f}_N(\lambda))$ and $\sigma^2(f_N(\lambda))$ have the same asymptotic behavior as (30) with $m(N) = N$ as $N \to \infty$. Also it is clear that if $m(N) = o(N)$ and $k(N)\,m(N) = N$, then

$$k(N)\,h_N(m(N)) \cong h_N(N)$$

by the summability of the cumulant functions for $k = 2, 3, 4$.

To apply the central limit theorem here condition (16) of Chapter III with $\delta = 2$ will be verified, that is, we shall show that

$$\sigma^{-4}\left(\sum_{u=1}^{m} Y_u^{(N)}\right) E\left|\sum_{u=1}^{m} (Y_u^{(N)} - E\,Y_u^{(N)})\right|^4 = O(1) \tag{31}$$

with $Y_u^{(N)}$ given by (26) for $m = m(N)$ as $N \to \infty$. Notice that

$$E\left|\sum_{u=1}^{m} (Y_u^{(N)} - E\,Y_u^{(N)})\right|^4 = \sigma^4\left(\sum_{u=1}^{m} Y_u^{(N)}\right) + \mathrm{cum}_4\left(\sum_{u=1}^{m} Y_u^{(N)}\right).$$

The multilinear character of the fourth cumulant implies that

$$\mathrm{cum}_4\left(\sum_{u=1}^{m} Y_u^{(N)}\right) = \sum_{u_1,u_2,u_3,u_4=1} \mathrm{cum}_4\,(Y_{u_1}^{(N)}, Y_{u_2}^{(N)}, Y_{u_3}^{(N)}, Y_{u_4}^{(N)})$$

and

$$\mathrm{cum}_4\,(Y_{u_1}^{(N)}, Y_{u_2}^{(N)}, Y_{u_3}^{(N)}, Y_{u_4}^{(N)}) = \prod_{i=1}^{4} w_{k_i}^{(N)} \cos k_i\lambda \sum_{\nu} \mathrm{cum}(X_s,\ s\in\nu_1) \ldots$$
$$\mathrm{cum}(X_s,\ s\in\nu_p)$$

where the summation $\nu = \nu_1 \cup \cdots \cup \nu_p$ is over all indecomposable partitions of the table (see Theorem 2 of section II. 2).

$$\begin{array}{l} X_{u_1}\ X_{u_1+k_1} \\ X_{u_2}\ X_{u_2+k_2} \\ X_{u_3}\ X_{u_3+k_3} \\ X_{u_4}\ X_{u_4+k_4}\,. \end{array}$$

One of the many indecomposable partitions consisting only of pairs leads to the sum

(32) $$\sum_{k_i} \sum_{u_i} r_{u_1-u_2} r_{u_3-u_4} r_{u_3-u_1+k_3-k_1} r_{u_4-u_2+k_4-k_2}$$
$$w_{k_1}^{(N)} \cos k_1 \lambda \, w_{k_2}^{(N)} \cos k_2 \lambda \, w_{k_3}^{(N)} \cos k_3 \lambda \, w_{k_4}^{(N)} \cos k_4 \lambda \, .$$

Let $a = u_1 - u_2$, $b = u_3 - u_4$, $\alpha = k_3 - k_1$, $\beta = k_4 - k_2$. The expression (32) is bounded by

$$\sum_{\substack{a,\, b,\, u_1,\, u_3,\\ k_1,\, k_2,\, \alpha,\, \beta}} | r_a | \, | r_b | \, | r_{u_3-u_1+\alpha} | \, | r_{u_3-u_1-b+\alpha+\beta} |$$
$$| w_{k_1}^{(N)} | \, | w_{k_1+\alpha}^{(N)} | \, | w_{k_2}^{(N)} | \, | w_{k_2+\beta}^{(N)} | \, .$$

On first summing over k_1, k_2 a constant times

$$b_N^{-2} \sum_{\substack{a,\, b,\, u_1,\, u_3,\\ \alpha,\, \beta}} | r_a | \, | r_b | \, | r_{u_3-u_1+\alpha} | \, | r_{u_3-u_1-b+\alpha+\beta} |$$

is obtained as a bound. By summing over α, β one is led to a constant times

$$b_N^{-2} \sum_{a,\, b,\, u_1,\, u_3} | r_a | \, | r_b | \, .$$

The sum over a, b and u_1, u_3 gives us a final bound which is

$$O(b_N^{-2} m^2) \, .$$

The other indecomposable partitions that consist entirely of pairs lead to the same bound. Indecomposable partitions that do not consist entirely of pairs will yield smaller bounds. As an example consider the partition leading to the sum

(33) $$\sum_{k_i} \sum_{u_i} c(u_2 - u_1, u_3 - u_1, u_4 - u_1) \, r_{k_2-k_1+u_2-u_1} r_{k_4-k_3+u_4-u_3}$$
$$\prod_{i=1}^{4} (w_{k_i}^{(N)} \cos k_i \lambda) \, .$$

Let $u_2 - u_1 = a$, $u_3 - u_1 = b$, $u_4 - u_1 = s$, $\alpha = k_2 - k_1$, $\beta = k_4 - k_3$. The expression (33) is bounded by

$$\sum_{\substack{u_1,\, a,\, b,\, s\\ \alpha,\, \beta,\, k_1,\, k_2}} | c(a, b, s) | \, | r_{\alpha+a} | \, | r_{\beta+b} | \, | w_{k_1}^{(N)} | \, | w_{k_1+\alpha}^{(N)} | \, | w_{k_2}^{(N)} | \, | w_{k_2+\beta}^{(N)} | \, .$$

By summing first over k_1, k_2 one is led to a constant times

$$b_N^{-2} \sum_{u_1,\, a,\, b,\, s,\, \alpha,\, \beta} | c(a, b, s) | \, | r_{\alpha+a} | \, | r_{\beta+b} |$$

as a bound. By summing over α, β and then over a, b, s

$$O\left(b_N^{-2} \sum_{u_1} 1\right) = O(m\, b_N^{-2})$$

is obtained as a bound. These estimates indicate that (31) is valid and so Theorem III. 4 can be applied to obtain asymptotic normality of these spectral estimates.

An additional set of estimates enables one to show that *Theorem 4 is still valid if the condition that $a(x)$ be of finite support* (retaining the *other conditions on $a(x)$*) *is replaced by the condition that $a(x)$ be square integrable with $x\, a(x)$ bounded.*

4. Prewhitening and Tapering

It is clear that if one has a spectral density to estimate that has a number of sharp peaks or a widely varying range of magnitudes, there will be difficulties in estimation. This is strongly suggested by some of the approximate expressions obtained for the bias of estimates such as (5) and (24). From these, it is clear that the bias will be small if the spectral density is close in some appropriate sense to a flat of "white spectrum". The use of "prewhitening" has been suggested by Press and Tukey [1956]. The object is to estimate the spectral density by the following procedure. First pass the original time series (reduced to mean zero) through a filter so as to flatten the spectral density. Estimate the flattened spectral density and derive from it an estimate of the original spectral density by dividing by the absolute square of the transfer function of prewhitening filter. One should note that an initial rough estimate of the spectral density will typically be required so as to make a reasonable choice of the prewhitening filter.

Another technique called tapering has been suggested by Cooley and Tukey [1965]. The object is to decrease the bias due to the periodogram. This can be accomplished by multiplying the time series by appropriate weights, then computing a finite Fourier transform, renormalizing and proceeding as in the usual spectral analysis. Suppose the weights are generated from a piecewise continuous real function $h(u)$ that vanishes outside (0, 1). We take the weights as $h(j/N)$ and compute the finite Fourier transform of the weighted time series

$$d^{(N)}(\lambda) = \sum_{j=0}^{N-1} h(j/N)\, X_j \exp\{-i\,\lambda\, j\}\,.$$

The analogue of the periodogram is

$$I^{(N)}(\lambda) = \left(2\,\pi \sum_{j} h(j/N)^2\right)^{-1} |\, d^{(N)}(\lambda)\,|^2\,.$$

Set

$$H(\alpha) = \int h(u) \exp\{-i\alpha u\}\, du\,,$$

$$H^{(N)}(\alpha) = \sum_j h(j/N) \exp\{-i\alpha j\}\,,$$

and

$$K^{(N)}(\alpha) = |\, H^{(N)}(\alpha)\,|^2 \Big/ \int_{-\pi}^{\pi} |\, H^{(N)}(\alpha)\,|^2\, d\alpha\,.$$

It is then clear that

$$E\, I^{(N)}(\lambda) = \int_{-\pi}^{\pi} K^{(N)}(\alpha) f(\lambda - \alpha)\, d\alpha\,.$$

Assume that f is twice continuously differentiable and that $H(\alpha)$ decreases to zero sufficiently rapidly as $|\,\alpha\,| \to \infty$. We then have

$$E\, I^{(N)}(\lambda) = f(\lambda) + k_2 \left(\frac{\lambda}{N}\right)^2 f''(\lambda) + o(N^{-2}) \tag{34}$$

as $N \to \infty$ with

$$k_2 = \int \alpha^2\, |\, H(\alpha)\,|^2\, d\alpha \Big/ \int |\, H(\alpha)\,|^2\, d\alpha\,. \tag{35}$$

The linear term does not appear in (34) because $|\, H(\alpha)\,|^2$ is symmetric about zero. Notice that the bias of this modified periodogram is $O(T^{-2})$ (under the assumptions made on H) as contrasted with the bias $O(T^{-1})$ for the classical periodogram. We require that $|\, H(\alpha)\,| = O(|\,\alpha\,|^{-3/2-\varepsilon})$ for some $\varepsilon > 0$ so that the second moment (35) is finite. In the case of the classical periodogram $H(\alpha)$ is the Dirichlet kernel and if condition (16) is satisfied

$$E\, I^{(N)}(\lambda) - f(\lambda) = \frac{1}{N}\frac{1}{2\pi} \sum f_k\, |\, k\,|\, e^{-ik\lambda} + O(N^{-1})\,.$$

5. *Spectral Density Estimates Using Blocks*

We shall informally discuss the construction of spectral density estimates by using blocks of data, a procedure suggested by Bartlett [1950]. Suppose we have $N = k M$ observations consisting of $k = k(N)$ blocks of $M = M(N)$ observations. Assume that the stationary process observed has finite fourth order moments, mean zero and short range dependence. It is clear from earlier discussion that the periodogram is an ineffective estimate of the spectral density because of lack of consistency. If the covariances are absolutely summable, the spectral density of the process will exist and be continuous. Suppose however that each of the $k(N)$ blocks of size $M(N) \to \infty$ as $N \to \infty$ and that $k(N) \to \infty$ as $N \to \infty$. Assume that a periodogram ${}_jI_M(\lambda)$, $j = 1, \ldots, k(N)$, is computed for each of the blocks of size $M(N)$. If the decay of dependence is sufficiently rapid, the periodograms computed from different blocks will be asymptotically independent and an estimate of the spectral density $f_N(\lambda)$ can be constructed by taking the average of the periodograms computed from the different blocks of data

$$f_N(\lambda) = \frac{1}{k(N)} \sum_{j=1}^{k(N)} {}_jI_M(\lambda) .$$

The estimate is clearly asymptotically unbiased because the periodogram is asymptotically unbiased. Further, the variance of the estimate tends to zero as $N \to \infty$ if the periodograms computed from different blocks are asymptotically independent. In particular, we'd expect that if $0 < \lambda < \pi$

$$\sigma^2\big(f_N(\lambda)\big) \cong \frac{f^2(\lambda)}{k(N)} \to 0$$

as $N \to \infty$ and this certainly contrasts with the case of the periodogram. Further, this procedure provides one with the option of gauging the stationarity of the process observed. One can construct two different estimates of the spectral density (at λ) by using the initial blocks for one and the final blocks for the other. If these estimates of the spectral density differ too greatly from each other, this could be interpreted either as a sign of nonstationarity or possibly of long range dependence. Under the assumption of short range dependence the two estimates could be regarded as approximately independent if the sample size is sufficiently large. Let the two spectral density estimates $f_N^{(1)}(\lambda)$ and $f_N^{(2)}(\lambda)$ be constructed from $k_1(N)$ and $k_2(N)$ blocks respectively. Given stationarity and short range dependence the variance of the difference

$$\sigma^2\big(f_N^{(1)}(\lambda) - f_N^{(2)}(\lambda)\big)$$
$$\cong f^2(\lambda)\,\big(k_1(N)^{-1} + k_2(N)^{-1}\big)$$

as $N \to \infty$ if $0 < \lambda < \pi$. Further, if the spectral density f is twice continuously differentiable we have

$$E\big(f_N^{(1)}(\lambda) - f_N^{(2)}(\lambda)\big) = 0\big(M(N)^{-2}\big)$$

as $N \to \infty$. Mixing and moment conditions of the type discussed in Chapters V and VI are enough to imply asymptotic normality of these estimates.

Kolmogorov and Zurbenko [1980] have considered estimates that employ tapering in the construction of periodograms from blocks of data. These periodograms from different blocks are then averaged to obtain density estimates. Let $h_M(t)$, $t = \ldots, -1, 0, 1, \ldots$ be nonnegative values that are equal to zero outside the range $t = 0, 1, \ldots, M$. Given the values $x_Q, \ldots, x_{Q+M}$ construct

$$d_M^Q(\lambda) = \sum_{t=-\infty}^{\infty} h_M(t-Q)\, e^{it\lambda}\, x_t$$

with

$$(36) \qquad H_M(u) = \sum_{t=-\infty}^{\infty} h_M(t)\, e^{itu}\,, \qquad \int_{-\pi}^{\pi} |\, H_M(u)\,|^2\, du = 1\,.$$

The estimate of the spectral density is

$$(37) \qquad f_N(\lambda) = \frac{1}{T} \sum_{k=0}^{T-1} |\, d_M^{Lk}(\lambda)\,|^2$$

and $N = T\,L + M + 1 - L$ sample values of x_t are used. One should note that in the case of estimate (37) the blocks used are of length M but they are not disjoint. In the case of the specific estimate suggested by Kolmogorov $M = K(P-1)$ with

$$h_M(t) = a(K, P)\, l_{K,P}(t)$$

where $a(K, P)$ is determined by condition (36) and the coefficients $l_{K,P}(t)$ by

$$\sum_{t=0}^{K(P-1)} z^t\, l_{K,P}(t) = (1 + z + \ldots + z^{P-1})^K$$

$$= \left(\frac{1-z^P}{1-z}\right)^K .$$

With appropriate choices of the parameters of the estimate as functions of the sample size N, it is clear that one can set some bounds on the size of spectral leakage from frequencies that are not close to each other. Also, as already remarked procedures of this type allow one to check the stationarity assumption. However, it is also clear that breaking the data into blocks of size M and Fourier transforming sets a limit on resolving the difference between frequencies that are closer to each other than $O(M^{-1})$.

6. A Lower Bound for the Precision of Spectral Density Estimates

We now consider a result that gives a lower bound for the precision of spectral density estimates, at least in the context of stationary Gaussian sequences. The result and its derivation are due to Samarov [1977]. Let $\{X_k, k = \ldots, -1, 0, 1, \ldots\}$ be a real stationary Gaussian sequence with mean zero and spectral density $f(\lambda)$, $-\pi \leq \lambda \leq \pi$. Let $W_{r,\alpha}(K)$ be the set of spectral densities which when continued periodically onto the entire real line are r times differentiable, $r = 0, 1, 2, \ldots$, with

$$|\, f^{(k)}(\lambda)\,| \leq K\,, \quad k = 0, 1, \ldots, r\,,$$

and such that $f^{(r)}(\lambda)$ satisfies a Lipschitz condition with exponent α, $0 < \alpha \leq 1$, that is,

$$|\, f^{(r)}(\lambda') - f^{(r)}(\lambda'')\,| \leq K\,|\, \lambda' - \lambda''\,|^{\alpha}$$

for all $\lambda, \cdot \lambda'$. Set $\beta = r + \alpha$. Notice that if r is positive, this implies that $f^{(k)}(-\pi) = f^{(k)}(\pi) = 0$, $k = 1, \ldots, r$.

$\mathcal{M}_n$ denotes the class of all possible estimates $\hat{f}_n$ of $f(\lambda)$ at $\lambda = 0$ based on the observations $X_1, \ldots, X_n$. The value $\lambda = 0$ is chosen for convenience. Any other value λ, $|\lambda| < \pi$ could have been chosen. Consider the norm

$$(38) \qquad \| g \|_{r,\alpha'} = \sum_{k=1}^{r} \sup_{\lambda \in [-\pi, \pi]} | g^{(k)}(\lambda) | + \sup_{\lambda_1, \lambda_2 \in [-\pi, \pi]} \frac{| g^{(r)}(\lambda_1) - g^{(r)}(\lambda_2) |}{| \lambda_1 - \lambda_2 |^{\alpha'}}$$

for some $\alpha' < \alpha$. Let $f_0 \in W_{r,\alpha}(K)$ and $U_\delta(f_0)$ a δ-neighborhood of f_0 using the norm (38). L is the set of nonnegative symmetric loss functions l that are non-decreasing for $x \geq 0$ and increase strictly in an interval with left endpoint 0. The object is to prove the following Theorem.

Theorem 5. *Let $f_0 \in W_{r,\alpha}(K - \sigma)$ for some $\sigma > 0$. Assume that $f_0(\lambda) > 0$ for all $\lambda \in [-\pi, \pi]$. Then for any loss function $l \in L$ one has*

$$\lim_{\delta \to 0} \mathit{lim}_{n \to \infty} \inf_{\hat{f}_n \in \mathcal{M}_n} \sup_{f \in U_\delta(f_0)} E_f\, l(n^{\beta/(2\beta+1)} | \hat{f}_n - f(0) |) > 0$$

Let $l_0(\cdot)$ be the loss function

$$l_0(x) = \begin{cases} 1 & \text{for } |x| \geq 1 \\ 0 & \text{for } |x| < 1 . \end{cases}$$

Give any loss function $l \in L$ and any number $c > 0$ there is a number $d > 0$ such that $l(x) > d\, l_0(x/c)$. For any such estimate

$$(39) \quad \sup_{f \in U_\delta(f_0)} E_f\, l(n^{\beta/(2\beta+1)} | \hat{f}_n - f(0) |) \geq d \sup_{f \in U_\delta(f_0)} E_f\, l_0(n^{\beta/(2\beta+1)} | \hat{f}_n - f(0) | c^{-1})$$

$$= d \sup_{f \in U_\delta(f_0)} P_f\{ n^{\beta/(2\beta+1)} | \hat{f}_n - f(0) | c^{-1} \geq 1 \}$$

$$= d \left(1 - \inf_{f \in U_\delta(f_0)} P_f\{ n^{\beta/(2\beta+1)} | \hat{f}_n - f(0) | c^{-1} < 1 \} \right) .$$

Notice that (39) implies that to prove the theorem it is enough to show that for some $c > 0$

$$(40) \quad \lim_{\delta \to 0} \lim_{n \to \infty} \sup_{\hat{f}_n \in \mathcal{M}_n} \inf_{f \in U_\delta(f_0)} P_f\{ n^{\beta/(2\beta+1)} | \hat{f}_n - f(0) | c^{-1} < 1 \} < 1 .$$

Consider a function $h(\cdot)$ with the following properties:

(i) $h(\cdot) \in W_{r,\alpha}(1)$

(ii) $\max_{\lambda \in [-\pi, \pi]} | h(\lambda) | = h(0) = k_1, \quad 0 < k_1 < 1,$

(iii) $\int_{-\pi}^{\pi} h^2(\lambda)\, d\lambda = k_2 > 0$

(iv) $h(\lambda) = 0$ outside $[-\pi, \pi]$.

Let $h_n(\lambda) = n^{-\beta/(2\beta+1)} h(n^{1/(2\beta+1)} \lambda)$. It then directly follows that

(i) $h_n(\cdot) \in W_{r,\alpha}(1)$,

(ii) $\max_{\lambda \in [-\pi, \pi]} | h_n(\lambda) | = h_n(0) = n^{-\beta/(2\beta+1)} k_1$,

(iii) $\int_{-\pi}^{\pi} h_n^2(\lambda)\, d\lambda = k_2/n$,

(iv) $h_n(\lambda) = 0$ outside $[-\pi\, n^{-1/(2\beta+1)}, \pi\, n^{-1/(2\beta+1)}]$.

We consider the spectral density $f_0(\cdot) \in W_{r,\alpha}(K - \sigma)$ and the sequence of spectral densities

$$g_n(\lambda) = f_0(\lambda) + \varepsilon\, h_n(\lambda)\,.$$

$g_n(\lambda) \in U_\delta(f_0)$ if n is sufficiently large and $\varepsilon < \sigma$. Let the probability distributions of the segment $X_1, \ldots, X_n$ of the Gaussian stationary sequence with mean zero and spectral densities $g_n(\lambda)$ and $f_0(\lambda)$ respectively be denoted by Q_n and P_n. Λ_n is the logarithm of the likelihood ratio $\Lambda_n = \log(dQ_n/dP_n)$. m_n and σ_n^2 are the mean and variance of Λ_n with respect to the measure Q_n.

A number of lemmas will be required to prove the theorem.

Lemma 1. Under the conditions specified above $m_n = E_{g_n} \Lambda_n \leq \gamma_1 \varepsilon^2$ and $\sigma_n^2 = E_{g_n}(\Lambda_n - m_n)^2 \leq \gamma_2 \varepsilon^2$ where the constants $\gamma_1, \gamma_2 > 0$ only depend on f_0, k_2 and K.

Let A_n and B_n be the $n \times n$ covariance matrices corresponding to the probability distributions P_n and Q_n. Since

$$\Lambda_n = \log\left(\frac{dQ_n}{dP_n}\right) = \frac{1}{2} \log | A_n | - \frac{1}{2} \log | B_n | - \frac{1}{2} \{x'(A_n^{-1} - B_n^{-1})\, x\}$$

it follows that

(41) $$m_n = E_{g_n} \Lambda_n = \frac{1}{2} \{\mathrm{tr}[B_n A_n^{-1} - I_n] + \log | A_n | - \log | B_n |\}\,.$$

Consider the spectral densities $g_{n,\theta} = f_0(\lambda) + \varepsilon\, \theta\, h_n(\lambda)$, $0 \leq \theta \leq 1$, with the corresponding covariance matrices $B_{n,\theta} = A_n + \varepsilon\, \theta\, D_n$ with the (j, k)th element of D_n

$$d_{j-k}^{(n)} = \frac{1}{2\pi} \int_{-\pi}^{\pi} h_n(\lambda)\, e^{i(j-k)\lambda}\, d\lambda\,.$$

Since $B_{n,0} = A_n$ and $B_{n,1} = B_n$ formula (41) can be written as

$$m_n = \frac{1}{2}\{\operatorname{tr}[B_{n,1}(B_{n,0}^{-1} - B_{n,1}^{-1})] + \log | B_{n,0} | - \log | B_{n,1} |\}.$$

By applying Taylor's formula to the function $\varphi(\theta) = \operatorname{tr}(B_{n,1}\, B_{n,\theta}^{-1}) + \log | B_{n,\theta} |$ and using the identities

$$\frac{d}{d\theta} A^{-1}(\theta) = - A^{-1}(\theta) \left(\frac{d}{d\theta} A(\theta)\right) A^{-1}(\theta)$$

$$\frac{d}{d\theta} \log | A(\theta) | = \operatorname{tr}\left(A^{-1}(\theta) \frac{d}{d\theta} A(\theta)\right)$$

one can replace (41) by

$$\begin{aligned} m_n &= \frac{1}{2}\, \varepsilon \operatorname{tr}[B_{n,1}\, B_{n,\xi}^{-1}\, D_n\, B_{n,\xi}^{-1} - D_n\, B_{n,\xi}^{-1}] \\ &= \frac{1}{2}\, \varepsilon \operatorname{tr}[(B_{n,1} - B_{n,\xi})\, B_{n,\xi}^{-1}\, D_n\, B_{n,\xi}^{-1}] \end{aligned}$$

for some $\xi \in (0, 1)$. Let $\| A \|_1$ be the norm

$$\| A \|_1 = \left(\sum_{i,j} a_{ij}^2\right)^{1/2}$$

and $\| A \|_2$ the norm

$$\| A \|_2 = \sup\{\| A\,\mathbf{x} \|_1;\ \| \mathbf{x} \|_1 = 1,\ \mathbf{x} \in R^n\}$$

with $\| \mathbf{x} \|_1$ the Euclidean length of the n-vector $\mathbf{x}$. First of all

$$(42) \qquad | \operatorname{tr}(A\,B) | = \left| \sum_{i,j} a_{ij}\, b_{ji} \right| \le \left\{ \sum_{i,j} a_{ij}^2 \sum_{i,j} b_{ij}^2 \right\}^{1/2} \le \| A \|_1 \cdot \| B_1 \| .$$

Let ${}_1A, \ldots, {}_nA$ denote the columns of the matrix A. If A and B are symmetric matrices

$$(43) \qquad \begin{aligned} \| A\,B \|_1 = \| B\,A \|_1 &= \left\{ \sum_{j=1}^{n} \{ \| B_j\, A \|_1^2 \} \right\}^{1/2} \\ &\le \| B \|_2 \left\{ \sum_{j=1}^{n} \| {}_jA \|_1^2 \right\}^{1/2} = \| B \|_2 \| A \|_1 . \end{aligned}$$

The inequalities (42) and (43) imply that

$$m_n \le \frac{1}{2}\, \varepsilon \| B_{n,1} - B_{n,\xi} \| \; \| D_n \|_1 \| B_{n,\xi}^{-1} \|_2^2 \le \frac{1}{2}\, \varepsilon^2 \| D_n \|_1^2 \| B_{n,\xi}^{-1} \|_2^2 .$$

The following Lemma will be useful in completing the proof of Lemma 1.

*Lemma 2. Let a_j, $j = \ldots, -1, 0, 1, \ldots$, be a sequence of real numbers with $a_{-j} = a_j$ and such that $\Sigma\, a_j^2 < \infty$. Set $\alpha(\lambda) = \Sigma\, a_j\, e^{ij\lambda}$. A_n designates the $n \times n$ matrix with (j, k)*th *element a_{j-k}, $1 \leq j, k \leq n$. Then the following inequalities hold:*

(a) $\| A_n \|_2 \leq \sup_{\lambda \in [-\pi, \pi]} | \alpha(\lambda) |$.

(b) If A_n is positive definite

$$\| A_n^{-1} \|_2 \leq \sup_{\lambda \in [-\pi, \pi]} | \alpha(\lambda)^{-1} |$$

(c) $n^{-1} \| A_n \|_1^2 \leq \sum_j a_j^2 = \frac{1}{2\pi} \int_{-\pi}^{\pi} \alpha(\lambda)^2\, d\lambda$.

$\| A \|_2$ equals the largest eigenvalue in absolute value of A. But this equals

$$\sup_{\| x \|_1 = 1} \left| \sum_{j, k=1}^{n} x_j\, a_{j-k}\, x_k \right| = \sup_{\| x \|_1 = 1} \left| \frac{1}{2\pi} \int_{-\pi}^{\pi} \left| \sum_{j=1}^{n} x_j\, e^{ij\lambda} \right|^2 \alpha(\lambda)\, d\lambda \right|$$

$$\leq \sup_{\lambda \in [-\pi, \pi]} | \alpha(\lambda) | .$$

If A_n is positive definite, $\| A_n^{-1} \|_2$ is the reciprocal of the minimal eigenvalue of A_n. Since the minimal eigenvalue of A_n is

$$\inf_{\| x \|_1 = 1} \sum_{j, k=1}^{n} x_j\, a_{j-k}\, x_k = \inf_{\| x \|_1 = 1} \frac{1}{2\pi} \int_{-\pi}^{\pi} \left| \sum_{j=1}^{n} x_j\, e^{ij\lambda} \right|^2 | \alpha(\lambda) |\, d\lambda$$

$$\geq \inf_{\lambda \in [-\pi, \pi]} | \alpha(\lambda) | ,$$

it follows that

$$\| A_n^{-1} \|_2 \leq \sup_{\lambda \in [-\pi, \pi]} | 1/\alpha(\lambda) | .$$

Finally

$$n^{-1} \| A_n \|_1^2 = n^{-1} \sum_{j, k=1}^{n} a_{j-k}^2 = \sum_{s=-n}^{n} a_s^2 \left(1 - \frac{|s|}{n}\right) \leq \sum_s a_s^2 = \frac{1}{2\pi} \int_{-\pi}^{\pi} \alpha(\lambda)^2\, d\lambda .$$

Lemma 2 and property (iii) of the functions $h_n(\lambda)$ directly imply that

$$m_n \leq \gamma_1\, \varepsilon^2\, n \int_{-\pi}^{\pi} h_n^2(\lambda)\, d\lambda = \gamma_1\, \varepsilon^2 . \tag{44}$$

This yields the first result of Lemma 1. Now

$$\sigma_n^2 = E_{g_n}(\Lambda_n - m_n)^2 = \frac{1}{4}\int \{y' B_n^{1/2}(A_n^{-1} - B_n^{-1}) B_n^{1/2} y$$

$$- \operatorname{tr}(B_n A_n^{-1} - I_n)\}^2 (2\pi)^{-n/2} \exp\left\{-\frac{1}{2} y' y\right\} dy$$

$$= \frac{1}{2} \| B_n(A_n^{-1} - B_n^{-1}) \|_1^2 .$$

Lemma 2, inequality (44) and property (iii) imply that

$$E_{g_n}(\Lambda_n - m_n)^2 \leq \frac{1}{2} \| A_n^{-1} \|_2^2 \| B_n - A_n \|_1^2$$

$$= \frac{1}{2} \varepsilon^2 \| A_n^{-1} \|_2^2 \| D_n \|_1^2 \leq \frac{1}{2} \varepsilon^2 \| A_n^{-1} \|_2^2 \frac{n}{2\pi} \int_{-\pi}^{\pi} h_n^2(\lambda)\, d\lambda \leq \alpha_2 \varepsilon^2 .$$

The proof of Lemma 1 is complete.

Lemma 3. Let P and Q be two probability distributions on a probability space. Assume that Q is absolutely continuous with respect to P and that $E_Q[\log(dQ/dP)]^2 \leq M$. Then given any number $\alpha > 0$ and any event A, it follows that

$$Q(A) \leq e^\alpha P(A) + M \alpha^{-2} .$$

Let $\Lambda = \log(dQ/dP)$. Then

$$(45) \qquad Q(A) = \int_A e^\Lambda \, dP = \int_{A \cap \{\Lambda \leq \alpha\}} e^\Lambda \, dP + \int_{\{\Lambda > \alpha\}} dQ$$

$$\leq e^\alpha P(A) + E_Q\left[\log \frac{dQ}{dP}\right]^2 \alpha^{-2} \leq e^\alpha P(A) + M \alpha^{-2} .$$

We now complete the proof of the theorem. Now $g_n \in U_\delta(f_0)$. Given d_1, $d_2 > 0$ with $d_1 + d_2 = 1$ it follows that

$$\inf_{f \in U_\delta(f_0)} P_f\{n^{\beta/(2\beta+1)} | \hat{f}_n - f(0) | c^{-1} < 1\} \leq d_1 P_{f_0}\{n^{\beta/(2\beta+1)} | \hat{f}_n - f_0(0) | c^{-1} < 1\}$$

$$+ d_2 P_{g_n}\{n^{\beta/(2\beta+1)} | \hat{f}_n - g_n(0) | c^{-1} < 1\} .$$

Let $d_1 = e^\alpha(e^\alpha + 1)^{-1}$ and $d_2 = (e^\alpha + 1)^{-1}$. Lemmas 1 and 3 and formula (45) imply that

$$\inf_{f \in U_\delta(f_0)} P_f\{n^{\beta/(2\beta+1)} | \hat{f}_n - f(0) | c^{-1} < 1 \}$$

$$\leq e^\alpha(e^\alpha + 1)^{-1} [P_{f_0}\{n^{\beta/(2\beta+1)} | \hat{f}_n - f_0(0) | c^{-1} < 1 \}$$

$$+ P_{f_0}\{n^{\beta/(2\beta+1)} | \hat{f}_n - g_n(0) | c^{-1} < 1 \}] + c_3 \alpha^{-2} \varepsilon^2 (e^\alpha + 1)^{-1} .$$

Let

$$A = \{n^{\beta/(2\beta+1)} \mid \hat{f}_n - f_0(0) \mid c^{-1} < 1, n^{\beta/(2\beta+1)} \mid \hat{f}_n - g_n(0) \mid c^{-1} < 1\}.$$

Notice that if A holds then

$$n^{\beta/(2\beta+1)} \mid g_n(0) - f_0(0) \mid c^{-1} < 2. \tag{46}$$

We shall show that for small enough $c > 0$ the set A is the empty set. If one takes $c > \varepsilon/2$, it follows from condition (ii) that inequality (46) is not satisfied and so A is empty. Then

$$\inf_{f \in U_\delta(f_0)} P_f \{n^{\beta/(2\beta+1)} \mid \hat{f}_n - f(0) \mid c^{-1} < 1\}$$
$$\leq e^\alpha(e^\alpha + 1)^{-1} + c_3\, a^{-2}\, \varepsilon^2 (e^\alpha + 1)^{-1} = (e^\alpha + c_3\, a^{-2}\, \varepsilon^2)\,(e^\alpha + 1)^{-1}.$$

If $a > \varepsilon\, c_3^{1/2}$ inequality (40) is obtained.

7. *Turbulence and the Kolmogorov Spectrum*

The term turbulence as used in fluid mechanics refers to a flow whose characteristics are so chaotic that statistical and probabilistic methods are used to study and describe many of its aspects. In many studies the velocity field $\mathbf{v}(\mathbf{x}, t)$ of the flow is considered a function of location $\mathbf{x} = (x_1, x_2, x_3)$ and time t that is a random solution in some sense of the Navier-Stokes equation

$$\frac{\partial \mathbf{v}(\mathbf{x}, t)}{\partial t} + (\mathbf{v} \cdot \boldsymbol{\nabla})\,\mathbf{v} = -\frac{1}{\varrho}\nabla p + \nu\,\nabla^2\,\mathbf{v} \tag{47}$$

and the continuity equation

$$\nabla \cdot \mathbf{v} = 0.$$

In the equations just written ∇ is the gradient operator, ϱ the density, p the thermodynamic pressure, ν the kinematic viscosity and ∇^2 the Laplacian. The dimensionless quantity $R = u\,l/\nu$, with u a characteristic velocity and l a characteristic length (u length/unit time, l length, ν length^2/unit time), represents an important aspect of the flow and it is for large Reynold's number R that the flow takes on the erratic and unstable character of turbulence. The Reynold's number represents the ratio of inertial to viscous forces. The study of onset of instability for nonlinear systems has a long history and recently there has been a great deal of interest in and progress in resolving such questions for simpler nonlinear systems. An engaging discussion of this recent work can be found in the book of Guckenheimer and Holmes [1983]. However, there are still many open questions relative to fully developed turbulence. An idealized model is that of homogeneous turbulence in which the process $\mathbf{v}(\mathbf{x}, t)$ is considered a stationary random process in $\mathbf{x}$ with finite second order moments. It is thought that the turbulence generated downstream in a wind tunnel by passing a uniform fluid flow through a regular grid of bars held at right angles to the flow is approximated reasonably by such a model locally. For a more detailed discussion of the historical background the book of Batchelor [1953] is useful. There are open questions concerning existence and uniqueness of deterministic solutions of the system (47) for low viscosity ν (or high Reynold's number). A classical result of E. Hopf confirms the existence of a weak deterministic solution for all $\nu > 0$ (see the book of Temam [1977] for a discussion). This is a weak solution in the sense that it satisfies a corresponding integral equation. There has been recent mathematical research (see Foias and Temam [1980] or Vishik, Komech and Fursikov [1979]) establishing the existence of weak homogeneous statistical solutions of the Navier-Stokes system by adapting ideas like those of Hopf.

There are a number of ideas that have come to play a role in a "theory of turbulence". One of these, as we have already noted, is that a turbulent velocity field can be described statistically as a random process. Another has been that the motion can be regarded qualitatively as a summation of turbulent eddies (corresponding to harmonics). Idealized models like that of homogeneous turbulence have been proposed where the random velocity field is assumed to be stationary with respect to spatial displacement. This by virtue of the appropriate multidimensional extension of Bochner's theorem and Cramér's theorem leads to a Fourier representation of the covariance function and the random field in homogeneous turbulence. We shall return to the discussion of a Fourier representation later on. Let us now consider some of the heuristics with which turbulence is often analyzed. One introduces the concept of Reynolds' numbers for turbulent eddies of different sizes. Let λ be the magnitude of a given eddy and v_λ its velocity. The corresponding Reynolds' number is taken to be $R_\lambda \sim v_\lambda \lambda/\nu$. In the case of large eddies, R_λ is large and the viscosity is thought to be unimportant. The viscosity is important for small eddies of magnitude λ_0. Energy transfer and dissipation is thought to be of the following character in turbulent flow. The energy passes from large eddies to smaller eddies with essentially no dissipation. Energy is dissipated in the smallest eddies and there the kinetic energy is transformed into heat. Many arguments in fluid mechanics use dimensional analysis. We give a very brief sketch of such an argument due to Kolmogorov [1941]. The object is in part to specify the parameters relevant in turbulent flow over regions small relative to the overall scale l of the turbulence, but large relative to the distance λ_0 at which viscosity is important. In the model of homogeneous turbulence with the implied spatial harmonic analysis, this corresponds to wave numbers k in the range $l^{-1} \ll k \ll \lambda_0^{-1}$, the inertial subrange. One argues that the relevant parameters are wave-number and the mean dissipation of energy per unit time per unit mass of fluid ε. Consider estimating the order of magnitude v_λ of turbulent velocity variation over distances of the order of λ in the range $l > \lambda > \lambda_0$. The only obvious quantity in terms of ϱ, ε and λ having the dimensions of velocity is $(\varepsilon\, \lambda)^{1/3}$. Notice that ε could be interpreted as the energy flux passing from larger to smaller eddies. In the model of homogeneous turbulence this suggests an approximate form for the spectrum in the inertial range. The estimate $v_\lambda \sim (\varepsilon\,\lambda)^{1/3}$ suggests $E\{v(x)\,v(x+\lambda)\} \sim (\varepsilon\,\lambda)^{2/3}$. This implies that the spectral density (averaged about a sphere of radius $|k|$ in wave number space) should have the form $c\,|k|^{-5/3}$, the Kolmogorov spectrum.

The frontspiece of the book shows the generation of turbulence by passing a laminar flow through a grid. Measurements are usually made downstream by a hot-wire anemometer. The measurements are temporal and are interpreted as spatial measurements by making use of a "frozen-flow" approximation suggested by Taylor.

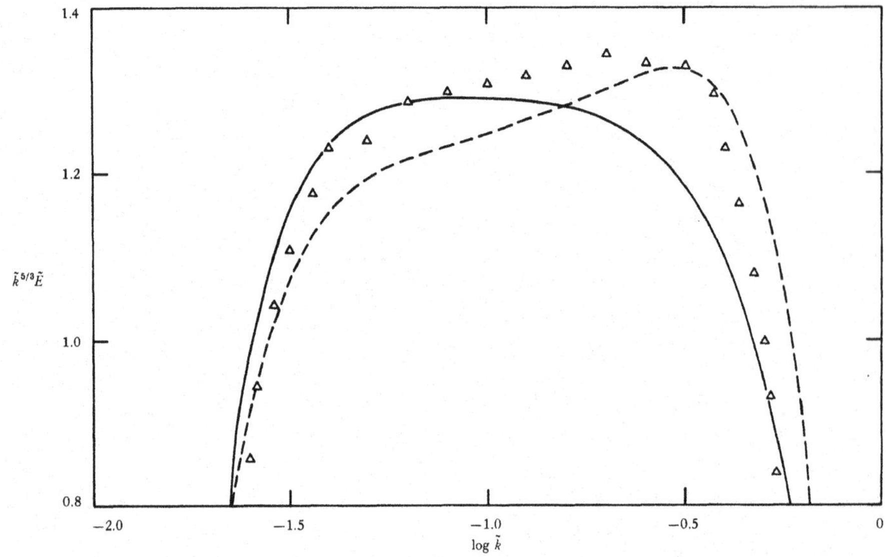

Figure 2. Inertial subrange for measured and empirical three-dimensional spectra. △, USCD jet, and — — — theoretical fits. (Figure courtesy of Dr. Kenneth Helland, University of California, San Diego).

The Figure 2 graphs the energy spectrum of a jet flow. The internal diameter of the jet was 29.2 cm and the turbulent Reynold's number of the flow 950. The experiment points are given by triangles and the two smooth curves are theoretical fits. Notice that the Kolmogorov spectrum would give only a crude approximation to the data. The readings were made at UCSD in 1970. One difficulty with such an approximation may be due to the fact that the Taylor approximation is not valid for such flows.

Figures 3 and 4 represent the results of measurements of velocity and temperature at small space and time scales using sensors and electronic equipment mounted on an NCAR Electra in flight. Here we can see that the Kolmogorov spectrum provides a very good fit to the data. For details see the article of Friehe and LaRue [1975].

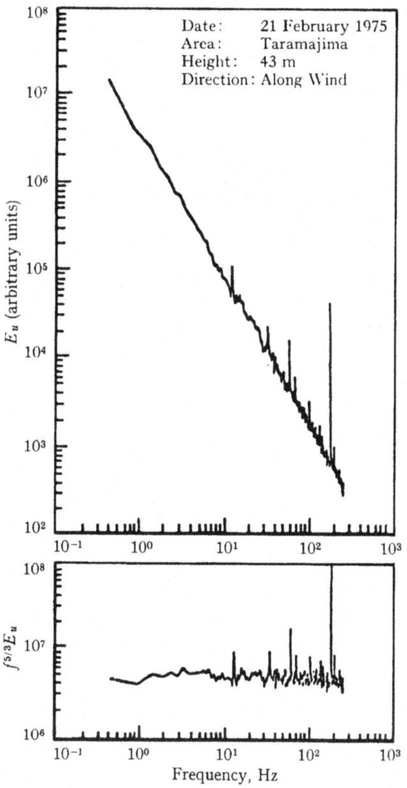

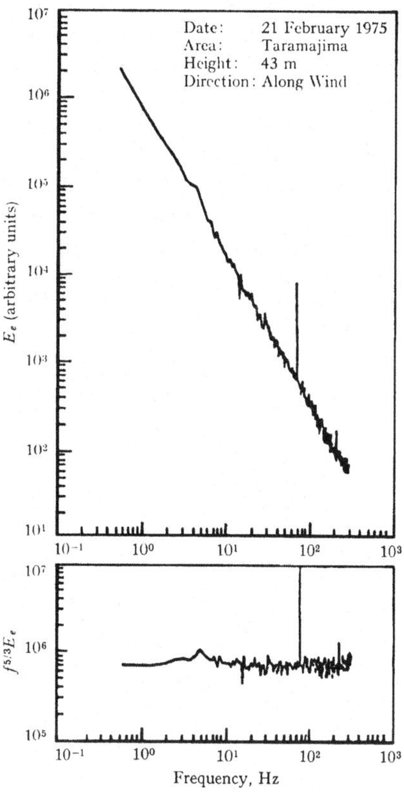

Figure 3a. Power spectrum of velocity (upper left-hand figure).
Figure 3b. Frequency to the 5/3 power times power spectrum value (lower left-hand figure).

Figure 4a. Power spectrum of temperature (upper right-hand figure).
Figure 4b. Frequency to the 5/3 power times power spectrum value (lower right-hand figure). (Figures 3 and 4, courtesy of Professor Carl Friehe, University of California, Irvine).

8. Spectral Density Estimates for Random Fields

In section 6 of Chapter IV asymptotic normality was indicated for a class of smoothed periodograms of random fields. Here, in a similar manner we shall describe conditions that are sufficient for the asymptotic normality of spectral density estimates of a random field. The following theorem is an analogue of Theorem 4 for random fields.

Theorem 6. *Let $\{X_{\mathbf{j}}\}$ be a strictly stationary strongly mixing random field Consider strictly stationary random fields $Y^{(n)} = \{Y_{\mathbf{j}}^{(n)}\}$, $E\, Y_{\mathbf{j}}^{(n)} \equiv 0$, where $Y_{\mathbf{j}}^{(n)}$ is measurable with respect to the Borel field $\mathscr{B}(\mathbf{j}, c(n))$ generated by $X_{\mathbf{s}}$ with $|\mathbf{s} - \mathbf{j}| \le c(n)$ and $c(n) \to \infty$, $c(n) = o(n)$ as $n \to \infty$. Let*

$$h_n(\mathbf{b} - \mathbf{a}) = E \left| \sum_{\substack{j_i = a_i \\ i = 1, \ldots, k}}^{b_i} Y_{\mathbf{j}}^{(n)} \right|^2 .$$

Assume that for any two k-vector sequences $\mathbf{s}(n)$, $\mathbf{m}(n)$ with $m_i(n) \le n$, $i = 1, \ldots, k$, $s_i(n) = O(m_i(n))$ for $i = 1, \ldots, k$ and for some i, $s_i(n) = o(m_i(n))$, $c(n) = o(m_i(n))$ for $i = 1, \ldots, k$ it follows that

$$h_n(\mathbf{s}(n)) / h_n(\mathbf{m}(n)) \to 0 .$$

Further let

$$\{h_n(\mathbf{m}(n))\}^{-(2+\delta)/2} E \left| \sum_{\substack{j_i = 1 \\ i = 1, \ldots, k}}^{m_i} Y_{\mathbf{j}}^{(n)} \right|^{2+\delta} = o(1)$$

for $\mathbf{m}(n)$ and some $\delta > 0$. There are then sequences $r(n)$, $p(n) \to \infty$ with $r(n)\, p(n) \cong n$ such that

$$\sum_{\substack{j_i = 1 \\ i = 1, \ldots, k}}^{n} Y_{\mathbf{j}}^{(n)} / \{r(n)^k\, h_n(p(n), \ldots, p(n))\}^{1/2} \tag{48}$$

is asymptotically normally distributed with mean zero and variance one. If $r(n)^k\, h_n(p(n), \ldots, p(n)) \cong h_n(n, \ldots, n)$ the normalization in (48) can be replaced by $\{h_n(n, \ldots, n)\}^{1/2}$.

Let $\{X_{\mathbf{j}}\}$ be a random field with $E\, X_{\mathbf{j}} \equiv 0$ and cumulant functions up to order eight are absolutely summable. The spectral density of the random field

$$f(\boldsymbol{\lambda}) = \frac{1}{(2\pi)^k} \sum r_{\mathbf{j}}\, e^{-i \mathbf{j} \cdot \boldsymbol{\lambda}} .$$

Let I_N be the hypercube (in k-space) of lattice points $\mathbf{j}$ with all components integers j_s, $1 \le j_s \le N$. An estimate $r_{\mathbf{j}}^{(N)}$ of $r_{\mathbf{j}}$ is given by

$$r_{\mathbf{j}}^{(N)} = N^{-k} \sum_{\mathbf{u}, \mathbf{u}+\mathbf{j} \in I_N} X_{\mathbf{u}}\, X_{\mathbf{u}+\mathbf{j}}$$

assuming that one observes the random field on the index set I_N. An estimate $f_N(\boldsymbol{\lambda})$ of $f(\boldsymbol{\lambda})$ is then given by

$$f_N(\boldsymbol{\lambda}) = (2\pi)^{-k} \sum_{|j_1|, \ldots, |j_k| \le N} r_{\mathbf{j}}^{(N)}\, w_{\mathbf{j}}^{(N)}\, e^{-i \mathbf{j} \cdot \boldsymbol{\lambda}}$$

where

$$w_{\mathbf{j}}^{(N)} = a(\mathbf{j}\, b_N),\ a(\mathbf{0}) = 1\ ,$$

and $a(\mathbf{x})$ *is assumed to be an even* $(a(\mathbf{x}) = a(-\mathbf{x}))$ *continuous function of bounded support.* Here $b_N \to 0$ with $N\, b_N \to \infty$ as $N \to \infty$. The following Theorem can be proved with an argument like that of Theorem 4.

Theorem 7. *Let* $X = \{X_{\mathbf{j}}\}$ *be a strictly stationary strongly mixing random field with* $E\, X_{\mathbf{j}} \equiv 0$. *Assume that the cumulant functions up to eighth order are absolutely summable. Also let the spectral density estimate* $f_N(\boldsymbol{\lambda})$ *have weights* $w_{\mathbf{j}}^{(N)}$ *defined in terms of a function* $a(\cdot)$ *satisfying the conditions specified above. It then follows that* $(N\, b_N)^{k/2}\, [f_N(\boldsymbol{\lambda}) - E\, f_N(\boldsymbol{\lambda})]$ *is asymptotically normally distributed with mean zero and variance*

$$(2\,\pi)^k \left\{1 + \eta(2\,\lambda_1) \ldots \eta(2\,\lambda_k)\right\} f^2(\boldsymbol{\lambda}) \int W^2(\boldsymbol{\alpha})\, d\boldsymbol{\alpha}$$

where

$$W(\boldsymbol{\alpha}) = \frac{1}{(2\,\pi)^k} \int a(\mathbf{u})\, e^{-i\mathbf{u}\cdot\boldsymbol{\alpha}}\, d\mathbf{u}\ .$$

The conclusion of this Theorem will still hold if we allow $a(\cdot)$ to have a jump discontinuity on a smooth bounded surface of finite area.

Problems

1. Let $X_1(n)$, $X_2(n)$, $n = 1, \ldots, N$ be two components of a weakly stationary vector-valued process with mean zero, and an absolutely continuous spectral distribution function with continuous spectral density. The cross-spectral density of X with Y

$$f_{12}(\lambda) = \frac{1}{2\,\pi} \sum_k \operatorname{cov}\big(X_1(n),\, X_2(n-k)\big)\, e^{-ik\lambda}$$

is generally complex-valued even if the processes X_1 and X_2 are real. The real and imaginary parts of $g(\lambda)$ are often called the co and quad spectral density functions (see example 8 of Chapter I for an interpretation). An analogue of the periodogram is given by

$$I_{12}^{(N)}(\lambda) = \frac{1}{2\,\pi\, N}\, d_{X_1}^{(N)}(\lambda)\, \overline{d_{X_2}^{(N)}(\lambda)}\ .$$

Show that

$$E\, I_{12}^{(N)}(\lambda) \to f_{12}(\lambda)$$

as $N \to \infty$.

2. Let X_1, X_2, X_3, X_4 be four components of a vector-valued process stationary up to fourth order moments. Under the assumption of summability of cumulants up to fourth order show that

$$\lim_{N\to\infty} \operatorname{cov}\{I_{12}^{(N)}(\lambda), I_{34}^{(N)}(\mu)\}$$
$$= \eta(\lambda - \mu) f_{13}(\lambda) f_{24}(-\lambda) + \eta(\lambda + \mu) f_{14}(\lambda) f_{23}(-\lambda) .$$

Notice that here the covariance of two complex-valued random variables U, V is understood to be

$$E\{(U - E\,U)\,\overline{(V - E\,V)}\}$$

and $\eta(\lambda) = \Sigma\, \delta(\lambda + 2\,k\,\pi)$.

3. Show that the spectral estimate of section 3 can be written as a smoothed periodogram

$$f_N(\mu) = \int_{-\pi}^{\pi} W_N(\mu - \lambda)\, I_N(\lambda)\, d\lambda .$$

Prove that the estimates $f_N(\mu_j)$ at distinct values μ_j, $0 \le \mu_j \le \pi$, are asymptotically uncorrelated if $f(\mu_j) > 0$ and one has summability of cumulants up to fourth order.

4. Prove that the estimates $f_N(\mu_j)$ at distinct values μ_j, $0 \le \mu_j \le \pi$, are asymptotically jointly normal under the conditions of Theorem 4.

5. Consider cross-spectral estimates

$$f_{12}^{(N)}(\mu) = \int_{-\pi}^{\pi} W_N(\mu - \lambda)\, I_{12}^{(N)}(\lambda)\, d\lambda$$

$$f_{34}^{(N)}(\mu) = \int_{-\pi}^{\pi} W_N(\mu - \lambda)\, I_{34}^{(N)}(\lambda)\, d\lambda .$$

The weight function $W_N(u)$ is given in terms of weights $w_k^{(N)}$ of the type specified in section 3. Under the assumptions of problem 2 derive the asymptotic covariance properties of these estimates.

6. Determine the joint asymptotic distribution of estimates $f_{12}^{(N)}(\mu)$, $f_{11}^{(N)}(\mu)$, $f_{22}^{(N)}(\mu)$ under assumptions of the type specified in Theorem 4.

7. Is $f_{12}^{(N)}(\mu)$ asymptotically a complex normal random variable under the assumptions of problem 6.

8. Determine the asymptotic distribution of

$$f_{12}^{(N)}(\mu)\,\{f_{11}^{(N)}(\mu)\, f_{22}^{(N)}(\mu)\}^{-1/2}$$

under the assumptions of problem 6.

Notes

5.1 In the case of an observational model

$$X_n = A_0 \cos(w_0 n + \theta_0) + V_n$$

with V_n a strongly mixing residual, a least squares procedure minimizing

$$\sum_{n=1}^{N} | X_n - A \cos(w n + \theta) |^2$$

for the estimation of the unknown parameters A_0, w_0 and θ_0 has been proposed. Under appropriate conditions, it can be shown that the least squares estimates $\hat{A}, \hat{w}, \hat{\theta}$ of A_0, w_0, θ_0 are such that $N^{1/2}(\hat{A} - A_0)$, $N^{3/2}(\hat{w} - w_0)$, $N^{1/2}(\hat{\theta} - \theta_0)$ are asymptotically normally distributed with means zero and covariance matrix

$$4 \pi f(w_0) \begin{bmatrix} 1 & 0 & 0 \\ 0 & \theta_0^2/3 & \theta_0^2/2 \\ 0 & \theta_0^2/2 & \theta_0^2 \end{bmatrix}^{-1} .$$

Here $f(w_0)$ is the spectral density of the process $\{V_n\}$ at w_0. An early discussion can be found in the paper Whittle [1952] for independent residuals. A later discussion for stationary residuals is given in Hannan [1973]. A recent treatment of related problems has been laid out in Hasan [1982].

Bartlett [1963] introduced the periodogram for a point process. This has been used as a tool in estimating the parameters of a Poisson process with sinusoidally varying intensity function

$$\lambda(t) = A \exp\{\varrho \cos(w t + \phi)\}$$

in Cox and Lewis [1966] and Lewis [1970]. This problem has been discussed at some length in Vere-Jones [1982].

5.2 Spectral estimates were initially introduced in the work of Bartlett [1948], Daniell [1946], and Tukey [1949]. Perhaps the earliest book discussing the properties of spectral estimates at some length is that of Grenander and Rosenblatt [1957]. Since then extended discussions have appeared in many books on time series analysis, among them Blackman and Tukey [1959], Hannan [1970], Anderson [1971], Brillinger [1975], and Priestley [1981] among others.

Spectral estimates have been used in the study of deterministic systems whose trajectories develop a quasi-random character (see Farmer [1982]). Initially there are discrete spectral peaks. As the chaotic character develops the spectral estimate contains less pronounced peaks and becomes flatter.

5.3 Brillinger [1972] obtained results on the asymptotic normality of second order spectral estimates for processes with stationary increments under conditions of the type considered in our Chapter VI. The details of the proof follow the lines given in Brillinger and Rosenblatt [1967] and require summability conditions on all cumulants. These processes include a large class of point processes with stationary increments. The proof we give of the asymptotic normality of spectral estimates in this section is essentially that of Rosenblatt [1984] and has the advantage that it only requires a limited number of moment conditions. It can clearly be adapted to the case of processes with stationary increments. The proof is given in more detail in the paper referred to. Also it is shown how the limited number of moment conditions can be replaced by a Lindeberg condition.

5.4 A procedure for spectral estimation based on tapering with a discrete prolate spheroidal wave function has been proposed (see Thomson [1982]). The discrete prolate spheroidal wave functions $U_k(N, w; \lambda)$ (with associated eigenvalue $\mu_k(N, w)$) are solutions of the integral equation

$$\int_{-w}^{w} \frac{\sin N\pi(\lambda - \lambda')}{\sin \pi(\lambda - \lambda')} U_k(N, w; \lambda')\, d\lambda'$$
$$= \mu_k(N, w)\, U_k(N, w; \lambda) ,$$

$-\infty < \lambda < \infty$, $k = 0, 1, \ldots, N-1$. These functions are useful in studying the extent to which sequences and their spectra can be simultaneously concentrated. See Slepian [1978] for a detailed discussion of the discrete prolate spheroidal wave functions. It is claimed in Thomson's paper that spectral estimates based on such tapering of the data are more effective than classical estimates in controlling bias when there is a very large variation in the magnitude of the spectral density estimated.

Problems of robustness in time series models are discussed in the paper of Martin and Thomson [1982]. Two models are discussed. In the first model

$$y_t = x_t + v_t$$

with y_t the observed process, x_t the process of interest and v_t a process of additive outliers. The second model is that of a linear process

$$y_t = x_t = \sum_{i=0}^{\infty} a_i\, v_{t-i}$$

with v_t a heavy-tailed nonGaussian sequence of independent, identically distributed random variables. Notions of robust prewhitening arise in data processing procedures they suggest.

There has been interest in symmetric α stable processes, $0 < \alpha < 2$, recently. A process X is symmetric α stable if all linear combinations $a_1 X(t_1) + \ldots + a_n X(t_n)$ are symmetric α stable, that is, with a characteristic function of the form $\exp(-c \mid t \mid^{\alpha})$ for a fixed α, $0 < \alpha < 2$. The second moments of these processes are infinite. The class of stationary symmetric α stable processes is considered in the paper of Cambanis, Hardin and Weran [1984]. They determine conditions under which these processes are ergodic or mixing. They also consider complex symmetric α stable processes X with a harmonic representation

$$X(t) = \int_{-\infty}^{\infty} e^{it\lambda}\, dW(\lambda), \quad -\infty < t < \infty,$$

with W a complex symmetric α stable process having independent increments (for $\lambda > 0$ since $dW(\lambda) = \overline{dW(-\lambda)}$). Stationary processes of this type with such a harmonic representation (as contrasted with normal processes) are not ergodic. A nonsymmetric counterpart of the inner product called the covariation is introduced. When $1 < \alpha < 2$ the covariation of such a process with a harmonic representation has a harmonic representation

$$\text{covariation}(X(t), X(s)) = \int_{-\infty}^{\infty} e^{i(t-s)\lambda}\, \phi(\lambda)\, d\lambda$$

under appropriate conditions. The function $\phi(\lambda)$ can be considered the analogue of a spectral density. Masry and Cambanis [1984] construct finite tapered Fourier transforms $d_T(\lambda)$ of the data $X(t)$, $|t| \leq T$. They introduce a modified periodogram

$$I_T(\lambda) = C_{p,\alpha} \,|\, d_T(\lambda) \,|^p, \quad 0 < p < \alpha/2\,.$$

A consistent estimate $\phi_T(\lambda)$ of $\phi(\lambda)$ (under certain conditions) is constructed by first smoothing $I_T(\lambda)$

$$g_T(\lambda) = \int_{-\infty}^{\infty} W_T(\lambda - u)\, I_T(u)\, du$$

and then setting $\phi_T(\lambda) = \{g_T(\lambda)\}^{\alpha/p}$. The asymptotic distribution of these estimates is not determined.

5.5 The idea of using blocks of data to construct spectral estimates is especially appealing if one wishes to test stationarity by using disjoint stretches of data to compute different spectral estimates and compare them.

5.6 The first lower bounds of this type were given for probability density estimates by Farrell [1972]. Since then results of this type have appeared for estimates of the probability density, regression, and the spectral density. An interesting paper of Kiefer [1982] discusses results of this type.

5.7 A discussion of the theoretical and experimental background of the Kolmogorov spectrum and allied questions can be found in Champagne [1978]. An extended development of the theory and problems of turbulence is presented in the books of Hinze [1975] and Monin and Yaglom [1971].

5.8 Aspects of spectral estimation in a multidimensional context are considered in the paper of McClellan [1982].

Chapter VI

Cumulant Spectral Estimates

1. Introduction

In Chapter V a class of spectral density estimates were introduced and their asymptotic properties were analyzed. There two alternative representations as given by (V. 18) and (V. 19) were given. In the early days of spectral analysis, it was common to use representation (V. 19) so that covariances were initially estimated and then Fourier transformed (with weights) so as to obtain spectral density estimates. In recent years, it has been more usual to employ the alternative representation (V. 18) of the spectral density estimate as a smoothed periodogram. Actually a Riemann sum as a discrete approximation to (V. 18) is considered. A discrete Fourier transform is used to compute the periodogram at the frequencies $2\pi j/N$, $j = 0, 1, \ldots, N-1$, and the periodogram is then smoothed to produce the spectral density estimate. For composite numbers N there is a computational advantage in making use of the finite Fourier transform and carrying out the computation by employing the fast Fourier transform. For this reason, there will be an introductory discussion of the finite Fourier transform and fast Fourier transform in section 2.

The spectral analysis dealt with in Chapter V is a spectral analysis of second order moments. We have already seen in Section III.2 that one can introduce a Fourier analysis of higher order cumulants. It is natural to speak of the results of such a harmonic analysis as cumulant spectra. A more detailed discussion of such cumulant spectra and their interpretation will be given in section 3. In Chapter V the asymptotic properties of spectral estimates were derived. Specifically, asymptotic normality was demonstrated using strong mixing and only a few auxiliary moment conditions. In principle, such a technique can also be used for the proof of asymptotic normality of appropriately designed cumulant spectral estimates (of third or higher order). A discussion of cumulant spectral estimates based on a discussion of higher order periodograms and the finite Fourier transform will be given. Then asymptotic normality of the estimates will be determined in terms of an alternative set of conditions, the use of a full set of cumulant summability conditions like those mentioned in section III.4. These cumulant summability conditions do have certain advantages but on the other hand, require the existence of all moments. Also a number of the results obtained are in a form appropriate for vector-valued processes.

2. The Discrete Fourier Transform and the Fast Fourier Transform

Suppose we are given the finite sequence of complex numbers $x(j), j = 0, 1, \ldots, N-1$. The finite Fourier transform of this sequence is given by

$$\hat{x}(k) = \sum_{j=0}^{N-1} x(j) \exp(-2\pi i j k / N)$$

where $k = 0, 1, \ldots, N-1$. The inverse transform is then

$$x(j) = \frac{1}{N} \sum_{k=0}^{N-1} \hat{x}(k) \exp(2\pi i j k / N) \tag{1}$$

with $j = 0, 1, \ldots, N-1$. The relation (1) can be seen to follow directly from the elementary identity

$$\sum_{k=0}^{N-1} \exp(-2\pi i j k / N) \exp(2\pi i l k / N) = N \delta_{j,l}$$

for $j, l = 0, 1, \ldots, N-1$. Consider the number of computations required to evaluate $\hat{x}(k)$ from the sequence $x(\cdot)$. It's clear that N multiplications and additions are required. This indicates that N^2 computations are required for the computation of $\hat{x}(k)$ at all the integers $k = 0, 1, \ldots, N-1$.

We shall at this point explain the basic idea of the fast Fourier transform. Suppose N is a composite number $N = r_1 r_2$ with two factors $r_1, r_2 > 1$. There are then symmetries of the complex exponential weights that allow one to cut down the number of computations required in the finite Fourier transform if it is laid out properly. If j and k run through the integers $0, 1, \ldots, N-1$, they can also be represented in the form

$$k = s_2 r_1 + s_1 \tag{2}$$

with $s_1 = 0, 1, \ldots, r_1 - 1$ and $s_2 = 0, 1, \ldots, r_2 - 1$,

$$j = t_1 r_2 + t_2$$

with $t_1 = 0, 1, \ldots, r_1 - 1$ and $t_2 = 0, 1, \ldots, r_2 - 1$. Now

$$\begin{aligned} x(k) &= \sum_{t_2} \sum_{t_1} x(t_1 r_2 + t_2) \exp(2\pi k \{t_1 r_2 + t_2\} / N) \\ &= \sum_{t_2} \exp(2\pi i k t_2 / N) \sum_{t_1} x(t_1 r_2 + t_2) \exp(2\pi i k t_1 / r_1) . \end{aligned} \tag{3}$$

From (2) it follows that

$$\exp(2\pi i k t_1 / r_1) = \exp(2\pi i s_1 t_1 / r_1) .$$

This implies that the inner summation on the right of (3)

$$\sum_{t_1} x(t_1 r_2 + t_2) \exp(2\pi i k t_1 / r_1) = y(s_1, t_2)$$

depends only on s_1 and t_2. The computation of $\hat{x}(k)$ can be carried out in two stages. First compute $y(s_1, t_2)$. There are $r_1 r_2$ pairs (s_1, t_2). Further r_1 multiplications and summations are required for the computation of $y(s_1, t_2)$. The $r_1 r_2$ pairs imply that $r_1^2 r_2$ such operations are required for the full computation of $y(\cdot, \cdot)$. Then the computation of $\hat{x}(k)$ in terms of $y(\cdot, \cdot)$ requires r_2 further operations. Since there are N values of k, the additional operations are $r_2 N$ in number. All together

$$r_1^2 r_2 + r_2 N = N(r_1 + r_2)$$

operations are required and this is a good deal smaller than N^2 operations. If N has a factorization $N = r_1 r_2 \cdots r_p$ with $p > 2$, the technique used above can be used in a nested manner to reduce the computation to one of $N(r_1 + r_2 + \cdots + r_p)$ computations. If N is of the form $N = 2^p$, the number of operations required is $2\, p\, N$ which is $O(N \log N)$ as contrasted with N^2 operations.

The fast Fourier transform was propesed in a paper of Cooley and Tukey [1965]. Later it was discovered that related ideas had been presented as early as 1924 in a work of Runge and König [1924]. For a much more detailed discussion of the fast Fourier transform and an interesting digression on history see the paper of Cooley, Lewis, and Welch [1977].

3. Vector Valued Processes

Up to this point we have generally dealt with real or complexvalued processes. From this point on, vector-valued processes will be dealt with often. For convenience, at this point the notation will be changed. Let

$$\mathbf{X}(t) = \begin{pmatrix} X_1(t) \\ \vdots \\ X_k(t) \end{pmatrix}, \quad E\,\mathbf{X}(t) \equiv \mathbf{0}\,,$$

$t = \ldots, -1, 0, 1, \ldots$ be a k-vector valued process. We also assume from this point on in the chapter that all moments exist. It has already been noted in problem 7 of Chapter I that $\mathbf{X}(t)$ has a vector-valued Fourier representation

$$\mathbf{X}(t) = \int_{-\pi}^{\pi} \exp(i\,t\,\lambda)\, d\mathbf{Z}(\lambda)$$

with $Z(\lambda) = \{Z_a(\lambda);\, a = 1, \ldots, k\}$ (a column vector) a process of orthogonal increments, that is,

$$E\{d\mathbf{Z}(\lambda)\, d\mathbf{Z}(\mu)'\} = \delta(\lambda - \mu)\, dG(\lambda)$$

with $\delta(\lambda)$ the Dirac delta function and $G(\lambda)$ the $k \times k$ Hermitian matrix-valued spectral distribution function of the process $\mathbf{X}(t)$. The covariance sequence $E\,\mathbf{X}(t)\,\mathbf{X}(\tau)' = r(t - \tau)$ is a sequence of $k \times k$ matrices having the Fourier representation

$$r(t) = \int_{-\pi}^{\pi} \exp\{i\,t\,\lambda\}\, dG(\lambda)$$

in terms of the spectral distribution function G. The spectral distribution function is nonnegative in the sense that each increment $G(\lambda + h) - G(\lambda)$, $h > 0$, is a positive semidefinite matrix. All these remarks on the spectral representation of a vector-valued weakly stationary process follow from the univariate results derived in Chapter I by considering the univariate processes $\boldsymbol{\alpha} \cdot \mathbf{X}(t)$ obtained by taking inner products of $\mathbf{X}(t)$ with fixed k-vectors $\boldsymbol{\alpha}$. If G is absolutely continuous, it's derivative $g(\lambda) = G'(\lambda)$ is called the spectral density of the process. $g(\lambda)$ is a $k \times k$ *matrix-valued nonnegative definite function.* The diagonal elements $g_{aa}(\lambda)$ are the spectral densities of the components $X_a(t)$, $a = 1, \ldots, k$, while the off-diagonal entries $g_{a,b}(\lambda)$, $a \neq b$, are the *cross-spectral densities* of the components $X_a(t)$ and $X_b(t)$. The cross-spectral densities are generally complex-valued even if the process $\mathbf{X}(t)$ has real components. The real and imaginary parts of $g_{ab}(\lambda)$ are called the *co-spectral* and *quadrature spectral densities* of $X_a(t)$ and $X_b(t)$. Problem 8 of Chapter I gives a natural and interesting interpretation of the co- and quadrature spectrum. *We shall assume that the process* $\mathbf{X}(t)$ *has real-valued components.* In that case, as indicated in example 8 of Chapter I

$$dG(\lambda) = \overline{dG(-\lambda)} \tag{4}$$

and

$$d\mathbf{Z}(\lambda) = \overline{d\mathbf{Z}(-\lambda)}\,.$$

The existence of moments and stationarity imply that the moments of the process satisfy

$$\begin{aligned} m_{a_1,\ldots,a_s}(t_1,\ldots,t_s) &= E\{X_{a_1}(t_1)\cdots X_{a_s}(t_s)\} \\ &= m_{a_1,\ldots,a_s}(t+t_1,\ldots,t+t_s) \end{aligned}$$

for all integers t. It is convenient to introduce a periodic form of the delta function. At this point we shall assume that the moments m have a Fourier representation in terms of functions G of bounded variation

$$\begin{aligned} m_{a_1,\ldots,a_s}&(t_1,\ldots,t_s) \\ &= \int_{-\pi}^{\pi}\cdots\int_{-\pi}^{\pi} \exp\left\{i\sum_1^s t_j\,\omega_j\right\} dG_{a_1,\ldots,a_s}(\omega_1,\ldots,\omega_s) \end{aligned} \tag{5}$$

with

$$E\left\{\prod_1^s dZ_{a_j}(\omega_j)\right\} = \eta\left(\sum_1^s \omega_j\right) dG_{a_1,\ldots,a_s}(\omega_1,\ldots,\omega_s)\,.$$

Notice that $dG_{a_1,\ldots,a_s}$ must be zero unless

$$\sum_1^s \omega_j \equiv 0 \text{ modulo } 2\pi\,.$$

This follows from the assumption of stationarity for the process. The assumption of such a Fourier representation (5) for moments higher than the second is not valid for all stationary processes with finite moments (see Sinai [1963]). However, it will be shown to be valid for a reasonably broad class.

Let the joint cumulant of $X_{a_1}(t_1), \ldots, X_{a_s}(t_s)$ be denoted by $c_{a_1, \ldots, a_s}(t_1, \ldots, t_s)$. The assumption (5) for all moments is equivalent to the representation

$$(6) \qquad c_{a_1, \ldots, a_s}(t_1, \ldots, t_s) = c_{a_1, \ldots, a_s}(t + t_1, \ldots, t + t_s)$$

$$= \int_{-\pi}^{\pi} \cdots \int_{-\pi}^{\pi} \exp\left\{i \sum_1^s t_j \omega_j\right\} c\{dZ_{a_j}(\omega_j);\ j = 1, \ldots, s\}$$

with the cumulant

$$c\{dZ_{a_j}(\omega_j);\ j = 1, \ldots, s\} = \eta\left(\sum_1^s \omega_j\right) dF_{a_1, \ldots, a_s}(\omega_1, \ldots, \omega_s)$$

and $F_{a_1, \ldots, a_s}$ of bounded variation with $dF_{a_1, \ldots, a_s} = 0$ unless $\Sigma_1^s \omega_j \equiv 0$ modulo 2π. Because of the stationarity of $X(t)$ the cumulants $c_{a_1, \ldots, a_s}(t_1, \ldots, t_s)$ (as well as the corresponding moments) can be regarded as functions of any set of $s-1$ of the t variables. *We shall assume that the cumulants $c_{a_1, \ldots, a_s}$ are absolutely summable as functions of $s-1$ of the t variables.* It then follows that

$$(7) \qquad dF_{a_1, \ldots, a_s}(\omega_1, \ldots, \omega_s)\, \eta\left(\sum_1^s \omega_j\right)$$

$$= f_{a_1, \ldots, a_s}(\omega_1, \ldots, \omega_s)\, \eta\left(\sum_1^s \omega_j\right) d\omega_1 \ldots d\omega_s .$$

Here the function f is written as a function of s variables even though it is zero off the manifold $\Sigma_1^s \omega_j \equiv 0$ modulo 2π. One can also show that f is a continuous function on $\Sigma_1^s \omega_j \equiv 0$ modulo 2π. We shall say that the s-tuple $(\omega_1, \ldots, \omega_s)$ of frequencies corresponding to $c_{a_1, \ldots, a_s}$ lies on a *proper submanifold* of

$$(8) \qquad \sum_1^s \omega_j \equiv 0 \text{ modulo } 2\pi$$

if it not only lies on (8) but also on

$$\sum_{j \in J} \omega_j \equiv 0 \text{ modulo } 2\pi$$

with J a proper subset of the set of integers $1, 2, \ldots, s$. One can then verify that

$$dG_{a_1, \ldots, a_s}(\omega_1, \ldots, \omega_s)\, \eta\left(\sum_1^s \omega_j\right)$$

$$= f_{a_1, \ldots, a_s}(\omega_1, \ldots, \omega_s)\, \eta\left(\sum_1^s \omega_j\right) d\omega_1 \cdots d\omega_s$$

if $(\omega_1, \ldots, \omega_s)$ does not lie on a proper submanifold of $\Sigma_1^s \omega_j \equiv 0$ modulo 2π. Actually one deals with cumulants and their transforms rather than moments and their transforms so as to avoid difficulties associated with proper submanifolds. Notice that in the case $s = 2$ the cumulants are just covariances. It is reasonable to call $F_{a_1, \ldots, a_s}$ an sth *order cumulant spectral distribution function* and $f_{a_1, \ldots, a_s}$ an sth *order cumulant spectral density* of the process $X(t)$.

If the (second order) spectral distribution function is absolutely continuous (differentiable), relation (4) implies that the cross-spectral density g_{a_1, a_2} satisfies

$$g_{a_1, a_2}(\lambda) = \overline{g_{a_1, a_2}(-\lambda)}\,. \tag{9}$$

The fact that the components of $X(t)$ are assumed to be real-valued implies that the following analogue of (4) for sth order cumulant spectral densities is valid

$$f_{a_1, \ldots, a_s}(\omega_1, \ldots, \omega_s) = \bar{f}_{a_1, \ldots, a_s}(-\omega_1, \ldots, -\omega_s)\,. \tag{10}$$

It has already been noted that the moments and cumulants of order s depend on only $s - 1$ t variables. For convenience, we shall often write these as functions of the $s - 1$ variables v as follows

$$\begin{aligned} m'_{a_1, \ldots, a_s}(v_1, \ldots, v_{s-1}) &= m_{a_1, \ldots, a_s}(t + v_1, \ldots, t + v_{k-1}, t) \\ c'_{a_1, \ldots, a_s}(v_1, \ldots, v_{s-1}) &= c_{a_1, \ldots, a_s}(t + v_1, \ldots, t + v_{k-1}, t) \end{aligned} \tag{11}$$

even though this leads to an asymmetry. Also since $f_{a_1, \ldots, a_s}(\omega_1, \ldots, \omega_s)$ is defined only for s-tuples $(\omega_1, \ldots, \omega_s)$ satisfying (8) it is convenient to introduce a contracted form of the spectral density depending on only $s - 1$ frequencies

$$f'_{a_1, \ldots, a_s}(\lambda_1, \ldots, \lambda_{s-1}) = f_{a_1, \ldots, a_s}(\lambda_1, \ldots, \lambda_{s-1}, \lambda_s) \tag{12}$$

where it is understood that $\Sigma_1^s \lambda_j \equiv 0$ modulo 2π. The representation (12) also implies that $|\lambda_j| \leq \pi$, $j = 1, \ldots, s$. We can now also see why the contracted form for cumulants (11) was chosen. Both (11) and (12) imply that

$$\begin{aligned} &f'_{a_1, \ldots, a_s}(\lambda_1, \ldots, \lambda_{s-1}) \\ &\quad = (2\pi)^{-s+1} \sum_{v_1, \ldots, v_{s-1}} \exp\left\{-i \sum_1^{s-1} \lambda_j v_j\right\} c'_{a_1, \ldots, a_s}(v_1, \ldots, v_{k-1})\,. \end{aligned} \tag{13}$$

The assumption of absolute summability of the cumulants $c'_{a_1, \ldots, a_s}(v_1, \ldots, v_{s-1})$ implies that $f'_{a_1, \ldots, a_s}(\lambda_1, \ldots, \lambda_{s-1})$ can be considered as a continuous function on the $(s-1)$-dimensional torus $|\lambda_j| \leq \pi$, $j = 1, \ldots, s-1$. The points $\lambda_j = \pm\pi$ are identified with each other. When $s = 2$ we need only consider $\lambda_1 + \lambda_2 = 0$ and $f'_{a_1, a_2}(\lambda_1) = g_{a_1, a_2}(\lambda_1)$ for $0 \leq \lambda_1 \leq \pi$ because of (5). In the case of third order spectra, $s = 3$, the 2-dimensional torus is sliced into three sections corresponding to $\lambda_1 + \lambda_2 + \lambda_3 = 0$, $\lambda_1 + \lambda_2 + \lambda_3 = \pi$ and $\lambda_1 + \lambda_2 + \lambda_3 = -\pi$. The three sections are given in the accompanying figures.

$\lambda_1 + \lambda_2 + \lambda_3 = 0$

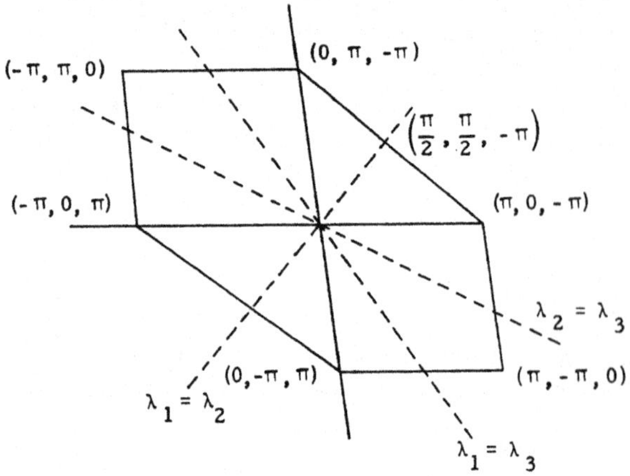

$\lambda_1 + \lambda_2 + \lambda_3 = -2\pi$

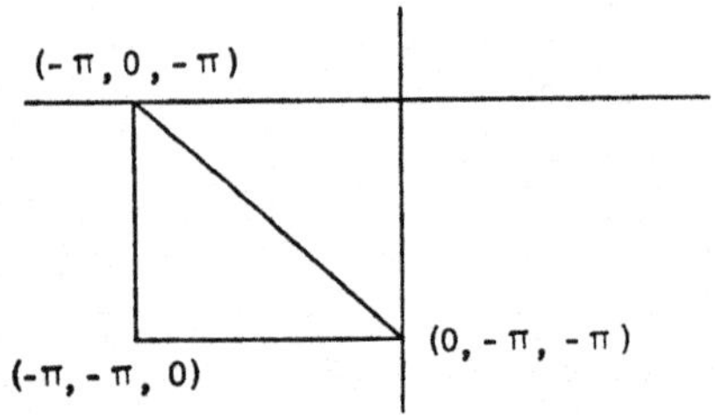

$\lambda_1 + \lambda_2 + \lambda_3 = 2\pi$

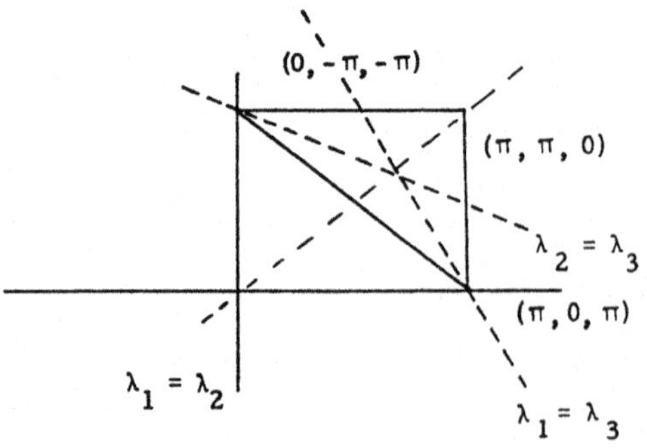

In the case of $f'_{a_1, a_2, a_3}(\lambda_1, \lambda_2)$ with the a_i distinct, only the part of the square $|\lambda_1|, |\lambda_2| \leq \pi$ to the right of $\lambda_1 = 0$ has to be considered because of relation (10). If some of the a_i's coincide, additional symmetries are introduced and one needs to consider only a smaller part of the region. As an example, let us consider the case in which $a_i = a$, $i = 1, 2, 3$. For the region $\lambda_1 + \lambda_2 + \lambda_3 = 0$ it is clear from the additional symmetries obtained by interchanging the subscripts and variables consistently that

$$f'_{a,a,a}(\lambda_1, \lambda_2) = f'_{a,a,a}(\lambda_2, \lambda_1) = f'_{a,a,a}(-\lambda_1 - \lambda_2, \lambda_2) . \tag{14}$$

The dotted lines represent the lines of symmetry $\lambda_1 = \lambda_2$, $\lambda_1 = \lambda_3 = -\lambda_1 - \lambda_2$, $\lambda_2 = \lambda_3 = -\lambda_1 - \lambda_2$. The relations (10) and (14) imply that one need only consider the triangle with vertices $(0, 0, 0)$, $\left(\frac{\pi}{2}, \frac{\pi}{2}, -\pi\right)$ and $(\pi, 0, -\pi)$ in the plane $\lambda_1 + \lambda_2 + \lambda_3 = 0$. Corresponding symmetries in the remaining regions indicate that one only needs in addition the triangle with vertices $\left(\frac{\pi}{2}, \frac{\pi}{2}, \pi\right)$, $\left(\frac{2\pi}{3}, \frac{2\pi}{3}, \frac{2\pi}{3}\right)$ and $(\pi, 0, \pi)$ in the plane $\lambda_1 + \lambda_2 + \lambda_3 = 2\pi$. Further discussion of third order as well as fourth order spectra can be found in Brillinger and Rosenblatt [1967].

At this point we introduce the following *assumption I: Let the strictly stationary k-vector valued process $X(t)$ satisfy*

$$\sum_{v_1, \ldots, v_{s-1} = -\infty}^{\infty} (1 + |v_j|) \, | c'_{a_1, \ldots, a_s}(v_1, \ldots, v_{s-1}) | < \infty \tag{15}$$

for $j = 1, \ldots, s-1$ and any s-tuple $a_1, \ldots, a_s$ with $s = 2, 3, \ldots$.

Since the cumulant spectral densities are given by (13), it is clear that the preceding assumption implies that the spectral densities have bounded uniformly continuous gradients.

Let us now introduce the finite Fourier transform $d_a^{(T)}(\lambda)$ based on observations $X_a(t)$, $t = 0, 1, \ldots, T-1$

$$d_a^{(T)}(\lambda) = \sum_{t=0}^{T-1} X_a(t) \exp(-i \lambda t) .$$

The second order periodogram of the univariate series $X_a(t)$ is given by

$$I_a^{(T)}(\lambda) = (2\pi T)^{-1} | d_a^{(T)}(\lambda) |^2 .$$

A higher order analogue of the second order periodogram is

$$I_{a_1, \ldots, a_s}^{(T)}(\lambda_1, \ldots, \lambda_s) = (2\pi)^{-s+1} T^{-1} \prod_{j=1}^{s} d_{a_j}^{(T)}(\lambda_j)$$

with $\Sigma_1^s \lambda_j \equiv 0$ modulo 2π. Let us set

$$\Delta_T(\lambda) = \sum_{t=0}^{T-1} \exp(-i \lambda t) = \exp\left(i \lambda (T-1)/2\right) \sin(\lambda T/2) / \sin(\lambda/2) .$$

This is essentially a Dirichlet kernel. Notice that $\Delta_T(\lambda) = T$ if $\lambda \equiv 0$ modulo 2π while $\Delta_T(\lambda) = 0$ if $\lambda = 2\pi n/T$ with n an integer that is not a multiple of T.

We shall first estimate cumulants of the finite Fourier transforms $d_a^{(T)}(\lambda)$ in the following lemma.

Lemma 1. Assume that (15) *holds for* $j = 1, \ldots, s-1$. *Then the cumulant*

$$c\{d_{a_1}^{(T)}(\lambda_1), \ldots, d_{a_s}^{(T)}(\lambda_s)\} = (2\pi)^{s-1} \Delta_T\left(\sum_1^s \lambda_j\right) f'_{a_1,\ldots,a_s}(\lambda_1, \ldots, \lambda_{s-1}) + O(1)$$

with the error term $O(1)$ *uniform for all* $\lambda_1, \ldots, \lambda_s$.

Because of multilinearity the cumulant

$$(16) \qquad c\{d_{a_1}^{(T)}(\lambda_1), \ldots, d_{a_s}^{(T)}(\lambda_s)\} = \sum_{t_1=0}^{T-1} \cdots \sum_{t_s=0}^{T-1} \exp\left\{-i\sum_1^s \lambda_j t_j\right\} c_{a_1,\ldots,a_s}(t_1, \ldots, t_s).$$

Set $u_i = t_i - t_s$, $i = 1, \ldots, s-1$ and let $\lambda = \Sigma_1^s \lambda_j$,

$$t_\alpha = -\min(u_1, \ldots, u_{s-1}, 0) \quad \text{and} \quad t_\beta = T - 1 - \max(u_1, \ldots, u_{s-1}, 0).$$

First of all it is clear that $|u_j| \le T-1$ for $j = 1, \ldots, s-1$. For values of the u_j's in this range we have $0 \le t_\alpha \le t_\beta \le T-1$. The expression on the right of (16) can be rewritten as

$$(17) \qquad \sum_{u_1=-T+1}^{T-1} \cdots \sum_{u_{s-1}=-T+1}^{T-1} \exp\left\{-i\sum_1^{s-1} \lambda_j u_j\right\} c'_{a_1,\ldots,a_s}(u_1, \ldots, u_{k-1}) \sum_{t=t_\alpha}^{t_\beta} \exp\{-i\lambda t\}.$$

If $t_\beta < t_\alpha$ the sum over t in (17) is understood to be zero. Notice that

$$\sum_{t=0}^{t_\alpha-1} \exp\{-i\lambda t\} \le |t_\alpha| \le |u_1| + \cdots + |u_{s-1}|,$$

$$\left|\sum_{t=t_\beta+1}^{T-1} \exp\{-i\lambda t\}\right| \le |T-1-t_\beta| \le |u_1| + \cdots + |u_{s-1}|.$$

These observations imply that

$$\left|(2\pi)^{s-1} f'_{a_1,\ldots,a_s}(\lambda_1, \ldots, \lambda_{s-1}) \Delta_T\left(\sum_1^s \lambda_j\right) - c\left\{d_{a_1}^{(T)}(\lambda_1), \ldots, d_{a_s}^{(T)}(\lambda_s)\right\}\right|$$

$$\le 2\sum_{-\infty}^{\infty} \cdots \sum_{-\infty}^{\infty} (|u_1| + \cdots + |u_{s-1}|)\,|c'_{a_1,\ldots,a_s}(u_1, \ldots, u_{s-1})| = O(1).$$

The estimate for cumulants of finite Fourier transforms obtained in the preceding Lemma allows us to get estimates for the mean and covariance of the higher order periodograms.

Lemma 2. Let $X(t)$ be a k-vector strictly stationary process satisfying assumption I. The mean of the sth *order periodogram is such that*

$$E\, I^{(T)}_{a_1, \ldots, a_s}(\lambda_1, \ldots, \lambda_s) = f_{a_1, \ldots, a_s}(\lambda_1, \ldots, \lambda_s) + O(T^{-1}) \tag{18}$$

provided $\Sigma^s_{=1} \lambda_j \equiv 0 \bmod 2\pi$ but $\lambda_1, \ldots, \lambda_s$ do not lie in a proper submanifold. The error term $O(T^{-1})$ is not uniform unless all the λ_j's are of the form $2\pi m T^{-1}$ with m an integer. Further the second order central moment

$$\begin{aligned} T^{-s+2} \{ & E[I^{(T)}_{a_1, \ldots, a_s}(\lambda_1, \ldots, \lambda_s)\, I^{(T)}_{b_1, \ldots, b_s}(\mu_1, \ldots, \mu_s)] \\ & - E\, I^{(T)}_{a_1, \ldots, a_s}(\lambda_1, \ldots, \lambda_s)\, E\, I^{(T)}_{b_1, \ldots, b_s}(\mu_1, \ldots, \mu_s)\} \\ & = (2\pi)^{2-s} \sum \prod_{j=1}^{s} \frac{\sin \frac{1}{2} T(\lambda_j + \mu_{Pj})}{T \sin \frac{1}{2}(\lambda_j + \mu_{Pj})} f_{a_j, b_{P_j}}(\lambda_j) + O(T^{-1}) \end{aligned} \tag{19}$$

where the summation extends over all permutations P of 1, 2, ..., s and it is assumed that $(\lambda_1, \ldots, \lambda_s)$ and $(\mu_1, \ldots, \mu_s)$ do not lie on proper submanifolds but do satisfy

$$\sum_{j=1}^{s} \lambda_j, \sum_{j=1}^{s} \mu_j \equiv 0 \text{ modulo } 2\pi .$$

Let us first deal with (18). Consider all partitions $(\nu_1, \ldots, \nu_q)$ of $(1, \ldots, s)$. Let $a_\nu = \{a_j : j \in \nu\}$, $\tilde{\lambda}_\nu = \Sigma_{j \in \nu} \lambda_j$ and $\lambda'_\nu = \{\lambda_j : j \in \nu$ with the last λ suppressed$\}$. One can then show that

$$\begin{aligned} & E\, I^{(T)}_{a_1, \ldots, a_s}(\lambda_1, \ldots, \lambda_s) \\ & = (2\pi)^{-s+1} T^{-1} \sum_{\nu} (2\pi)^{s-q} [\Delta_T(\tilde{\lambda}_{\nu_1}) f'_{a_{\nu_1}}(\lambda'_{\nu_1}) + O(1)] \\ & \qquad \cdots [\Delta_T(\tilde{\lambda}_{\nu_q}) f'_{a_{\nu_q}}(\lambda'_{\nu_p}) + O(1)] \end{aligned} \tag{20}$$

by using Lemma 1. First notice that $\Delta_T(\lambda)$ is bounded as $T \to \infty$ if λ is not a multiple of 2π. The main term of (20) arises from the partition consisting of one set $\nu = (1, \ldots, s)$ and is

$$\Delta_T(\tilde{\lambda}_\nu) f'_{a_\nu}(\lambda'_\nu) = f'_{a_1, \ldots, a_s}(\lambda_1, \ldots, \lambda_{s-1})$$

since $(\lambda_1, \ldots, \lambda_{s-1})$ does not lie in a proper submanifold. The estimate (18) then follows directly.

Now consider (19). Given the random variables $y_1, \ldots, y_{2s}$

$$E\left(\prod_{j=1}^{2s} y_j\right) - E\left(\prod_{j=1}^{s} y_j\right) E\left(\prod_{j=s+1}^{2s} y_j\right) = \sum C_{\nu_1} \cdots C_{\nu_p} \tag{21}$$

where the sum on the right is over the indecomposable partitions $\{\nu_1, \ldots, \nu_q\}$ of the table

$$\begin{array}{c} 1 \cdots s \\ \\ s+1 \cdots 2s \end{array}$$

and C_ν is the cumulant of the y's with subscripts in ν. If $y_j = d_{a_j}^{(T)}(\lambda_j)$, $j = 1, \ldots, 2s$ with $\lambda_{s+j} = \mu_j$ for $j = 1, \ldots, s$, by making use of Lemma we find that the term for partition ν on the right of (21) can be written

$$\sum_{\nu} (2\pi)^{2s-q} [\Delta_T(\tilde{\lambda}_{\nu_1}) f'_{a_{\lambda_1}}(\lambda'_{\nu_1}) + O(1)] \cdots [\Delta_T(\tilde{\lambda}_{\nu_q}) f'_{a_{\nu_q}}(\lambda'_{\nu_q}) + O(1)] . \tag{22}$$

From (22) it is clear that the principal terms are those with $q = s$ and the partition sets of frequencies $(\lambda_1, \mu_{P1}), \ldots, (\lambda_s, \mu_{P_s})$ with P a permutation of the integers $1, \ldots, s$. The expression of largest order of magnitude will be obtained when $\mu_{P1} = -\lambda_1, \ldots, \mu_{P_s} = -\lambda_s$. The result (19) follows directly from these observations. Notice that the covariance of the periodogram (as a complex-valued random variable) is obtained by setting $\mu_1 = -\lambda_1, \ldots, \mu_s = -\lambda_s$.

The first result (18) of Lemma 2 indicates that the periodogram of order s is an asymptotically unbiased estimate of the corresponding cumulant spectral density as $T \to \infty$ as long as $(\lambda_1, \ldots, \lambda_s)$ does not lie in a proper submanifold. However, if it does lie in a proper submanifold, the expected value of the periodogram will typically diverge. In Chapter V we noticed that the second order periodogram was not a reasonable estimate of the second order spectral density at a frequency λ since its variance tended to a positive value as the sample size T increased. However, the periodogram values at distinct nonnegative frequencies are asymptotically uncorrelated and this suggested using smoothed versions of the periodogram as effective estimates of the spectral density. As indicated in Chapter V, this is a viable and meaningful program. The second result (19) of Lemma 2 tells us that the variance of a periodogram of order s (≥ 2) is of order T^{s-2} as $T \to \infty$. Thus the periodogram of order $s > 2$ has even worse properties than that of order 2. Nonetheless this result suggests that periodogram values at distinct frequency s-tuples are asymptotically uncorrelated as long as they are off proper submanifolds.

4. Smoothed Periodograms

The asymptotic properties of the mean and covariance of periodograms were examined in section 3. Our object is now to show that certain weighted or smoothed periodograms will serve as reasonable estimates of the corresponding spectra of order s. Let $W(u_1, \ldots, u_s)$ be a bounded continuous weight function on $\Sigma_1^s u_j = 0$ such that

$$\int_{-\infty}^{\infty} \cdots \int W(u_1, \ldots, u_s) \, \delta \left(\sum_1^s u_j \right) du_1 \cdots du_s = 1$$

and

$$W(-u_1, \ldots, -u_s) = W(u_1, \ldots, u_s) .$$

The following assumption is made concerning these weight functions.

Assumption II. There are positive constants A and ε such that

$$\left| W \left(u_1, \ldots, u_{s-1}, -\sum_1^{s-1} u_j \right) \right| \leq A \left(1 + \left[\sum_1^{s-1} u_j^2 \right]^{1/2} \right)^{-(k+\varepsilon-1)}$$

and

$$\left| \frac{\partial}{\partial u_l} W \left(u_1, \ldots, u_{s-1}, -\sum_1^{s-1} u_j \right) \right| \leq A \left(1 + \left[\sum_1^{s-1} u_j^2 \right]^{1/2} \right)^{-(k+\varepsilon-1)}$$

for $l = 1, \ldots, s-1$.

Let b_T be a bandwidth parameter tending to zero as $T \to \infty$. Set

$$W_T(u_1, \ldots, u_s) = b_T^{-s+1} W(b_T^{-1} u_1, \ldots, b_T^{-1} u_s) .$$

b_T will be chosen so that $b_T^{s-1} T \to \infty$ as $T \to \infty$ also. A plausible estimate of $f_{a_1, \ldots, a_s}(\lambda_1, \ldots, \lambda_s)$ is given by

(23) $$f_{a_1, \ldots, a_s}^{(T)}(\lambda_1, \ldots, \lambda_s)$$

$$= (2\pi)^{s-1} T^{-s+1} \sum_{k_i=-\infty}^{\infty} W_T \left(\lambda_1 - \frac{2\pi k_1}{T}, \ldots, \lambda_s - \frac{2\pi k_s}{T} \right)$$

$$\Phi \left(\frac{2\pi k_1}{T}, \ldots, \frac{2\pi k_s}{T} \right) I_{a_1, \ldots, a_s}^{(T)} \left(\frac{2\pi k_1}{T}, \ldots, \frac{2\pi k_s}{T} \right)$$

where $\Sigma_1^s \lambda_u \equiv 0$ modulo 2π and

$$\Phi(u_1, \ldots, u_s) = 1 \quad \text{if} \quad \sum_1^s u_j \equiv 0 \text{ modulo } 2\pi$$

but $\Sigma_{j \in J} u_j \not\equiv 0$ modulo 2π where J is any nonvacuous proper subset of $1, \ldots, s$ and $\Phi(u_1, \ldots, u_s) = 0$ otherwise.

Bounds of the type given in Assumption II are useful in that they allow us to approximate integrals by Riemann sums effectively. This is explicitly noted in the following lemma.

Lemma 3. Let $h(u_1, \ldots, u_s)$ be a function that satisfies

$$| h(u_1, \ldots, u_s) | , \left| \frac{\partial}{\partial u_l} h(u_1, \ldots, u_s) \right| \leq A \left(1 + \left[\sum_1^s u_j^2 \right]^{1/2} \right)^{-(k+\varepsilon)},$$

$l = 1, \ldots, s$, with A, ε positive constants. Assume that $A_j \leq B_j$ are finite constants and that $d_j > 0, j = 1, \ldots, s$. Suppose that $N_j = (B_j - A_j)/d_j, j = 1, \ldots, s$, are integers. It then follows that

$$d_1 \cdots d_s \sum_{n_1 = 0}^{N_1} \cdots \sum_{n_s = 0}^{N_s} h(A_1 + d_1 n_1, \ldots, A_s + d_s n_s)$$

$$= \int_{A_1}^{B_1} \cdots \int_{A_s}^{B_s} h(u_1, \ldots, u_s)\, du_1 \cdots du_s + R$$

and the remainder is such that $| R | \leq (\Sigma_1^s d_j) K$ with K an absolute constant.

The lemma follows on applying a Taylor formula with remainder. A direct application of Lemma 3 leads to the following remark on the mean of cumulant spectral estimates of type (23).

Theorem 1. *Let $X(t)$ be a strictly stationary k-vector valued process satisfying Assumption I. Assume that $f_{a_1, \ldots, a_s}^{(T)}(\lambda_1, \ldots, \lambda_s)$ is of type (23) with a weight function satisfying Assumption II. If $b_T \to 0$, $b_T T \to \infty$ as $T \to \infty$ it follows that*

$$E f_{a_1, \ldots, a_s}^{(T)}(\lambda_1, \ldots, \lambda_s)$$

$$= \int_{-\pi}^{\pi} \cdots \int W_T(\lambda_1 - \alpha_1, \ldots, \lambda_s - \alpha_s)\, f_{a_1, \ldots, a_s}(\alpha_1, \ldots, \alpha_s)\, \eta \left(\sum_1^s \alpha_j \right) d\alpha_1 \cdots d\alpha_s$$

$$+ O(b_T^{-1} T^{-1}) .$$

The following result estimates the covariance of spectral estimates as $T \to \infty$.

Theorem 2. *Assume that $X(t)$ satisfies the conditions specified in Theorem 1. Let $f_{a_1, \ldots, a_s}^{(T)}(\lambda_1, \ldots, \lambda_s)$ and $f_{a_1', \ldots, a_s'}^{(T)}(\mu_1, \ldots, \mu_s)$ be estimates of*

$$f_{a_1, \ldots, a_s}(\lambda_1, \ldots, \lambda_s) \text{ and } f_{a_1', \ldots, a_s'}(\mu_1, \ldots, \mu_s)$$

of the type specified in formula (23) with weight function satisfying Assumption II. It is assumed that

$$\sum_1^s \lambda_j,\ \sum_1^s \mu_j \equiv 0 \text{ modulo } 2\pi$$

and that $b_T^{s-1}\,T \to \infty$ as $b_T \to 0$ and $T \to \infty$. Then it follows that the covariance

$$\operatorname{cov}\left[f^{(T)}_{a_1,\ldots,a_s}(\lambda_1,\ldots,\lambda_s), f^{(T)}_{a_1',\ldots,a_s'}(\mu_1,\ldots,\mu_s)\right]$$

$$= 2\pi T^{-1} \sum_P \int_{-\pi}^{\pi}\cdots\int W_T(\lambda_1-\alpha_1,\ldots,\lambda_s-\alpha_s)\, W_T(\mu_1+\beta_1,\ldots,\mu_s+\beta_s)$$

$$\eta\left(\sum_1^s \alpha_j\right) \prod_1^s \{\eta(\alpha_j+\beta_{P_j})\, f_{a_j,a'_{P_j}}(\alpha_j)\}\, d\alpha_1\cdots d\alpha_s\, d\beta_1\cdots d\beta_s$$

$$+ O(b_T^{-s+2}\,T^{-1})$$

with the summation over all permutations on the integers 1, ..., s. The error term is uniform in λ's and μ's of the form $2\pi m/T$ with m an integer.

The covariance of the spectral estimates at $(\lambda_1,\ldots,\lambda_s)$ and $(\mu_1,\ldots,\mu_s)$ can be written

$$(24)\qquad (2\pi/T)^{2s-2} \sum_{r_i}\sum_{g_j} W_T\left(\lambda_1-\frac{2\pi r_1}{T},\ldots,\lambda_s-\frac{2\pi r_s}{T}\right)$$

$$W_T\left(-\mu_1-\frac{2\pi q_1}{T},\ldots,-\mu_s-\frac{2\pi q_s}{T}\right) \Phi\left(\frac{2\pi r_1}{T},\ldots,\frac{2\pi r_s}{T}\right)$$

$$\Phi\left(\frac{2\pi q_1}{T},\ldots,\frac{2\pi q_s}{T}\right)(2\pi)^{-2s+2}\,T^{-2}$$

$$\left[E\left\{\prod_1^s d^{(T)}_{a_j}\left(\frac{2\pi r_j}{T}\right)\prod_1^s d^{(T)}_{a_j'}\left(\frac{2\pi q_j}{T}\right)\right\}\right.$$

$$\left. - E\left\{\prod_1^s d^{(T)}_{a_j}\left(\frac{2\pi r_j}{T}\right)\right\} E\left\{\prod_1^s d^{(T)}_{a_j'}\left(\frac{2\pi q_j}{T}\right)\right\}\right].$$

The expression in the square brackets above can be written as

$$(25)\qquad \sum_\nu (2\pi)^{2s-p}\left[\Delta_T\left(\frac{2\pi\tilde{u}_{\nu_1}}{T}\right) f'_{b_{\nu_1}}\left(\frac{2\pi u'_{\nu_1}}{T}\right)+O(1)\right]$$

$$\cdots\left[\Delta_T\left(\frac{2\pi\tilde{u}_{\nu_p}}{T}\right) f'_{b_{\nu_p}}\left(\frac{2\pi u'_{\nu_p}}{T}\right)+O(1)\right]$$

where it is understood that one sums over all indecomposable partitions $\nu = (\nu_1, \ldots, \nu_p)$ of the table

$$\begin{matrix} 1 \cdots s \\ s+1 \cdots 2s . \end{matrix}$$

In (25) u denotes a set of r's and q's while b denotes a set of a's and a''s. On expanding (25) and inserting the result in (24), we obtain a number of terms of the form

$$(26) \qquad (2\pi/T)^{2s-2} \sum_{r_i, q_j} W_T\left(\lambda_i - \frac{2\pi r_i}{T} ;\ i = 1, \ldots, s\right) W_T\left(-\mu_j - \frac{2\pi q_j}{T} ;\ j = 1, \ldots, s\right)$$

$$\Phi\left(\frac{2\pi r_i}{T} ;\ i = 1, \ldots, s\right) \Phi\left(\frac{2\pi q_j}{T} ;\ j = 1, \ldots, s\right) (2\pi)^{-2s+2}\, T^{-2}\, (2\pi)^{2s-p}$$

$$\Delta_T\left(\frac{2\pi \tilde{u}_{\nu_1}}{T}\right) \cdots \Delta_T\left(\frac{2\pi \tilde{u}_{\nu_l}}{T}\right) f'_{b_{\nu_1}}\left(\frac{2\pi u'_{\nu_1}}{T}\right) \cdots f'_{b_{\nu_l}}\left(\frac{2\pi u'_{\nu_l}}{T}\right) O(1)$$

with $l \le p$. The sum of the $2\pi r_i/T$ as well as the sum of the $2\pi q_j/T$ is congruent to zero modulo 2π. This means that 2 of the $2s$ variables can be expressed essentially in terms of the other variables. Let us now see how many more of the variables can be essentially eliminated by using the restraints indicated by the product.

$$\Delta_T\left(\frac{2\pi \tilde{u}_{\nu_1}}{T}\right) \cdots \Delta_T\left(\frac{2\pi \tilde{u}_{\nu_l}}{T}\right).$$

If all of the $2s$ frequencies are involved in these l restraints, the number of variables can be reduced further by $l-1$. However, if all the $2s$ frequencies are not involved in the restraints, the number of variables can be reduced further by l. Set $\delta = 1$ or 0 according as to whether all the $2s$ frequencies are involved in the restraints or not. Lemma 3 implies that (26) can be approximated to the first order by

$$T^{-2s}\, (2\pi)^{-2s+2+l-\delta}\, T^{2s-2-l+\delta}\, (2\pi)^{s-p}$$

$$\int_{-\pi}^{\pi} \cdots \int W_T(\lambda_1 - r_1, \ldots, \lambda_s - r_s)\, W_T(-\mu_1 - q_1, \ldots, -\mu_s - q_s)$$

$$T^l\, \eta(\tilde{u}_{\nu_1}) \cdots \eta(\tilde{u}_{\nu_l})\, \eta\left(\sum_1^s r_j\right) \eta\left(\sum_1^s q_j\right)$$

$$f'_{b_{\nu_1}}(u'_{\nu_1}) \cdots f'_{b_{\nu_l}}(u'_{\nu_l})\, d\alpha_1 \cdots d\alpha_s\, d\beta_1 \cdots d\beta_s .$$

Since this expression is $O(T^{-2+\delta}\, b_T^{-l+\delta})$ the terms of largest order occur for $l = s$ and in that case $\delta = 1$. The result follows from this estimate.

The result of this theorem can be shown to lead to the following simpler form (27) which is stated as a corollary.

Corollary 1: Under the assumptions of Theorem 2:

$$\lim_{T\to\infty} b_T^{s-1}\, T \operatorname{cov}[f^{(T)}_{a_1,\ldots,a_s}(\lambda_1,\ldots,\lambda_s), f^{(T)}_{a'_1,\ldots,a'_s}(\mu_1,\ldots,\mu_s) \tag{27}$$

$$= 2\pi \sum_P \eta(\lambda_1 - \mu_{P_1}) \cdots \eta(\lambda_s - \mu_{P_s})\, f_{a_1 a'_{P_1}}(\lambda_1) \cdots f_{a_s a'_{P_s}}(\lambda_s)$$

$$\int \cdots \int W(u_1, \ldots, u_s)\, W(u_{P_1}, \ldots, u_{P_s})\, \delta\left(\sum_1^s u_j\right) du_1 \cdots du_s$$

where the summation is over all permutations P of the integers $1, \ldots, s$.

The following theorem describes the asymptotic distribution of cumulant spectral estimates.

Theorem 3. *Let $X(t)$ be a strictly stationary k-vector valued process satisfying Assumption I. Let $f^{(T)}_{A_j}(\lambda^{(j)})$, $j = 1, \ldots, m$, be cumulant spectral estimates*

$$f^{(T)}_{A_j}(\lambda^{(j)}) = (2\pi/T)^{s_j-1} \sum W^{(j)}_T\left(\lambda^{(j)} - \frac{2\pi r^{(j)}}{T}\right) \Phi\left(\frac{2\pi r^{(j)}}{T}\right) I^{(T)}_{A_j}\left(\frac{2\pi r^{(j)}}{T}\right) \tag{28}$$

of orders $s_1 \le s_2 \le \cdots \le s_m$ with the weight functions $W^{(j)}$ satisfying Assumption II. Here A_j denotes the indices of the s_j series involved in the jth spectral estimate. It is assumed that the bandwidth $b^{(j)}_T$ of the estimates satisfy

$$b^{(1)}_T \le \cdots \le b^{(m)}_T$$

as well as

$$b^{(j)}_T \to 0, \ \{b^{(j)}_T\}^{s_j-1}\, T \to \infty$$

as $T \to \infty$. Spectral estimates of the same order are given the same bandwidth. Consider

$$(\{b^{(j)}_T\}^{s_j-1}\, T)^{1/2}\, (f^{(T)}_{A_j}(\lambda^{(j)}) - f^{(T)}_{A_j}(\lambda^{(j)}))\,, \tag{29}$$

$j = 1, \ldots, m$. The assumptions made imply that the normalized and centered estimates (29) are asymptotically jointly normally distributed as $T \to \infty$ and estimates of different orders are asymptotically independent. The limiting covariance structure of expressions (29) of the same order is given by (27).

We shall first show that the correlation of estimates of different order tends to zero. This together with joint asymptotic normality of estimates (to be demonstrated later) will imply asymptotic independence of estimates of different orders. Consider the covariance of two normalized and centered estimates (29) of orders s_1 and s_2 with $s_1 \leq s_2$. The typical teim of the covariance is of the form

$$T^{-s_1-s_2}\{b_T^{(1)}\}^{\frac{1}{2}(s_1-1)}\{b_T^{(2)}\}^{\frac{1}{2}(s_2-1)}(2\pi)^{s_1+s_2-p}$$

$$\sum_{r^{(1)},\, r^{(2)}}\{b_T^{(1)}\}^{-s_1+1}W^{(1)}\left(b_T^{(1)-1}\left\{\lambda^{(1)}-\frac{2\pi r^{(1)}}{T}\right\}\right)$$

$$\{b_T^{(2)}\}^{-s_2+1}W^{(2)}\left(b_T^{(2)-1}\left\{\lambda^{(2)}-\frac{2\pi r^{(2)}}{T}\right\}\right)$$

$$\Phi\left(\frac{2\pi r^{(1)}}{T}\right)\Phi\left(\frac{2\pi r^{(2)}}{T}\right)\Delta_T\left(\frac{2\pi\tilde{u}_{\nu_1}}{T}\right)\cdots\Delta_T\left(\frac{2\pi\tilde{u}_{\nu_l}}{T}\right)\cdot O(1)$$

and this arises from a partition of the table

$$\begin{matrix} 1\cdots s_1 \\ s_1+1\cdots s_1+s_2 \end{matrix}$$

with p sets so that $l \leq p$. There are two restraints arising from the factors Φ, namely

$$\sum_{r^{(i)}}\frac{2\pi r^{(i)}}{T} \equiv 0 \text{ modulo } 2\pi,\quad i = 1, 2\,.$$

This leads to a reduction to $s_1 - 1$ variables in the first row and $s_2 - 1$ in the second row. The Δ_T factors lead to l additional restraints linking the s_2 variables of the second row with the s_1 variables of the first row. This is due to the fact that the partition is irreducible and the row variables cannot lie in proper submanifolds. The l restraints are used to solve for a variable in the second row in terms of variables of the first row and possibly other variables of the second row. $l - 1$ or l additional variables are eliminated depending on whether $l = p$ (all variables of the table are involved) or $l < p$. The product of the Δ_T factors contributes T^l at most. Set $\delta^T = 1$ or 0 as to whether $l = p$ or not. The sum of the product of W terms multiplied by

$$\{b_T^{(1)}\}^{-s_1+1}\{b_T^{(2)}\}^{-s_2+1+l-s}\,T^{-s_1-s_2+2+l-\delta}$$

can be approximated by a finite integral since it is a Riemann sum. The product of the remaining factors is

$$T^{-s_1-s_2+1}\,T^{s_1+s_2-2-l+\delta}\,T^l\{b_T^{(2)}\}^{-l+\delta}\{b_T^{(1)}\}^{\frac{1}{2}(s_1-1)}\{b_T^{(2)}\}^{\frac{1}{2}(s_2-1)} \tag{30}$$

$$= T^{-1+\delta}\{b_T^{(1)}\}^{\frac{1}{2}(s_2-1)}\{b_T^{(2)}\}^{\frac{1}{2}(s_2-1)-l+\delta}$$

$$= T^{-1+\delta}\,(b_T^{(1)}/b_T^{(2)})^{\frac{1}{2}(s_1-1)}\{b_T^{(2)}\}^{\frac{1}{2}(s_1-1)+\frac{1}{2}(s_2-1)-l+\delta}\,.$$

When $\delta = 1$ since $l \leq s_1$ the expression (30) is $O(1)$ if $s_1 = s_2 = l$ and is $O(b_T^{(2)})^{1/2}$ otherwise. When $\delta = 0$ since $l \leq s_1 - 1$ the expression (30) is $O(T^{-1})$. Therefore all the terms tend to zero as $T \to \infty$ except for those for which $\delta = 1$ and $l = s_1 = s_2$. These types of terms arise only in the case of covariances of terms of the same order.

At this point we shall consider the asymptotic behavior of cumulants of $J \geq 3$ of the centered and normalized expressions (29) and will show that they tend to zero as $T \to \infty$. This is enough to imply the joint asymptotic normality of these expressions since it indicates that the moments of any given linear combination tend to those of a normal random variable. The typical term of such a cumulant has the form

$$T^{-s_1-\cdots-s_J}\, T^{J/2}\, \{b_T^{(1)}\}^{\frac{1}{2}(s_1-1)} \cdots \{b_T^{(J)}\}^{\frac{1}{2}(s_J-1)}$$

$$\sum_{r^{(i)}} \{b_T^{(1)}\}^{-s_1+1}\, W^{(1)}\left(\{b_T^{(1)}\}^{-1}\left\{\lambda^{(1)} - \frac{2\pi r^{(1)}}{T}\right\}\right) \cdots$$

$$\{b_T^{(J)}\}^{-s_J+1}\, W^{(J)}\left(\{b_T^{(J)}\}^{-1}\left\{\lambda^{(1)} - \frac{2\pi r^{(J)}}{T}\right\}\right)$$

$$\Phi\left(\frac{2\pi r^{(1)}}{T}\right)\cdots\Phi\left(\frac{2\pi r^{(J)}}{T}\right) \Delta_T\left(\frac{2\pi \tilde{u}_{\nu_1}}{T}\right)\cdots\Delta_T\left(\frac{2\pi \tilde{u}_{\nu_l}}{T}\right) O(1)\,.$$

This term arises from an irreducible partition P of the table

$$\begin{matrix} 1 \cdots s_1 \\ s_1 + 1 \cdots s_1 + s_2 \\ \cdots \\ s_1 + \cdots + s_{J-1} \cdots s_1 + \cdots + s_J \end{matrix}$$

with p sets so that $l \leq p$. The factors Φ imply a reduction to $s_i - 1$ variables in the ith row, $i = 1, \ldots, J$. The Δ_T factors lead to l additional restraints. If the additional l restraints involve all variables of the table, one can eliminate $l - 1$ additional variables. If the additional l restraints do not involve all the variables of the table, one can eliminate l additional variables. Let v_i be the remaining free variables in row i, $i = 1, \ldots, J$. Set $\delta = 1$ or 0 according as to whether the l additional restraints involve all the variables of the table or not. The sum of the product of W terms multiplied by

$$\left(\prod_i \{b_T^{(i)}\}^{v_i}\right) T^{-(s_1+\cdots+s_J)+J+l-\delta}$$

can be approximated by a finite integral. Notice that

$$v_1 + \cdots + v_J = s_1 + \cdots + s_J - J - l + \delta\,. \tag{31}$$

The product of the remaining factors is

$$T^{-s_1-\cdots-s_J}\, T^{J/2} \prod_i \{b_T^{(i)}\}^{\frac{1}{2}(s_i-1)}$$

$$\left(\prod_i \{b_T^{(i)}\}^{-s_i+1}\right) T^l \left(\prod_i \{b_T^{(i)}\}^{v_i}\right) T^{v_1+\cdots+v_J}$$

(32) $$= T^{-J/2+\delta} \prod_i \{b_T^{(i)}\}^{v_i-\frac{1}{2}(s_i-1)} .$$

The restraint (31) implies that

$$v_1+\cdots+v_J-\frac{1}{2}(s_1-1)-\cdots-\frac{1}{2}(s_J-1)=\frac{s_1+\cdots+s_J}{2}-\frac{J}{2}-l+\delta .$$

If $\delta=0$ then $l\leq(s_1+\cdots+s_J)/2$ while $\delta=1$ implies that $l<(s_1+\cdots+s_J)/2$. One can then conclude that if $\delta=0$, expression (32) is $O(T^{-J/2}\{b_T^{(1)}\}^{-J/2})\rightarrow 0$ as $T\rightarrow\infty$ while if $\delta=1$ the expression is $O(T^{-J/2+1}\{b_T^{(1)}\}^{-J/2+3/2})\rightarrow 0$ as $T\rightarrow\infty$. Thus all cumulants of order $J\geq 3$ tend to zero as $T\rightarrow\infty$. Of course, the estimates (23) are generally complex-valued. However, the multilinear property of the cumulants together with the remark just made on cumulants of order $J\geq 3$ implies that all cumulants of the real and imaginary parts of the random variables (23) of order $J\geq 3$ tend to zero. They are therefore asymptotically jointly normally distributed.

5. *Aliasing and Discretely Sampled Time Series*

In many applications the basic model is that of a continuous time parameter weakly stationary process $X(t)$. It is also plausible that the process is continuous in mean square, that is,

$$\lim_{t\rightarrow s} E\,|\,X(t)-X(s)\,|^2=0 .$$

In the problems of Chapter I, it was noted that the covariance function $r(t)$, which is continuous as a consequence of the continuity in mean square of the process, has the Fourier representation

$$r(t)=\int_{-\infty}^{\infty} e^{it\lambda}\, dF(\lambda)$$

in terms of the spectral distribution function F of the process. For convenience we assume that the mean $m\equiv E\,X(t)=0$. The process itself has a Fourier representation

(33) $$X(t)=\int_{-\infty}^{\infty} e^{it\lambda}\, dZ(\lambda)$$

in terms of a process of orthogonal increments $Z(\lambda)$

$$E\, Z(\lambda) \equiv 0$$

$$E\, dZ(\lambda)\, \overline{dZ(\mu)} = \delta(\lambda - \mu)\, dF(\lambda)\,.$$

If $X(t)$ is real-valued

$$dZ(\lambda) = \overline{dZ(-\lambda)}$$

$$dF(\lambda) = dF(-\lambda)\,.$$

We shall be interested in the case in which F is absolutely continuous with spectral density $f(\lambda) = F'(\lambda)$. Even though the basic model is that of a continuous time parameter process, suppose one observes the process only at the discrete times $t\,h$. The object is to see what can be said about the spectrum of the discretely sampled process and spectral estimates based on the discrete sampling. For the process $X(t)$ discretely sampled with sampling interval h one has

$$X(t\,h) = \int_{-\infty}^{\infty} e^{ith\lambda}\, dZ(\lambda) = \int_{-\pi/h}^{\pi/h} e^{ith\lambda}\, d\,{}_hZ(\lambda)$$

with

$$d\,{}_hZ(\lambda) = \sum_{j=-\infty}^{\infty} dZ\left(\lambda + \frac{2\pi j}{h}\right), \quad |\lambda| < \frac{\pi}{h}\,.$$

The second order covariance function r has the representation

$$r(t\,h) = \int_{-\infty}^{\infty} e^{ith\lambda} f(\lambda)\, d\lambda = \int_{-\pi/h}^{\pi/h} e^{ith\lambda}\, {}_hf(\lambda)\, d\lambda\,,$$

$${}_hf(\lambda) = \sum_{j=-\infty}^{\infty} f\left(\lambda + \frac{2\pi j}{h}\right),$$

$$E\,\{d_h\, Z(\lambda)\, d_h\, Z(\mu)'\} = \delta(\lambda + \mu)\, {}_hf(\lambda)\, d\lambda\,, \quad |\lambda| < \frac{\pi}{h}\,.$$

This folding effect on the spectrum or aliasing was noted in problem 4 of Chapter I.

There are also higher order effects of aliasing due to the discrete sampling. Suppose $X(t)$ is strictly stationary with moments of order $k > 2$ finite. The moment

$$m_k(t_1, \ldots, t_k) = E[X(t_1) \ldots X(t_k)] = r_k(t_2 - t_1, \ldots, t_k - t_1)$$

depends only on the time differences $t_j - t_1$, $j = 2, \ldots, k$. Existence of the moments up to order k is equivalent to existence of cumulants

$$c_k(t_2 - t_1, \ldots, t_k - t_1) = \operatorname{cum}\,(X(t_1), \ldots, X(t_k))$$

up to order k. A representation of the form

$$c_k(\tau_1, \ldots, \tau_{k-1}) = \int \exp\left\{ i \sum_{a=1}^{k-1} \tau_a \lambda_a \right\} dG_k(\lambda_1, \ldots, \lambda_{k-1})$$

with G_k of bounded variation is not generally valid. However, absolute integrability of the cumulant function $c_k(\tau_1, \ldots, \tau_k)$ (which can be regarded as a multinomial mixing of kth order) does imply such a representation with the cumulant spectral distribution function G_k absolutely continuous with density g_k. Further

$$g_k(\lambda_1, \ldots, \lambda_{k-1}) = (2\pi)^{-k+1} \int \exp\left\{ -i \sum_{a=1}^{k-1} \tau_a \lambda_a \right\} c_k(\tau_1, \ldots, \tau_{k-1})\, d\tau_1 \ldots d\tau_{k-1}\,.$$

These cumulant spectra give us a measure of the higher order interactions of the random spectral function $Z(\lambda)$ (in the Fourier representation (33)) at the various frequencies λ in that

$$\begin{aligned} \operatorname{cum}\big(dZ(\lambda_1), \ldots, dZ(\lambda_k)\big) &= \delta(\lambda_1 + \cdots + \lambda_k)\, dG_k(\lambda_1, \ldots, \lambda_{k-1}) \\ &= \delta(\lambda_1 + \cdots + \lambda_k)\, g_k(\lambda_1, \ldots, \lambda_{k-1})\, d\lambda_1 \cdots d\lambda_{k-1}\,. \end{aligned}$$

The cumulants and cumulant spectra of order $k > 2$ are zero for Gaussian processes. For this reason they are principal interest for non-Gaussian processes or nonlinear problems.

Suppose that $X(t) = \big(X_a(t),\, a = 1, \ldots, m\big)$ is a column vector-valued weakly stationary process. There is then still a vector-valued Fourier representation of the form (33) with

$$E\, dZ(\lambda)\, dZ(\mu)' = \delta(\lambda - \mu)\, dF(\lambda)\,.$$

Notice that the matrix A' denotes the conjugated transpose of A. The spectral distribution function $F(\lambda)$ is $m \times m$ matrix-valued and Hermitian. The function $F(\lambda)$ (problem 5 of Chapter I) is bounded and nondecreasing. The covariance function $r(t) = E\, X(\tau + t)\, X(\tau)'$ is an $m \times m$ matrix-valued function. If $X(t)$ has real-valued components one has

$$\overline{dF(-\lambda)} = dF(\lambda)'\,.$$

In this multivariate context we have for the higher order moments, cumulants and spectra

$$\begin{aligned} m_{a_1, \ldots, a_k}(t_1, \ldots, t_k) &= E\, X_{a_1}(t_1)\, X_{a_2}(t_2) \ldots X_{a_k}(t_k) \\ &= r_{a_1, \ldots, a_k}(t_2 - t_1, \ldots, t_k - t_1)\,, \end{aligned}$$

$$c_{a_1, \ldots, a_k}(t_2 - t_1, \ldots, t_k - t_1) = \operatorname{cum}\big(X_{a_1}(t_1), \ldots, X_{a_k}(t_k)\big)\,,$$

$$\begin{aligned} &\operatorname{cum}\big(dZ_{a_1}(\lambda_1), \ldots, dZ_{a_k}(\lambda_k)\big) \\ &\quad = \delta(\lambda_1 + \cdots + \lambda_k)\, dG_{a_1, \ldots, a_k}(\lambda_1, \ldots, \lambda_{k-1}) \\ &\quad = \delta(\lambda_1 + \cdots + \lambda_k)\, g_{a_1, \ldots, a_k}(\lambda_1, \ldots, \lambda_{k-1})\, d\lambda_1 \ldots d\lambda_{k-1}\,. \end{aligned}$$

The higher order effects of aliasing are shown by

$$c_{a_1, \ldots, a_k}(j_1 h, \ldots, j_{k-1} h)$$

$$= \int_{-\pi/h}^{\pi/h} \cdots \int_{-\pi/h}^{\pi/h} {}_h g_{a_1, \ldots, a_k}(\lambda_1, \ldots, \lambda_{k-1}) \exp\left\{ i \sum_{u=1}^{k-1} \lambda_u j_u h \right\} d\lambda_1 \ldots d\lambda_{k-1},$$

$j_1, \ldots, j_{k-1} = 0, \pm 1, \pm 2, \ldots$, with

$${}_h g_{a_1, \ldots, a_k}(\lambda_1, \ldots, \lambda_{k-1}) =$$

$$\sum_{j_1, \ldots, j_{k-1} = -\infty}^{\infty} g_{a_1, \ldots, a_k}\left(\lambda_1 + \frac{2\pi j_1}{h}, \ldots, \lambda_{k-1} + \frac{2\pi j_{k-1}}{h}\right),$$

$$|\lambda_1|, \ldots, |\lambda_{k-1}| < \frac{\pi}{h}.$$

We first discuss spectral estimates in the univariate second order case. The general multivariate kth order case will be considered after this preliminary discussion.

Suppose $X(t)$ is a univariate stationary process observed at $t = k h$, $k = 0, 1, \ldots, N - 1$. The periodogram ${}_h I^{(T)}(\lambda)$ is given by the formula

$${}_h I^{(T)}(\lambda) = \frac{h}{2\pi T} \left| \sum_{k=0}^{T-1} X(k h) e^{-ikh\lambda} \right|^2.$$

Continuity of the spectral density f implies that

$$E\, {}_h I^{(T)}(\lambda) \to {}_h f(\lambda), \quad |\lambda| < \frac{\pi}{h},$$

as $T \to \infty$. Under the assumption that the fourth order cumulants are summable one can show that

$$\operatorname{cov}\, [{}_h I^{(T)}(\lambda), {}_h I^{(T)}(\mu)] \to \eta((\lambda - \mu) h)\, [1 + \eta(2 \lambda h)]\, {}_h f^2(\lambda)$$

as $T \to \infty$ if $0 \leq \lambda, \mu$. Since the periodograms do not converge (in mean square) to the aliased spectral density, it is not a plausible estimate of the aliased spectral density. However, reasonable estimates can be constructed out of the periodogram in the following manner. Consider a bounded weight function $W(u)$ that is symmetric, with bounded support, and such that $\int W(u)\, du = 1$. Let

$$W_T(u) = B_T^{-1} W(B_T^{-1} u)$$

with B_T a bandwidth parameter that tends to zero as $T \to \infty$. The stringent conditions on $W(u)$ can be relaxed appreciably. Note they are adopted here simply for convenience. Let $B_T T \to \infty$ as $T \to \infty$. We consider the following estimate of ${}_h f(\lambda)$

$$f^{(T)}(\lambda) = \int_{-\pi/h}^{\pi/h} W_T(\lambda - \alpha)\, {}_h I^{(T)}(\alpha)\, d\alpha$$

$$= \frac{h}{2\pi} \sum_{k=-(T-1)}^{T-1} w_T(k\,h)\, {}_h m^{(T)}(k\,h) \exp(-\, i\,k\,h\,\lambda)$$

where

$${}_h m^{(T)}(k\,h) = \frac{1}{T} \sum_{0 \le k,\, k+j \le T-1} X(k\,h)\, X\big((k+j)\,h\big)$$

and

$${}_h w_T(k\,h) = \int_{-\pi/h}^{\pi/h} W_T(u) \exp(-\, i\,k\,h\,u)\, du\,.$$

Here the estimate $f^{(T)}(\lambda)$ is given as a smoothed version of a periodogram or alternatively as a Fourier transform of weighted estimates of the covariances of the process. Such dual representations also are valid for higher order spectral estimates. However, we will only discuss the representation of higher order spectral estimates obtained by smoothing higher order counterparts of the periodogram. Continuity of f and absolute summability of fourth order cumulants implies that

$$\lim_{T\to\infty} B_T\, T \operatorname{cov}\left[f^{(T)}(\lambda), f^{(T)}(\mu)\right]$$

$$= \frac{2\pi}{h}\, \eta\big((\lambda - \mu)\,h\big)\,[1 + \eta(2\,\lambda\,h)]\, {}_h f^2(\lambda) \int W^2(u)\, du$$

for $0 \le \lambda, \mu$ if $B_T \to 0$, $B_T\, T \to \infty$. It is apparent that

$$\lim_{T\to\infty} E\, f^{(T)}(\lambda) = {}_h f(\lambda)\,.$$

Strong mixing and the counterpart of the moment conditions up to eighth order detailed in Theorem 4 of section V.3 imply that the estimates $f^{(T)}(\lambda)$ at a finite number of values λ are jointly asymptotically normal with the mean and covariance properties noted above.

Let $X(t) = \big(X_a(t);\, a = 1, \ldots, m\big)$ be an m-vector strictly stationary process. Assume

$$\sum_{t_1, \ldots, t_{k-1}=-\infty}^{\infty} |\, t_j\, c_{a_1, \ldots, a_k}\, (t_1\, h, \ldots, t_{k-1}\, h)\,| < \infty$$

for $j = 1, \ldots, k-1$, any k-tuple $a_1, \ldots, a_k$ $(k = 2, 3, \ldots)$ and any $h > 0$. This is parallel to the cumulant assumptions made in Theorem 3 of section 4. The finite Fourier transform of the discretely sampled data for component $X_a(t)$ is

$$d_a^{(T)}(\lambda) = \sum_{t=0}^{T-1} X_a(t\,h) \exp(-\,i\,t\,h\,\lambda)\,.$$

kth order counterparts of the periodogram are given by

$$I^{(T)}_{a_1,\ldots,a_k}(\lambda_1,\ldots,\lambda_k) = \left(\frac{h}{2\,\pi}\right)^{k-1} T^{-1}\, d^{(T)}_{a_1}(\lambda_1)\,\ldots\, d^{(T)}_{a_k}(\lambda_k)\,.$$

The weight function $W(u_1, \ldots, u_k)$ is assumed to be bounded, piecewise continuous and defined on $\Sigma_1^k u_j = 0$ with bounded support. Further let

$$\int\cdots\int W(u_1,\ldots,u_k)\,\delta\left(\sum_1^k u_j\right) du_1 \ldots du_k = 1$$

and

$$W(-\,u_1,\ldots,-\,u_k) = W(u_1, u_2, \ldots, u_k)\,.$$

Set

$$W_T(u_1,\ldots,u_k) = B_T^{-k+1}\, W(B_T^{-1} u_1 \ldots B_T^{-1} u_k)\,.$$

An estimate $g^{(T)}_{a_1,\ldots,a_k}(\lambda_1,\ldots,\lambda_k)$, $\Sigma_1^k \lambda_j = 0$, of the kth order cumulant spectral density ${}_h g_{a_1,\ldots,a_k}(\lambda_1,\ldots,\lambda_{k-1})$, $|\,\lambda_i\,| \leq \pi/h$ (analogous to those considered in section 4) is given by

$$\begin{aligned} &g^{(T)}_{a_1,\ldots,a_k}(\lambda_1,\ldots,\lambda_k) \\ &= \left(\frac{2\,\pi}{h}\right)^{k-1} T^{-k+1} \sum_{s_1=-\infty}^{\infty}\cdots\sum_{s_k=-\infty}^{\infty} W_T\left(\lambda_1 - \frac{2\,\pi\,s_1}{T\,h},\ldots,\lambda_k - \frac{2\,\pi\,s_k}{T\,h}\right) \\ &\qquad \Phi\left(\frac{2\,\pi\,s_1}{T},\ldots,\frac{2\,\pi\,s_k}{T}\right) I^{(T)}_{a_1,\ldots,a_k}\left(\frac{2\,\pi\,s_1}{T\,h},\ldots,\frac{2\,\pi\,s_k}{T\,h}\right). \end{aligned}$$

If $B_T \to 0$ as $T \to \infty$ then

$$\lim_{T\to\infty} E\, g^{(T)}_{a_1,\ldots,a_k}(\lambda_1,\ldots,\lambda_k) = {}_h g_{a_1,\ldots,a_k}(\lambda_1,\ldots,\lambda_{k-1})\,.$$

Moreover if $B_T^{k-1}\,T \to \infty$ as $T \to \infty$, then

$$\begin{aligned} (34)\quad &\lim_{T\to\infty} B_T^{k-1}\, T \operatorname{cov}\left[g^{(T)}_{a_1,\ldots,a_k}(\lambda_1,\ldots,\lambda_k),\, g^{(T)}_{a'_1,\ldots,a'_k}(\mu_1,\ldots,\mu_k)\right] \\ &= \frac{2\,\pi}{h} \sum_P \eta((\lambda_1 - \mu_{P(1)})\,h)\,\ldots\,\eta((\lambda_k - \mu_{P(k)})\,h)\, f_{a_1, a'_{P(1)}}(\lambda_1)\,\ldots\, f_{a_k, a'_{P(k)}}(\lambda_k) \\ &\qquad \int\cdots\int W(\tau_1,\ldots,\tau_k)\, W(\tau_{P(1)},\ldots,\tau_{P(k)})\,\delta\left(\sum_1^k \tau_j\right) d\tau_1 \ldots d\tau_k\,. \end{aligned}$$

The sum in formula (34) is over all permutations P of the set $1, 2, \ldots, k$. Suppose all the estimates of order k have the same bandwidth B_T. The assumptions made imply all estimates of order k are asymptotically normal with the means and covariances specified above.

6. Turbulence and Third Order Spectra

Let us consider homogeneous turbulence. Assume that the mean velocity $E\,\mathbf{v}(\mathbf{x}, t) \equiv 0$. In the case of a homogeneous random field $\mathbf{v}(\mathbf{x}, t)$, because of stationarity and existence of second order moments, there is a Fourier representation

$$\mathbf{v}(\mathbf{x}, t) = \int e^{i\mathbf{k}\cdot\mathbf{x}}\, d\mathbf{Z}(\mathbf{k}, t)\,.$$

Notice that $\mathbf{k} = (k_1, k_2, k_3)$ is a spatial wavenumber vector. Further $\mathbf{Z}(\mathbf{k}, t) = \big(Z_1(\mathbf{k}, t), Z_2(\mathbf{k}, t), Z_3(\mathbf{k}, t)\big)$ is a random process with orthogonal increments. The spectral density matrix $f(\mathbf{k}, t) = \big(f_{\alpha,\beta}(\mathbf{k}, t)\,;\ \alpha, \beta = 1, 2, 3\big)$ describes the properties of the variance functions of $\mathbf{Z}(\mathbf{k}, t)$

$$\begin{aligned} E\{d\mathbf{Z}(\mathbf{k}, t)\, d\mathbf{Z}(\mathbf{k}', t)'\} &= (E\{dZ_\alpha(\mathbf{k}, t)\, \overline{dZ_\beta(\mathbf{k}', t)}\}\,;\ \alpha, \beta = 1, 2, 3) \\ &= f(\mathbf{k}, t)\, \delta_{\mathbf{k},\mathbf{k}'}\, dk\,. \end{aligned}$$

Even though we are discussing a time dependent context, quite often the dependence on t will not be explicitly given so as to simplify the notation. Also, existence of higher order moments will be assumed whenever necessary. Consider the third order spectra

$$\begin{aligned} Q_{\alpha,\beta}(\mathbf{k}, \mathbf{k}') &= k_\alpha \operatorname{Im} E\{dZ_\alpha(\mathbf{k} - \mathbf{k}')\, dZ_\beta(-\mathbf{k})\, dZ_\beta(\mathbf{k}')\} \\ &= k_\alpha\, B_{\alpha\beta\beta}(\mathbf{k}, \mathbf{k}')\, dk\, dk', \quad \alpha, \beta = 1, 2, 3\,. \end{aligned}$$

First it is clear that the energy density associated with the wavenumber vector $\mathbf{k}$ is

$$\sum_{i=1}^{3} f_{ii}(\mathbf{k})\,.$$

We shall derive an equation linking second and third order spectra that has an appealing physical interpretation. By taking the divergence of the nonlinear equation (47) of Chapter V one obtains the equality

$$\frac{1}{\varrho}\nabla^2 p = -\sum_{i,j=1}^{3} \frac{\partial^2 v_i v_j}{\partial x_i\, \partial x_j}\,.$$

A particular solution in R^3 is given by

$$\frac{1}{\varrho}\, p(\mathbf{x}) = \sum_{i,j} \frac{1}{4\pi} \int \frac{\partial^2 v_i' v_j'}{\partial x_i\, \partial x_j} \frac{d\mathbf{x}'}{|\mathbf{x}' - \mathbf{x}|}\,.$$

By substituting this expression back in equation (47) of Chapter V the pressure p can be eliminated and everything is given in terms of the velocity v. Of course we are formally proceeding as if one actually has a solution of the system of differential equations rather than a weak solution. However, this can be justified (see Višik et al. [1979]). Now take the ith component of the resulting equation, multiply by $v_i(\mathbf{x}', t)$ and take the expected value of the expression obtained. Sum over i and Fourier transform the resulting equation. The following equation is obtained

$$\frac{\partial}{\partial t}\left(\frac{1}{2}\sum_{i=1}^{3} f_{ii}(\mathbf{k})\right) = \int Q(\mathbf{k}, \mathbf{k}')\, dk' - \nu\,|\,\mathbf{k}\,|^2 \sum_{i=1}^{3} f_{ii}(\mathbf{k}) \tag{35}$$

with

$$Q(\mathbf{k}, \mathbf{k}') = \sum Q_{\alpha,\beta}(\mathbf{k}, \mathbf{k}') .$$

Here $Q(\mathbf{k}, \mathbf{k}')$ is considered the net mean rate of energy transfer from dk' to dk. The term of equation (35) involving $Q(\mathbf{k}, \mathbf{k}')$ couples the wavenumber vectors $\mathbf{k}$ and $\mathbf{k}'$ and arises from the nonlinearity of the Navier-Stokes system. If we had a linear system the wavenumber vectors $\mathbf{k}$ would be decoupled. The relation

$$Q(\mathbf{k}, \mathbf{k}') = -\, Q(\mathbf{k}', \mathbf{k})$$

follows from the continuity equation.

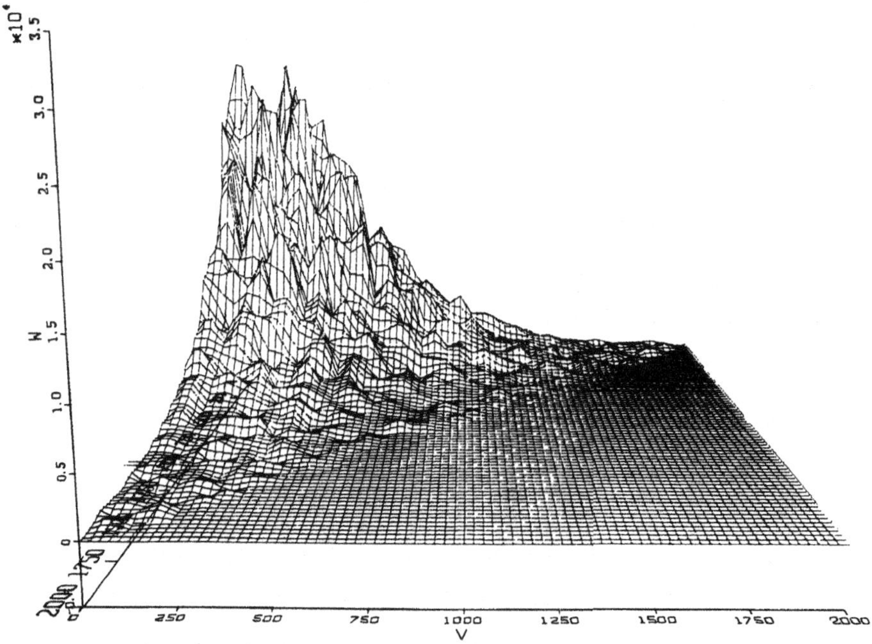

Figure 5. Bispectral estimate of $d\theta/dt$ for heated grid experiment. (Figure courtesy of Dr. Kenneth Helland, University of California, San Diego).

Readings were made of the temperature derivative in a low speed wind tunnel of the AMES Department at the University of California, San Diego. A heated grid with mesh size $M = 5$ cm was set up at one end of the tunnel. The mean temperature rise across the grid was 10 °C. The mean speed was 7.7 m/s. A cold-wire temperature sensor was mounted close to the x-wire velocity sensor at a distance 48 M downstream. The temperature derivative was formed with an electronic differentiator circuit. An estimate of the bispectrum (3rd order cumulant spectral density) of the time derivative was computed. A graph of this estimate is given in the accompanying figure.

Notes

6.1 An extensive bibliography of research on and applications of higher order spectra is given in Tryon [1981]. This lists much of the literature up to 1980.

6.2 The fast Fourier transform has been very useful in making spectral analysis a standard tool. In most cases it has speeded up computation very effectively. However, there are occasionally cases that arise in which the older method of Fourier transforming covariance estimates may be more effective and flexible.

6.3 One of the earliest discussions of higher moments with a Fourier-Stieltjes representation of the type given in formula (5) is to be found in Blanc-Lapierre and Fortet [1953]. One should note that one can construct processes for which higher moments exist but for which such a representation is invalid. The discussion given in sections 3 and 4 follow that given in Brillinger and Rosenblatt [1967].

6.4 An early application of bispectral estimates to obtain information about a nonlinear wave propagation problem is given in Hasselman, Munk and Mac Donald [1963]. The use of such estimates is plausible in problems of a non-Gaussian or nonlinear character. Recently bispectral estimates have been used to test for linearity (the model of a linear process) or departures from linearity. A discussion of such procedure can be found in Subba Rao and Gabr [1980].

6.5 Such a discussion of aliasing was inserted at this point so as to have a meaningful discussion of its effect on higher order cumulant spectral estimates. The discussion is similar to that given in Rosenblatt [1984].

6.6 The equation of heat conduction in a moving medium is

$$\frac{\partial\theta}{\partial t} + \sum_{i=1}^{3} v_i \frac{\partial\theta}{\partial x_i} = k \sum_{i=1}^{3} \frac{\partial^2\theta}{\partial x_i^2}$$

with θ the temperature, $\mathbf{v} = (v_i)$ the velocity vector and k the coefficient of thermal diffusivity.

Lii, Helland and Rosenblatt [1982] try to estimate $Q(k, k')$ in a three-dimensional spectral analysis.

Chapter VII

Density and Regression Estimates

1. Introduction. The Case of Independent Observations

Many of the methods considered thus far are intimately tied up with the harmonic analysis of stationary processes. In this chapter we shall consider procedures intuitively based on the model of a smoothed histogram. First we consider some elementary results for sequences of independent random variables. Our object is not that of generality, rather that of understanding. Later on, it will be seen that there are counterparts of these results for independent identically distributed random variables in the domain of suitably restricted stationary processes.

First the case of density function estimates is considered. Let $X_1, \ldots, X_n$ be independent, identically distributed random points in k-dimensional Euclidean space with common *continuous bounded* density function $f(x)$. Consider a bounded piece-wise continuous integrable weight function (or kernel function) $w(x)$ such that

$$\int w(x)\, dx = 1 .$$

A plausible estimate of the density function f at x is given by the kernel estimate

$$f_n(x) = \{n\, b(n)^k\}^{-1} \sum_{j=1}^{n} w\left(\frac{x - X_j}{b(n)}\right)$$

where $b(n)$ is understood to be a linear bandwidth parameter such that $b(n) \downarrow o$ and $n\, b(n)^k \to \infty$ as $n \to \infty$. Notice that

$$\int f_n(x)\, dx = 1$$

but typically $f_n(x)$ will not be positive for all x with probability one unless $w(x)$ is a nonnegative weight function. The mean of $f_n(x)$ is given by

$$\begin{aligned} E\, f_n(x) &= E\left\{b(n)^{-k}\, w\left(\frac{x - X}{b(n)}\right)\right\} \\ &= \int b(n)^{-k}\, w\left(\frac{x - u}{b(n)}\right) f(u)\, du \\ &= \int w(v)\, f\big(x - b(n)\, v\big)\, dv . \end{aligned}$$

Under the assumption that $b(n) \downarrow o$ as $n \to \infty$ it is clear that

$$E\, f_n(x) \to f(x) .$$

The covariance of $f_n(x)$ and $f_n(y)$ is given by

$$\text{(1)} \qquad \operatorname{cov}[f_n(x), f_n(y)] = n^{-1}\, b(n)^{-2k} \operatorname{cov}\left[w\left(\frac{x - X}{b(n)}\right), w\left(\frac{y - X}{b(n)}\right)\right]$$

and

(2) $$\operatorname{cov}\left[w\left(\frac{x-X}{b(n)}\right), w\left(\frac{y-X}{b(n)}\right)\right]$$

$$= b(n)^k \int w(v)\, w\left(\frac{y-x}{b(n)} + v\right) f\big(x - b(n)\, v\big)\, dv$$

$$- b(n)^{2k} \int w(v)\, f\big(x - b(n)\, v\big)\, dv \int w(v)\, f\big(y - b(n)\, v\big)\, dv\,.$$

It is clear from (1) and (2) that

(3) $$\lim_{n\to\infty} n\, b(n)^k \operatorname{cov}\left[f_n(x), f_n(y)\right]$$

$$= \delta_{x-y}\, f(x) \int w^2(v)\, dv$$

if $b(n) \downarrow o$, $n\, b(n)^k \to \infty$ as $n \to \infty$. If f is strictly positive, this implies that $f_n(x)$ and $f_n(y)$ are asymptotically uncorrelated as $n \to \infty$. A straightforward application of the Liapounov central limit theorem for independent random variables indicates that

$$\{n\, b(n)^k\}^{\frac{1}{2}} \left[f_n(x^{(i)}) - E\, f_n(x^{(i)})\right],$$

$i = 1, \ldots, m$, are asymptotically jointly normal and independent with the covariance structure given by (3) if $b(n) \downarrow o$, $n\, b(n)^k \to \infty$. The condition $n\, b(n)^k \to \infty$ indicates that the number of random points that lie in any fixed k-dimensional hypercube of positive volume diverges as $n \to \infty$.

Let us now consider the bias

$$E\, f_n(x) - f(x)$$

$$= \int w(v) \left\{f\big(x - b(n)\, v\big) - f(x)\right\} dv\,.$$

Assume that w is symmetric about 0 in the sense that

$$w(v) = w(-v)$$

and that

$$\int |\, v\,|^2\, |\, w(v)\,|\, dv < \infty\,.$$

A standard argument using Taylor's formula with remainder term then indicates that if f *is continuously differentiable up to second order with bounded derivatives*

(4) $$E f_n(x) - f(x) = \frac{1}{2} \sum_{i,j=1}^{k} (D_i D_j f)(x) \int u_i u_j w(u)\, du\, b(n)^2 + o\big(b(n)^2\big)$$

where D_i and D_j denote the partial derivatives of f with respect to the ith and jth components of x respectively.

The mean square error of $f_n(x)$ as an estimate of $f(x)$ is

$$E\,|\,f_n(x) - f(x)\,|^2 = \sigma^2\big(f_n(x)\big) + \big(E f_n(x) - f(x)\big)^2 .$$

The estimates of variance and bias given by (3) and (4) indicate that under the assumptions made, the most rapid decrease of mean square error is obtained by setting $b(n) = C\, n^{-1/(k+4)}$ with C an appropriately chosen constant. Then the mean square error will be $0(n^{-4/(k+4)})$ as $n \to \infty$.

Let us now consider regression estimates. Assume that $Z_1, Z_2, \ldots, Z_n$ are independent identically distributed $k+1$ dimensional random variables with common density function f. Further when we write

$$Z_j = (Y_j, X_j)$$

let Y_j denote the first component of Z_j and X_j the k-tuple of the remaining components. The common density function of the X_j's is $g(x)$. The conditional density function of Y_j given X_j is then

$$f(y \mid x) = f(y, x)/g(x) .$$

Assume that f and g are bounded continuous and positive. Further let the regression functions

$$r(x) = E(Y \mid X = x)$$
$$m(x) = E(Y^2 \mid X = x)$$

both be well-defined and integrable with weight function $g(x)$. The regression function $r(x)$ is assumed to be continuous. We should like to estimate the regression function $r(x)$ in terms of the observations $Z_1, \ldots, Z_n$. A possible estimate can be given in terms of a kernel function $w(x)$. For convenience assume that $w(x)$ is bounded and of bounded support. Take as the estimate

$$r_n(x) = n^{-1} b(n)^{-k} \sum_{j=1}^{n} Y_j\, w\left(\frac{x - X_j}{b(n)}\right) \left\{ n^{-1} b(n)^{-k} \sum_{j=1}^{n} w\left(\frac{x - X_j}{b(n)}\right)\right\}^{-1} = a_n(x)/g_n(x) .$$

The following expansion is carried out in order to gauge the asymptotic behaviour of the estimate as $n \to \infty$

$$\frac{a_n(x)}{g_n(x)} = \frac{a_n(x) - E\,a_n(x) + E\,a_n(x)}{g_n(x) - E\,g_n(x) + E\,g_n(x)}$$

$$= \{a_n(x) - E\,a_n(x) + E\,a_n(x)\}$$

$$\left(E\,g_n(x)\right)^{-1}\left\{1 - \frac{g_n(x) - E\,g_n(x)}{E\,g_n(x)}\right.$$

$$\left. + O\left(g_n(x) - E\,g_n(x)\right)^2\right\}$$

$$= \frac{E\,a_n(x)}{E\,g_n(x)} + \left(E\,g_n(x)\right)^{-1}\left(a_n(x) - E\,a_n(x)\right)$$

$$- \frac{E\,a_n(x)}{(E\,g_n(x))^2}\left(g_n(x) - E\,g_n(x)\right)$$

$$+ O\left(a_n(x) - E\,a_n(x)\right)^2 + O\left(g_n(x) - E\,g_n(x)\right)^2.$$

Now

(5)
$$E\,a_n(x) = E\left\{Y_j\,w\left(\frac{x - X_j}{b(n)}\right)\right\} b\,(n)^{-k}$$

$$= \int y\,w\left(\frac{x - u}{b(n)}\right) f(y \mid u)\,dy\,g(u)\,du\,b(n)^{-k}$$

$$= \int r\left(x - b(n)\,v\right) g\left(x - b(n)\,v\right) w(v)\,dv\,.$$

It is clear that $E\,a_n(x) \to r(x)\,g(x)$ as $n \to \infty$. If w is symmetric about zero and r and g are continuously differentiable up to second order, it follows from (4) and (5) that

$$\frac{E\,a_n(x)}{E\,g_n(x)} - r(x) = \frac{E\,a_n(x) - r(x)\,g(x)}{E\,g_n(x)}$$

$$- \frac{r(x)}{E\,g_n(x)}\{E\,g_n(x) - g(x)\}$$

$$= \left\{g(x)^{-1}\frac{1}{2}\sum_{i,j=1}^{k} D_i\,D_j\left(r(x)\,g(x)\right)\right.$$

$$\int u_i\,u_j\,w(u)\,du$$

$$- \frac{r(x)}{g(x)}\,\frac{1}{2}\sum_{i,j=1}^{k} D_i\,D_j\,g(x)$$

$$\left.\int u_i\,u_j\,w(u)\,du\right\} b_N^2$$

$$+ o(b_N^2)\,.$$

Earlier in (3), one had seen that $f_n(x)$ and $f_n(y)$ are asymptotically uncorrelated if $x \neq y$. Similarly one can show that $a_n(x)$ and $a_n(y)$ as well as $a_n(x)$ and $g_n(y)$ are asymptotically uncorrelated as $n \to \infty$ if $x \neq y$. The variance of $a_n(x)$ is

$$\begin{aligned}(6) \qquad \sigma^2\big(a_n(x)\big) &= n^{-1}\, b(n)^{-2k}\, \sigma^2 \left[Y\, w\left(\frac{x - X}{b(n)}\right)\right] \\ &= n^{-1}\, b(n)^{-k} \int y^2 f\big(y \mid x - b(n)\, v\big)\, dy \\ &\qquad\qquad g\big(x - b(n)\, v\big)\, w^2(v)\, dv \\ &\quad + O\big(n^{-1}\, b(n)^{-2k}\big) \\ &= n^{-1}\, b(n)^{-k}\, m(x)\, g(x) \int w^2(v)\, dv \\ &\quad + o\big(n^{-1}\, b(n)^{-k}\big)\,.\end{aligned}$$

A similar computation indicates that the covariance

$$\begin{aligned}(7) \qquad & \operatorname{cov}\big(a_n(x), g_n(x)\big) \\ &= n^{-1}\, b(n)^{-k}\, r(x)\, g(x) \int w^2(v)\, dv \\ &\quad + o\big(n^{-1}\, b(n)^{-k}\big)\,.\end{aligned}$$

A corresponding estimate for the variance of $g_n(x)$ is obtained directly from (3).

Let

$$\begin{aligned} h_n(x) &= \big(E\, g_n(x)\big)^{-1} \big(a_n(x) - E\, a_n(x)\big) \\ &\quad - \frac{E\, a_n(x)}{(E\, g_n(x))^2} \big(g_n(x) - E\, g_n(x)\big)\,.\end{aligned}$$

It follows from (3), (6) and (7) that

$$\begin{aligned}(8) \qquad & \lim_{n \to \infty} n\, b(n)^k\, \sigma^2\big(h_n(x)\big) \\ &= g(x)^{-1} \{m(x) - r(x)^2\} \int w(u)^2\, du\,.\end{aligned}$$

If $E \mid Y \mid^{2+\delta} < \infty$ for some $\delta > 0$, an application of the Liapounov central limit theorem shows that $\{n\, b(n)^k\}^{\frac{1}{2}} \big(h_n(x) - E\, h_n(x)\big)$ is asymptotically normally distributed with mean zero and variance given by (8).

2. *Density and Regression Estimates for Stationary Sequences*

Let $\{X_j\}$ be a strictly stationary sequence. We shall assume that joint distributions up to fourth order of the random variables are absolutely continuous with bounded continuous density functions. Assuming some additional conditions involving a version of short range dependence, it will be shown that the density estimates considered in section 1 have the same asymptotic distribution when sampling from the stationary sequence as they had in the case of independent random variables. The asymptotic distribution of regression and conditional probability estimates will be determined under similar assumptions when sampling from a stationary sequence.

In discussing the properties of a probability density estimate, we shall for ease and simplicity in notation consider the one dimensional case $k = 1$. The density estimate as before is

$$f_n(x) = \{n\, b(n)\}^{-1} \sum_{j=1}^{n} w\left(\frac{x - X_j}{b(n)}\right).$$

The weight function w is assumed to satisfy the same conditions as those specified in section 1. We shall first derive the following result.

Theorem 1. *Let $\{X_j\}$ be a strictly stationary sequence satisfying a strong mixing condition. Assume that joint distributions up to fourth order are absolutely continuous with uniformly bounded density functions. Let $_jf(x, y)$ be the joint density function of X_0 and X_j with*

$$\sum_j |\, _jf(x, y) - f(x)\, f(y)\,| < \infty$$

absolutely summable and bounded. If f is continuous, it follows that

$$\{n\, b(n)\}^{1/2}\, [f_n(x^{(i)}) - E\, f_n(x^{(i)})]\,,$$

$i = 1, \ldots, m$ are asymptotically jointly normal and independent as $n \to \infty$ with variances

$$f(x^{(i)}) \int w^2(v)\, dv\,,$$

$i = 1 \ldots, m$ if $b(n) \to 0$, $n\, b(n) \to \infty$.

As in the independent case

$$E\, w\big(b(n)^{-1}(x - X)\big) = b(n) \int w(v)\, f\big(x - b(n)\, v\big)\, dv\,, \tag{9}$$

$$E\, w^2\big(b(n)^{-1}(x - X)\big) = b(n) \int w^2(v)\, f\big(x - b(n)\, v\big)\, dv\,, \tag{10}$$

$$\begin{aligned} &E\{w\big(b(n)^{-1}(x - X)\big)\, w\big(b(n)^{-1}(y - X)\big)\} \\ &\quad = b(n) \int w\big(b(n)^{-1}(y - x) + v\big)\, w(v)\, f\big(x - b(n)\, v\big)\, dv\,. \end{aligned} \tag{11}$$

Also

$$(12)\quad \operatorname{cov}\{w(b(n)^{-1}(x-X_0)),\, w(b(n)^{-1}(y-X_j))\}$$
$$= b(n)^2 \int w(v)\, w(u)$$
$$\{{}_jf(x-b(n)\,u, y-b(n)\,v) - f(x-b(n)\,u)\, f(y-b(n)\,v)\}\, du\, dv\,,$$

$j = 1, 2, \ldots$. Relations (9), (10) and (12) imply that the variance

$$\operatorname{var}\left[\sum_{j=1}^{m} w(b(n)^{-1}(x-X_j))\right]$$
$$= \sum_{u=-m}^{m} (m-|u|) \operatorname{cov}\{w(b(n)^{-1}(x-X_0)),\, w(b(n)^{-1}(x-X_u))\}$$
$$= m\, b(n) \int f(x-b(n)\,v)\, w^2(v)\, dv + O(m\, b(n)^2)\,.$$

Let us now consider estimating the fourth central moment

$$(13)\quad E\left|\sum_{j=1}^{m} \{w(b(n)^{-1}(x-X_j)) - E\, w(b(n)^{-1}(x-X_j))\}\right|^4$$
$$= \sum_{j=1}^{m} E\,|g_j|^4 + 3 \sum_{j_1, j_2=1}^{m}{}' E\, g_{j_1}^2 g_{j_2}^2$$
$$+ 6 \sum_{j_1, j_2, j_3=1}^{m}{}' E\, g_{j_1}^2 g_{j_2} g_{j_3} + \sum_{j_1, j_2, j_3, j_4=1}^{m}{}' E\, g_{j_1} g_{j_2} g_{j_3} g_{j_4}\,.$$

In equation (13) g_j represents

$$g_j = w(b(n)^{-1}(x-X_j)) - E\, w(b(n)^{-1}(x-X_j))$$

and the primed summation indicates summation over distinct subscripts j. The four sums on the right of equation (13) are of order of magnitude $m\, b(n)$, $(m\, b(n))^2$, $(m\, b(n))^3$, $(m\, b(n))^4$ respectively because we have assumed that all joint distributions up to fourth order are absolutely continuous with bounded density functions. Now $b(n) \to 0$ and $n\, b(n) \to \infty$ as $n \to \infty$. One can specify the sequence $m(n)$ such that $m(n) = o(n)$ but still $m(n)\, b(n) \to \infty$ as $n \to \infty$. The Liapounov like condition in Theorem 4 of Chapter III is satisfied. A direct application of the theorem implies the asymptotic normality of

$$(14)\quad \{n\, b(n)\}^{1/2}\, [f_n(x) - E\, f_n(x)]$$

with limiting variance

$$f(x) \int w^2(v)\, dv\,.$$

Relation (11) implies that $f_n(x)$ and $f_n(y)$ are asymptotically uncorrelated if $f(x), f(y) > 0$ and $x \neq y$. By applying the same argument to any linear combination of the $f_n(x^{(i)})$ at distinct values $x^{(i)}$, $i = 1, \ldots, m$, the joint asymptotic normality and independence of the density estimates with variances (3) is demonstrated.

There are a number of regression functions that one might consider estimating in the case of a stationary sequence. A simple, interesting and perhaps fairly typical case is that of the regression function

$$r(x) = E(X_{n+1} \mid X_n = x) .$$

If we consider a regression estimate of the form dealt with in section 1, we are naturally led to

$$r_n(x) = \{n\, b(n)\}^{-1} \sum_{j=1}^{n} X_{j+1}\, w\big(b(n)^{-1}(x - X_j)\big)$$

$$\times \left[\big(n\, b(n)\big)^{-1} \sum_{j=1}^{n} w\big(b(n)^{-1}(x - X_j)\big) \right]^{-1} = a_n(x)/g_n(x) .$$

The asymptotic behavior of the density estimate $g_n(x)$ has already been resolved under appropriate conditions in Theorem 1. Most of the argument in Theorem 2 will be concerned with the numerator $a_n(x)$. Consider the family of functions

(15) $$h_{j_2-j_1, \ldots, j_a-j_{a-1}}(x_{j_1}, \ldots, x_{j_a})$$

$$= E(X_{j_1+1} \cdots X_{j_a+1} \mid X_{j_1} = x_1, \ldots, X_{j_a} = x_a)$$

$$f_{j_2-j_1, \ldots, j_a-j_{a-1}}(x_1, \ldots, x_a) ,$$

with $f_{j_2-j_1, \ldots, j_a-j_{a-1}}(x_1, \ldots, x_a)$ understood to be the joint density function of $X_{j_1}, \ldots, X_{j_a}$. The function

$$\sum_{j \neq 0} \{h_j(x_0, x_j) - h(x_0)\, h(x_j)\}$$

will also be of some interest.

Theorem 2. *Assume that the conditions of Theorem 1 are satisfied. Let the function* $h(x) = r(x)\, f(x)$ *be bounded and continuous. All the functions* $h_{j_2-j_1, \ldots, j_a-j_1}(x_1, \ldots, x_a)$, $a = 1, \ldots, 4$ *are assumed to be uniformly bounded. Further let*

(16) $$\sum_{j \neq 0} | h_j(x_0, x_j) - h(x_0)\, h(x_j) | < \infty$$

be absolutely summable and bounded. It then follows that

$$\{n\, b(n)\}^{1/2} \{r_n(x^{(i)}) - E\, a_n(x^{(i)}) / E\, g_n(x^{(i)})\} ,$$

$i = 1, \ldots, m$ are asymptotically jointly normal and independent as $n \to \infty$ with variances

$$[E(X_{j+1}^2 \mid X_j = x^{(i)}) - \{E(X_{j+1} \mid X_j = x^{(i)})\}^2]\, f(x^{(i)})^{-1} \int w^2(v)\, dv\,,$$

$i = 1, \ldots, m$ if $b(n) \to 0$, $n\, b(n) \to \infty$ and $f(x^{(i)}) > 0$.

Just as in the formula of section 1

$$(17)\quad r_n(x) = \frac{a_n(x)}{g_n(x)} = \frac{E\, a_n(x)}{E\, g_n(x)} + (E\, g_n(x))^{-1}\,(a_n(x) - E\, a_n(x)) - \frac{E\, a_n(x)}{(E\, g_n(x))^2}\,(g_n(x) - E\, g_n(x)) + O(a_n(x) - E\, a_n(x))^2 + O(g_n(x) - E\, g_n(x))^2\,.$$

Now

$$(18)\quad E\{X_{j+1}\, w(b(n)^{-1}\,(x - X_j))\} = b(n) \int w(\alpha)\, h(x - b(n)\, \alpha)\, d\alpha\,,$$

$$(19)\quad E\{X_{j+1}^2\, w^2(b(n)^{-1}\,(x - X_j))\} = b(n) \int w^2(\alpha)\, m_2(x - b(n)\, \alpha)\, d\alpha$$

with

$$m_2(x) = E\{X_{j+1}^2 \mid X_j = x\}\, f(x)\,.$$

Further

$$(20)\quad E\{X_1\, X_{j+1}\, w(b(n)^{-1}\,(x - X_0))\, w(b(n)^{-1}\,(x' - X_j))\} = b(n)^2 \int w(\alpha_0)\, w(\alpha_1)\, h_j(x - \alpha_0\, b(n),\, x' - \alpha_1\, b(n))\, d\alpha_0\, d\alpha_1\,,$$

$j = 1, 2, \ldots$. Relations (16), (18, (19) and (20) imply that

$$\operatorname{var}\left[\sum_{j=1}^{m} X_{j+1}\, w(b(n)^{-1}\,(x - X_j))\right] = m\, b(n) \int w^2(v)\, m_2(x - b(n)\, v)\, dv + O(m\, b(n)^2)\,.$$

Also the fourth central moment

$$(21)\quad E\left|\sum_{j=1}^{m} \{X_{j+1}\, w(b(n)^{-1}\,(x - X_j)) - E[X_{j+1}\, w(b(n)^{-1}\,(x - X_j))]\}\right|^4 = \sum_{j=1}^{m} E\,|l_j|^4 + 3 \sum_{j_1, j_2=1}^{m}{}' E\, l_{j_1}^2\, l_{j_2}^2 + 6 \sum_{j_1, j_2, j_3=1}^{m}{}' E\, l_{j_1}^2\, l_{j_2}\, l_{j_3} + \sum_{j_1, j_2, j_3, j_4=1}^{m}{}' E\, l_{j_1}\, l_{j_2}\, l_{j_3}\, l_{j_4}$$

with

$$l_j = X_{j+1}\, w\big(b(n)^{-1}(x - X_j)\big) - E\,[X_{j+1}\, w\big(b(n)^{-1}(x - X_j)\big)]\,.$$

As before the primed summation denotes a summation over distinct subscripts j. The four sums on the right of (21) are of order of magnitude $m\,b(n)$, $(m\,b(n))^2$, $(m\,b(n))^3$, $(m\,b(n))^4$ respectively because of the assumption that all the h functions up to order 4 as specified in (15) are uniformly bounded. Since this implies that the Liapounov condition (16) of Theorem 4 of Chapter III is satisfied, that theorem implies that

$$\{n\,b(n)\}^{1/2}\,[a_n(x) - E\,a_n(x)]$$

is asymptotically normal with limiting variance

$$m_2(x) \int w^2(v)\,dv\,. \tag{22}$$

Notice that

$$\begin{aligned} &E\,\{X_1^2\, w\big(b(n)^{-1}(x - X_0)\big)\, w\big(b(n)^{-1}(x' - X_0)\big)\} \\ &\quad = b(n) \int m_2\big(x - b(n)\,a\big)\, w(a)\, w\big(b(n)^{-1}(x' - x) + a\big)\,da\,. \end{aligned}$$

Theorem 4 of Chapter III as applied to linear combinations of the $a_n(x^{(i)})$ implies their joint asymptotic normality. Relations (18) and (20) imply that the $a_n(x^{(i)})$ at distinct values $x^{(i)}$, $i = 1, \ldots, m$, are asymptotically independent as $n \to \infty$. The fact that

$$\begin{aligned} &E[X_1\, w\big(b(n)^{-1}(x - X_0)\big)\, w\big(b(n)^{-1}(x' - X_0)\big)] \\ &\quad = b(n) \int h\big(x - b(n)\,a\big)\, w(a)\, w\big(b(n)^{-1}(x' - x) + a\big)\,da \end{aligned}$$

implies that

$$\lim_{n\to\infty} n\,b(n)\,\mathrm{cov}[a_n(x), g_n(x)] = h(x) \int w^2(a)\,da\,. \tag{23}$$

The $a_n(x^{(i)})$ and $g_n(x^{(i)})$, $i = 1, \ldots, m$ are jointly asymptotically normal by the type of argument indicated above. The representation (17) together with the estimates (14), (22), and (23) directly yield the conclusion of the theorem.

The derivations of Theorems 1 and 2 are quite similar. We now consider the asymptotic distribution of a plausible estimate of the conditional probability density

$$f(y \mid x) = \frac{f(x, y)}{g(x)}$$

of X_{n+1} given X_n. An obvious estimate is given by

$$f_n(y \mid x) = \frac{f_n(x, y)}{g_n(x)}$$

with

$$f_n(x, y) = n^{-1} b_1(n)^{-2} \sum_{j=1}^{n} k\big(b_1(n)^{-1}(x - X_j),\, b_1(n)^{-1}(y - X_{j+1})\big)$$

and

$$g_n(x) = n^{-1} b_2(n)^{-1} \sum_{j=1}^{n} w\big(b_2(n)^{-1}(x - X_j)\big) .$$

The weight functions $k(\cdot, \cdot)$ and $w(\cdot)$ are assumed to satisfy the conditions of boundedness and integrability mentioned in section 1. Notice that

$$\begin{aligned}(24)\quad \frac{f_n(x, y)}{g_n(x)} = \frac{E f_n(x, y)}{E g_n(x)} &+ \{E g_n(x)\}^{-1} \{f_n(x, y) - E f_n(x, y)\} \\ &- \frac{E f_n(x, y)}{(E g_n(x))^2} (g_n(x) - E g_n(x)) \\ &+ O\big(f_n(x, y) - E f_n(x, y)\big)^2 + O\big(g_n(x) - E g_n(x)\big)^2 .\end{aligned}$$

Here we have let $f(x, x')$ denote the joint density function of X_j, X_{j+1}. Also let

$${}_k f(x, x'; y, y')$$

be the joint density function of $X_j, X_{j+1}, Y_{j+k}, X_{j+k+1}$ for $k > 1$. The following theorem describes the asymptotic behavior of the estimate $f_n(y \mid x)$ as $n \to \infty$ under appropriate conditions.

Theorem 3. *Let $\{X_j\}$ be a strictly stationary sequence satisfying a strong mixing condition. Assume that joint distributions up to eight order are absolutely continuous with uniformly bounded density functions. Further let*

$$\sum_j | {}_j f(x, x'; y, y') - f(x, x') f(y, y') | < \infty$$

be absolutely summable and bounded. Also assume that $f(x, x')$ and $g(x)$ are continuous functions with $g(x) > 0$. Then if $b_1(n)^2 = o\big(b_2(n)\big)$ with $b_2(n) \to 0$, $n\, b_1(n)^2 \to \infty$, it follows that

$$(25)\qquad n^{1/2} b_1(n) \{f_n(y \mid x) - E f_n(x, y) / E g_n(x)\}$$

is asymptotically normal with variance

$$(26)\qquad g(x)^{-2} f(x, y) \int k^2(\alpha_1, \alpha_2)\, d\alpha_1\, d\alpha_2 .$$

If $b_2(n) = a\, b_1(n)^2$, $a > 0$, we still have asymptotic normality of (25) but now with variance

$$(27)\qquad g(x)^{-2} f(x, y) \int k^2(\alpha_1, \alpha_2)\, d\alpha_1\, d\alpha_2 + a^{-1/2} g(x)^{-3} f(x, y)^2 \int w^2(\alpha)\, d\alpha .$$

Under the conditions assumed, an argument paralleling that of Theorem 1 shows that

$$n^{1/2}\, b_1(n)\, \{f_n(x,y) - E\, f_n(x,y)\}$$

is asymptotically normal with mean zero and variance

$$f(x,y) \int k^2(\alpha_1, \alpha_2)\, d\alpha_1\, d\alpha_2 \,.$$

If $b_1(n)^2 = o(b_2(n))$, $g_n(x) - E\, g_n(x) = o(f_n(x,y) - E\, f_n(x,y))$ and so the second term on the right of (24) is the dominating one in

$$f_n(y \mid x) - E\, f_n(x,y) \,/\, E\, g_n(x) \,.$$

This implies that (25) is asymptotically normal with mean zero and variance (26) as $n \to \infty$. If $b_2(n) = a\, b_1(n)^2$, $a > 0$, both the second and third terms on the right of (24) are of the same order of magnitude. A simple estimate shows that $f_n(x,y)$ and $g_n(x)$ are asymptotically uncorrelated. Since $f_n(x,y)$ and $g_n(x)$ are asymptotically jointly normal, we see that (25) is now asymptotically normal with mean zero and variances (27).

Notes

7.1 There is at this time a rather large body of work in density and regression estimates of a nonparametric character. Early papers on kernel probability density estimates (in the case of independent identically distributed observations) are Rosenblatt [1956b] and Parzen [1962]. A question of some importance is the choice of the bandwidth. The paper of Silverman [1978] is of considerable interest relative to this problem.

7.2 Early papers on the case of dependent observations are Roussas [1967] and Rosenblatt [1970]. The papers of Bradley [1983] and Robinson [1983] discuss recent research.

Chapter VIII

Non-Gaussian Linear Processes

1. Estimates of Phase, Coefficients and Deconvolution for Non-Gaussian Linear Processes

In section 4 of Chapter II linear nonGaussian processes

$$X_t = \sum_{j=-\infty}^{\infty} a_j V_{t-j}$$

were considered. The sequence $\{V_t\}$ was one of independent, identically distributed nonGaussian random variables with $E V_t \equiv 0$, $E V_t^2 \equiv 1$. The sequence of coefficients $\{a_j\}$ is assumed real with $\Sigma a_j^2 < \infty$. As remarked there, in terms of observations on a Gaussian process $\{X_t\}$ only the absolute value $| a(e^{-i\lambda}) |$ of the transfer function

$$a(e^{-i\lambda}) = \sum a_j e^{-ij\lambda}$$

can be determined. However, under appropriate conditions, it was also noted that in the nonGaussian case, the transfer function $a(e^{-i\lambda})$ could be almost completely determined, in effect up to a factor $\pm e^{ik\lambda}$ with k an integer. The conditions are that

$$\sum |j| \, |a_j| < \infty , \tag{1}$$
$$a(e^{-i\lambda}) \neq 0 \quad \text{for all } \lambda ,$$

and that some cumulant α_k, $k > 2$, of V_t not be zero. The first condition (1) implies that $a(e^{-i\lambda})$ is continuously differentiable.

Consider the interpretation of some of these remarks when $a(z)$ is a rational function

$$a(z) = A(z) / B(z)$$

with $A(z)$ and $B(z)$ polynomials

$$A(z) = \sum_{k=0}^{q} a_k z^k, \; a_0 \neq 0, \; B(z) = \sum_{k=0} b_k z^k, \; b_0 = 1 ,$$

with no common factors. The process $\{X_t\}$ is then a finite parameter ARMA process. The possible indeterminate factor $e^{ik\lambda}$ with k an integer in $a(e^{-i\lambda})$ cannot occur here because of the normalization $a_0 \neq 0$, $b_0 = 1$ since it would correspond to a factor z^l, l a positive integer, in $A(z)$ or $B(z)$. The assumption that $\{X_t\}$ is a stationary process implies that $B(z)$ can have no zeros of absolute value one since otherwise

$$f(\lambda) = \frac{1}{2\pi} | a(e^{-i\lambda}) |^2 = \frac{1}{2\pi} | A(e^{-i\lambda}) / B(e^{-i\lambda}) |^2$$

would not be integrable. The assumption that $a(e^{-i\lambda}) \neq 0$ for all λ implies that $A(z)$ has no zeros of absolute value one. The statement made above that under the conditions cited $a(e^{-i\lambda})$ can be determined up to a factor $\pm e^{ik\lambda}$ with k an integer indicates that the *zeros of the polynomials* $A(z)$, $B(z)$ other than those of absolute value one or zero *can be determined if the ARMA process is non-Gaussian* with $\gamma_k \neq 0$ for some $k > 2$. Let us contrast this with what happens in the Gaussian case. Then any real root $z_j \neq 0$ of $A(z)$ or $B(z)$ can be replaced by its inverse and pairs of nonzero conjugate roots by their paired conjugated inverses $\bar{z}_j^{-1}$ without changing the probability structure of the process $\{X_t\}$ if the process is properly rescaled. This is a consequence of the fact that $|e^{i\lambda} - z_j| = |z_j|\ |e^{-i\lambda} - z_j^{-1}|$. Thus with real distinct roots $z_j \neq 0, \pm 1$ there are 2^{p+q} ways of specifying the roots without changing the probability structure of $\{X_t\}$. There is a different specification of the coefficients a_j, b_k corresponding to each of the possible root specifications leading to the same Gaussian probability structure. In the Gaussian case, to ensure a unique determination of the coefficients a_j, b_k it is usual to assume that all the roots of $A(z)$ and $B(z)$ are outside the unit circle $|z| \leq 1$ in the complex plane. In the nonGaussian case the actual location of the zeros $z_j \neq 0$ can be determined as contrasted with the Gaussian case. This means that the actual values of the a_j's and b_k's can be determined in the nonGaussian case up to a multiple of ± 1 for the a sequence as contrasted with the Gaussian case. Of course, all these remarks are made under the assumption that one can have access to a sample of arbitrary length. How to design estimates of $\alpha(e^{-i\lambda})$ on the basis of sample size n and the asymptotic behavior of such estimates are more detailed questions that will be taken up in this Chapter and section. We shall specifically deal with the cases in which γ_3 or γ_4 are nonzero. Estimates of $\{\alpha(e^{-i\lambda})\}^{-1}$ will also be considered relative to the problem of deconvolution, that is, determining the random quantities V_t from observations on the process $\{X_t\}$. Such questions arise in a geophysical context where the linear nonGaussian process serves as a model for seismic exploration. The constants a_j are considered the trace of a disturbance passing through a medium and the random weights V_j as the reflectivity of slabs in the layered medium. It is claimed that in many geophysical problems the data (the X_t's) that is observed is definitely nonGaussian and a primary aim is to deconvolve the data, estimating the a_j's and V_j's. The condition mentioned earlier on having the roots of the polynomials $A(z)$ and $B(z)$ outside the unit disc $|z| \leq 1$ is often referred to as a *minimum phase condition.* The classical procedures of parameter estimation for ARMA processes developed in Chapter IV all assume the minimum phase condition. The procedures discussed in this chapter are effective whether or not the minimum phase condition is satisfied.

Before describing the estimates of $\alpha(e^{-i\lambda})$, a simple example of two distinct moving average processes generated from exponential variables V_t is given having the same second order spectral structure. Consider the moving averages

$$X_t = 6\,V_t - 5\,V_{t-1} + V_{t-2}$$

and

$$Y_t = 3\,V_t - 7\,V_{t-1} + 2\,V_{t-2}$$

with the V_t independent, identically distributed random variables. The roots of $A(z)$ are 2 and 3 for the first process and 1/2 and 3 for the second process. The processes $\{X_t\}$ and $\{Y_t\}$ have the same second order spectral density but different marginal distributions even when centered so that they have mean value zero.

We shall consider constructing estimates of $a(e^{-i\lambda})$ when $\gamma_3 \neq 0$ and also when $\gamma_4 \neq 0$. Qualitatively the case $\gamma_3 \neq 0$ should be of interest when the process $\{X_t\}$ appears to have a nontrivial third moment.

The case $\gamma_4 \neq 0$ would be of importance when the fourth cumulant of the process $\{X_t\}$ is nontrivial but the third moment of the process is zero or close to zero. Now

$$a(e^{-i\lambda}) = \{2\pi f(\lambda)\}^{1/2} \exp\{i\, h(\lambda)\}\,.$$

$f(\lambda)$ can be estimated by using one of the spectral estimates discussed in Chapter V. Due to the indeterminacy of the factor $\pm e^{ik\lambda}$ (k integral) $h_1(\lambda) - \frac{h_1(\pi)}{\pi}\lambda$ or $h_1(\lambda)$ will be estimated rather than $h(\lambda)$. For this estimates of higher order cumulant spectral densities $b_k(\lambda_1, \ldots, \lambda_{k-1})$ will be required. The asymptotic properties of a large class of estimates of this type have been derived in Chapter VI. The approximation suggested in the following Lemma is useful in developing results on estimates of $a(e^{-i\lambda})$.

Lemma. Let the assumptions of Theorem 5 of Chapter II be satisfied. Consider an estimate ${}_nb(\lambda_1, \ldots, \lambda_{k-1})$ *of* $b_k(\lambda_1, \ldots, \lambda_{k-1})$ (as $n \to \infty$) *based on a sample of size* n. *Then*

$$\theta_n(\lambda_1, \ldots, \lambda_{k-1}) = \arctan\left(\operatorname{Im} {}_nb(\lambda_1, \ldots, \lambda_{k-1}) \,/\, \operatorname{Re} {}_nb(\lambda_1, \ldots, \lambda_{k-1})\right)$$

appears to be a plausible estimate of

$$\theta(\lambda_1, \ldots, \lambda_{k-1}) = \arg b_k(\lambda_1, \ldots, \lambda_{k-1})$$

and

(2)
$$\begin{aligned}
&\theta_n(\lambda_1, \ldots, \lambda_{k-1}) - \theta(\lambda_1, \ldots, \lambda_{k-1}) \\
&\quad = \frac{\operatorname{Im} b_k(\lambda_1, \ldots, \lambda_{k-1})}{|\, b_k(\lambda_1, \ldots, \lambda_{k-1})\,|^2} \{\operatorname{Re} {}_nb(\lambda_1, \ldots, \lambda_{k-1}) \\
&\qquad\qquad - \operatorname{Re} b_k(\lambda_1, \ldots, \lambda_{k-1})\} \\
&\quad + \frac{\operatorname{Re} b_k(\lambda_1, \ldots, \lambda_{k-1})}{|\, b_k(\lambda_1, \ldots, \lambda_{k-1})\,|^2} \{\operatorname{Im} {}_nb(\lambda_1, \ldots, \lambda_{k-1}) \\
&\qquad\qquad - \operatorname{Im} b_k(\lambda_1, \ldots, \lambda_{k-1})\} \\
&\quad + o_p\left({}_nb(\lambda_1, \ldots, \lambda_{k-1}) - b_k(\lambda_1, \ldots, \lambda_{k-1})\right).
\end{aligned}$$

The approximation suggested in the Lemma can be justified in the following manner. Notice that for a complex number

$$z = x + i\,y = r\,e^{i\theta}$$

with $r = |\,z\,|$ and $\theta = \arctan(y/x)$ a principal value determination, the relations

$$\frac{\partial\theta}{\partial y} = \frac{x}{r^2}\,, \quad \frac{\partial\theta}{\partial x} = -\frac{y}{r^2}\,, \quad \frac{\partial^2\theta}{\partial x^2} = \frac{2\,x\,y}{r^4}$$

$$\frac{\partial^2\theta}{\partial y^2} = -\frac{2\,x\,y}{r^4}$$

and

$$\frac{\partial^2\theta}{\partial x\,\partial y} = \frac{1}{r^2} - \frac{2\,x^2}{r^4} = -\frac{1}{r^2} + \frac{2\,y^2}{r^4}$$

hold. The approximation (2) is seen to be valid by making use of a Taylor expansion of the arctan function and using these relations.

At this point we shall consider estimates based on third order cumulant spectra (assuming $\gamma_3 \neq 0$). The discussion of estimates based on the fourth order cumulant is somewhat similar and will be given later on. Let $\Delta = \Delta(n)$, $k\,\Delta = \lambda$ and $\Delta = \Delta(n) \to 0$ as $n \to \infty$. First assume $b_3(0, 0)$ positive. Later on a simple modification of the procedure now given will show how to take care of the case in which $b_3(0, 0)$ is negative. Notice that

$$\begin{aligned} h_1(\lambda) &= h(\lambda) - h'(0)\,\lambda \cong h(k\,\Delta) - \frac{h(\Delta)}{\Delta}\,k\,\Delta \\ &= \sum_{j=1}^{k-1} \{h(j\,\Delta) + h(\Delta) - h((j+1)\,\Delta)\} \\ &= -\sum_{j=1}^{k-1} \arg b_3(j\,\Delta, \Delta)\,. \end{aligned}$$

This suggests that

$$H_n(\lambda) = -\sum_{j=1}^{k-1} \arg {}_n b(j\,\Delta, \Delta)$$

would be a reasonable estimate of $h_1(\lambda)$. We shall derive the following result for estimates of $h_1(\lambda)$ based on estimates of third order cumulant spectral estimates.

Theorem 1. *Let $b_3(0, 0)$ be positive and assume the linear process X_t satisfies the conditions of Theorem 3 of Chapter VI and $E\,X_t^6 < \infty$. Consider*

$$H_n(\lambda) = -\sum_{j=1}^{k-1} \arg {}_n b(j\,\Delta, \Delta)\,,$$

$k\,\Delta = \lambda$, as an estimate of

$$h_1(\lambda) = h(\lambda) - h'(0)\,.$$

The bispectral estimates ${}_nb(\lambda, \mu)$ *are understood to be weighted averages of third order periodogram values. Assume that* $b_3(\lambda, \mu) \in C^2$ *and that the weight function* W *of the bispectral estimates is symmetric and bandlimited with bandwidth* Δ. *Then*

$$H_n(\lambda) - h_1(\lambda) = R_n(\lambda) + o\big(H_n(\lambda) - h_1(\lambda)\big) \tag{3}$$

where

$$R_n(\lambda) = \sum_{j=1}^{k} \Big[\frac{\operatorname{Im} b(j\Delta, \Delta)}{|b(j\Delta, \Delta)|^2} \{\operatorname{Re} {}_nb(j\Delta, \Delta) - \operatorname{Re} b(j\Delta, \Delta)\} - \frac{\operatorname{Re} b(j\Delta, \Delta)}{|b(j\Delta, \Delta)|^2} \{\operatorname{Im} {}_nb(j\Delta, \Delta) - \operatorname{Im} b(j\Delta, \Delta)\}\Big]. \tag{4}$$

Further

$$E\,R_n = -\int_0^{\lambda} \frac{1}{2} \{b(u_1), 0\}^{-1} \sum_{j,k} A_{jk}(2 - \delta_{jk})\, D_{u_j} D_{u_k} \operatorname{Im} b(u_1, u_2)|_{u_2=0}\, du\, \Delta + o(\Delta) \tag{5}$$

where the A_{jk} *are the moments*

$$A_{jk} = \int u_j u_k W(u_1, u_2)\, du_1\, du_2$$

and D_{u_j} *is the partial derivative with respect to* u_j. *Also*

$$\operatorname{cov}\big(R_n(\lambda), R_n(\mu)\big) \cong \frac{2\pi^2}{\Delta^3 n \gamma_3^2} \min(\lambda, \mu) \int W^2(u_1, u_2)\, du_1\, du_2. \tag{6}$$

These results hold under the assumption that $\Delta(n) \to 0$, $\Delta^2 n \to \infty$ *as* $n \to \infty$.

The estimate (3) follows immediately from (2). Under the assumption made on the weight function of the bispectral estimates, it follows that the bias of a bispectral estimate is $0(\Delta^2)$ (see Section 4 of Chapter VI). Notice that

$$\operatorname{Im} b(\lambda, 0) = \frac{1}{(2\pi)^2} \sum_{u,v} c_{uv} \sin u\lambda.$$

However

$$\begin{aligned}\sum_v c_{uv} &= \sum_v E(X_0 X_u X_v) \\ &= \lim_{N\to\infty} E(X_0 X_u \sum_{v=-N}^{N} X_v) \\ &= \lim_{N\to\infty} E(X_0 X_{-u} \sum_{v=-N}^{N} X_v) = \sum_v c_{-u,v}.\end{aligned}$$

This implies that

$$\operatorname{Im} b(\lambda, 0) = 0 .$$

Consequently the first part of the sum in the expression (4) for $R_n(\lambda)$ is $0(\Delta^2)$. By using Theorem 1 of Chapter VI for the mean of a bispectral estimate and approximating a Riemann sum by an integral, the estimate (5) is obtained. The asymptotic expressions for variances and covariances of higher order cumulant spectral estimates given in Chapter VI, integral approximations to Riemann sums, and the special form of the bispectral density for the linear process

$$b(\lambda, \mu) = \frac{\gamma_3}{(2\pi)^2} \alpha(e^{-i\lambda}) \alpha(e^{-i\mu}) \alpha(e^{i(\lambda+\mu)})$$

lead to the simple approximation (6) for the covariance of the process $R_n(\lambda)$. The conditions on the linear process X_t (including the moment conditions) are enough to insure that the conclusion of Theorem VI on asymptotic estimates for covariances of cumulant spectral estimates are applicable here.

The Corollary below follows almost immediately.

Corollary. If all moments of the linear process $\{X_t\}$ *exist and the assumptions of Theorem 1 are satisfied, then* $H_n(\lambda)$ *converges to* $h_1(\lambda)$ *in probability and* $H_n(\lambda) - h_1(\lambda)$ *is asymptotically normal with the variance given by* (6). *The mean square error of* $R_n(\lambda)$ *is bounded by*

$$C_1 \Delta^2 + C_2(\Delta^3 n)^{-1} .$$

The optimal rate of convergence of this expression is $n^{-2/5}$ *and is attained when* $\Delta(n) \sim n^{-1/5}$. *Given* $b(\lambda, \mu) \in C^3$ *and a weight function bandlimited with first and second moments zero, the mean square error of* $R_n(\lambda)$ *is bounded by*

$$C_1 \Delta^4 + C_2(\Delta^3 n)^{-1} .$$

The optimal rate of convergence is now $n^{-4/7}$ *and is attained when* $\Delta(n) \sim n^{-1/7}$.

The existence of all moments for the linear process implies that the result on asymptotic normality of cumulant spectral estimates is applicable. By applying this result and using formula (3) the asymptotic normality of $H_n(\lambda)$ is demonstrated.

Generally the function $h_1(\lambda)$ will be estimated for a range of λ values. Since the sign of $b_3(0.0)$ may not be positive we estimate it by considering the real part of ${}_nb(0,0)$. If it is negative multiply all the values ${}_nb(j\Delta, \Delta)$ by minus one. The estimate $H_n(\lambda)$ is given then by

$$H_n(\lambda) = -\sum_{j=1}^{k-1} \arg\{-{}_nb(j\Delta, \Delta)\} .$$

We now make some remarks about the computation of the estimate of the phase of $a(e^{-i\lambda})$. Assume that one has a sample of $\{X_t\}$ of size $n = k N$. Center and normalize the sample so that it has mean zero and variance one. Then break the sample into k disjoint sections of length N so that the variance of the bispectral estimate from each section is not too large. This can be gauged by making use of second order spectral density estimates. Choose a grid of points $\lambda_j = j\Delta$ in $(0, 2\pi)$, $j = 1, \ldots, M$, $\Delta = 2\pi L/N$ where L is an appropriate integer. Form a bispectral estimate of the type mentioned above with a weight function of bandwidth Δ from each subsection. Then average the estimates from the different subsections to obtain a final estimate ${}_nb(j\Delta, \Delta)$. One should then compute $\theta_n(j\Delta) = \arg\{{}_nb(j\Delta, \Delta)\} + 2k\pi$ where the integer k is chosen to ensure continuity of

$$H_n(l\,\Delta) = H_n(\lambda_l) = -\sum_{j=1}^{l-1} \theta_n(j\,\Delta),\, l = 2, \ldots, M+1\,,$$

in the sense that neighboring values are as close to each other as possible. Because the upper index is $l - 1$ one starts with $l = 2$. Since $h(0) = 0$ set $H_n(0) = 0$ and estimate $H_n(\Delta) = H_n(\lambda_1)$ by an interpolation between 0 and $H_n(\lambda_2)$, $\lambda_2 = 2\,\Delta$. The value $H_n(\pi)$ is also computed by interpolation. Because

$$\alpha_k = \frac{1}{2\pi}\int_0^{2\pi} a(e^{-i\lambda})\, e^{ik\lambda}\, d\lambda$$

a plausible estimate $\hat{\alpha}_k$ of α_k is given by

$$\text{(7)} \qquad \hat{\alpha}_k = \frac{1}{2\pi}\int_0^{2\pi} \hat{\alpha}(e^{-i\lambda})\, e^{ik\lambda}\, d\lambda$$

$$\cong \frac{1}{M+2}\sum_{j=0}^{M+1} \{2\pi f_n(\lambda_j)\}^{\frac{1}{2}}$$

$$\exp\left\{ i\left(H_n(\lambda_j) - \frac{H_n(\pi)}{\pi}\,\lambda_j + k\,\lambda_j\right)\right\}$$

and this computation can be carried out by using the fast Fourier transform. The α_k's are real numbers but the computed $\hat{\alpha}_k$'s may or may not be real. If the symmetry of $f(\lambda)$ and odd property of $h(\lambda)$ about zero is used in an integration from $-\pi$ to π almost real values of the $\hat{\alpha}_k$'s will be obtained. The imaginary part of the $\hat{\alpha}_k$'s will be due to the rounding errors. In a computation there is usually no indication how good the estimates are apart from asymptotic results. In practice the sequence of points $\{j\Delta\}_{j=0}^{M+1}$ may not be symmetric about π. If the estimates $H_n(\lambda_j)$ are reasonably good the $\hat{\alpha}_j$'s using (7) should still be almost real. The size of the imaginary part reflects the noise level. In the case of estimates $H_n(\lambda_j)$ that are not good the imaginary part of the $\hat{\alpha}_j$'s becomes comparable to the real part. In this way one can get an indication of the quality of the estimation.

In the case of a one-sided linear process with a finite number of parameters one has a moving average of order q

$$X_t = \sum_{j=0}^{q} \alpha_j V_{t-j}, \alpha_0 \neq 0 .$$

The function $\alpha(z) = \sum_{j=0}^{q} \alpha_j z^j$ can be estimated by $\hat{\alpha}(z) = \sum_{j=0}^{q} \alpha_j z^j$.

In deconvolution one can try to recover the process $\{V_t\}$, $V_t = \{\alpha(B)\}^{-1} X_t$ (here B is the backward shift operator so that $B^j X_t = X_{t-j}$) by computing the approximation $\hat{V}_t = \{\hat{\alpha}(B)\}^{-1} X_t$. In case all the roots of $\alpha(z)$ $\big($and $\hat{\alpha}(z)\big)$ are outside the unit circle in the complex plane (the frequency function $\alpha(z)$ is minimum delay) the function $\hat{\alpha}^{-1}(z)$ has a one-sided expansion $\sum_{j=0}^{\infty} \alpha_j' B^j$. In computation, one will truncate the series after a certain number of terms. If some of the roots of $\hat{\alpha}(z)$ have modulus less than one, the expression $\hat{\alpha}^{-1}(B)$ can still be expanded with a Laurent series expansion by making use of the zeros of $\hat{\alpha}(z)$. To avoid finding a proper finite parameter model for $\{X_t\}$ and dealing with the sensitivity of root location in terms of their dependence on coefficients, notice that one can determine the deconvolution weights by inverting $\hat{\alpha}(e^{-i\lambda})$ directly. Set $b(e^{-i\lambda}) = \hat{\alpha}(e^{-i\lambda})$. The coefficient b_k in the Fourier expansion

$$b(e^{-i\lambda}) = \sum b_k e^{-ik\lambda}$$

can be computed by using

$$b_k = \frac{1}{2\pi} \int_0^{2\pi} [2\pi f_n(\lambda)]^{-\frac{1}{2}} \exp\left\{-i\left(H_n(\lambda) - \frac{H_n(\pi)}{\pi}\lambda + k\lambda\right)\right\} d\lambda$$

(8) $$\cong \frac{1}{M+2} \sum_{j=0}^{M+1} \{2\pi f_n(\lambda_j)\}^{-\frac{1}{2}} \exp\left\{-i\left(H_n(\lambda_j) - \frac{H_n(\pi)}{\pi}\lambda_j + k\lambda_j\right)\right\},$$

$k = \ldots, -1, 0, 1, \ldots$. Typically one determines suitable integers k_1 and k_2 and uses the real part of b_k for $k = k_1, \ldots, k_2$ as deconvolution weights (since we are dealing with a real-valued process).

We now give a sketch of an argument that enables us to get an asymptotic approximation for the covariances of the principal random part of the deconvolution weight estimates b_k. A similar argument can be used to derive a corresponding approximation for the covariances of the principal random part of the estimates $\hat{\alpha}_k$ (7). Expression (8) can be written as

$$b_k = \frac{2}{M+2} \sum_{j=0}^{M/2} \{2\pi f_n(\lambda_j)\}^{-\frac{1}{2}}$$
$$\cos\left\{-H_n(\lambda_j) + \frac{H_n(\pi)}{\pi}\lambda_j - k\lambda_j\right\}.$$

Notice that asymptotically

(9)
$$\begin{aligned}(f_n(\lambda_j))^{-\frac{1}{2}} &= (E f_n(\lambda_j))^{-\frac{1}{2}}\left(1 + \frac{f_n(\lambda_j) - E f_n(\lambda_j)}{E f_n(\lambda_j)}\right)^{-\frac{1}{2}} \\ &= (E f_n(\lambda_j))^{-\frac{1}{2}}\left(1 - \frac{1}{2}\frac{f_n(\lambda_j) - E f_n(\lambda_j)}{E f_n(\lambda_j)}\right. \\ &\qquad \left. + o\big(f_n(\lambda_j) - E f_n(\lambda_j)\big)\right).\end{aligned}$$

Also

(10)
$$\begin{aligned}&\cos\left\{-H_n(\lambda_j) + \frac{H_n(\pi)}{\pi}\lambda_j - k\lambda_j\right\} \\ &= \cos\left\{-E H_n(\lambda_j) + \frac{E H_n(\pi)}{\pi}\lambda_j - k\lambda_j\right. \\ &\qquad + \left[-H_n(\lambda_j) + E H_n(\lambda_j)\right. \\ &\qquad\qquad \left.\left. + \frac{H_n(\pi) - E H_n(\pi)}{\pi}\lambda_j\right]\right\} \\ &= \cos\left\{-E H_n(\lambda_j) + \frac{E H_n(\pi)}{\pi}\lambda_j - k\lambda_j\right\} \\ &\quad - \sin\left\{-E H_n(\lambda_j) + \frac{E H_n(\pi)}{\pi}\lambda_j - k\lambda_j\right\} \\ &\qquad \left[-H_n(\lambda_j) + E H_n(\lambda_j) + \frac{H_n(\pi) - E H_n(\pi)}{\pi}\lambda_j\right] \\ &\quad + o\left(-H_n(\lambda_j) + E H_n(\lambda_j) - \frac{H_n(\pi) - E H_n(\pi)}{\pi}\lambda_j\right).\end{aligned}$$

The second term on the right hand side of equation (9) is of smaller order than the second term on the right hand side of (10). This indicates that the principal random part of b_k (the deterministic mean is neglected here) can be approximated by

$$\frac{2}{M+2}\sum_{j=0}^{M/2} (2\pi f(\lambda_j))^{-\frac{1}{2}} \sin(-h_1(\lambda_j) + k\lambda_j)$$
$$\left[-H_n(\lambda_j) + E H_n(\lambda_j) + \frac{H_n(\pi) - E H_n(\pi)}{\pi}\lambda_j\right].$$

An argument like that leading to (6) indicates that the principal part of

$$-H_n(\lambda) + E\,H_N(\lambda) + \frac{H_n(\pi) - E\,H_n(\pi)}{\pi}\,\lambda\,,$$

$0 < \lambda < \pi$, asymptotically has the covariance

$$\frac{2\pi^2}{\Delta^3 n \gamma_3^2}\{\min(\lambda,\mu) - \lambda\mu\}\int W^2(u_1,u_2)\,du_1\,du_2$$

as $\Delta(n) \to 0$, $\Delta^2 n \to \infty$, $n \to \infty$. We conclude that the covariance of the principal random parts of b_j, b_k (j and k fixed) are

$$\frac{1}{\pi}\,\frac{1}{\Delta^3 n \gamma_3^2}\iint_0^\pi (f(\lambda)\,f(\mu))^{-\frac{1}{2}}\sin\big(h_1(\lambda) + k\lambda\big)$$

$$\sin\big(h_1(\mu) + j\mu\big)\{\min(\lambda,\mu) - \lambda\mu\}\,d\lambda\,d\mu$$

$$\int W^2(u,v)\,du\,dv\,.$$

We now describe one of several possible ways of estimating $h_1(\lambda)$ based on fourth order cumulant spectral estimates. Notice that

$$\sum_{j=1}^{k-1}\arg b_4(j\Delta,\Delta,\Delta)$$

$$= \sum_{j=1}^{k-1}\{h(j\Delta) + 2h(\Delta) - h\big((j+2)\Delta\big)\}$$

$$= 2[k\,h(\Delta) - h(k\Delta)] + B$$

with

$$B = h(2\Delta) - h(\Delta) + h(k\Delta) - h\big((k+1)\Delta\big)\,.$$

If $\lambda = k\Delta$ we have

$$h_1(\lambda) = h(\lambda) - h'(0)$$

$$\cong -\frac{1}{2}\sum_{j=1}^{k-1}\arg\{b_4(j\Delta,\Delta,\Delta)\} - \frac{1}{2}B$$

and if Δ is small the term B would also be small. A plausible estimate of $h_1(\lambda)$ could then be given by

$$G_n(\lambda) = -\frac{1}{2}\sum_{j=1}^{k-1}\arg\{{}_nb(j\Delta,\Delta,\Delta)\}$$

with ${}_nb(j\Delta,\Delta,\Delta)$ an estimate of the fourth order cumulant spectral density $b_4(j\Delta,\Delta,\Delta)$.

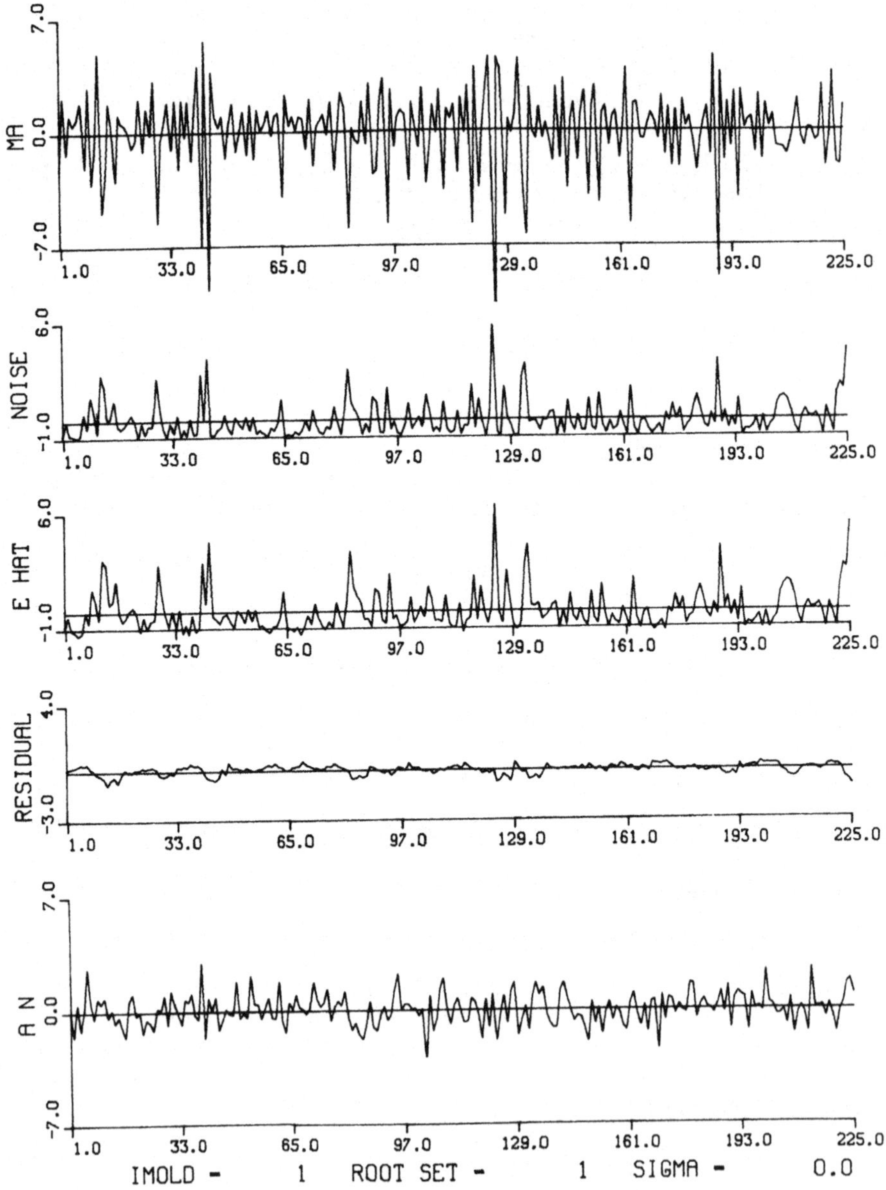

Figure 6. Exponential V_t.

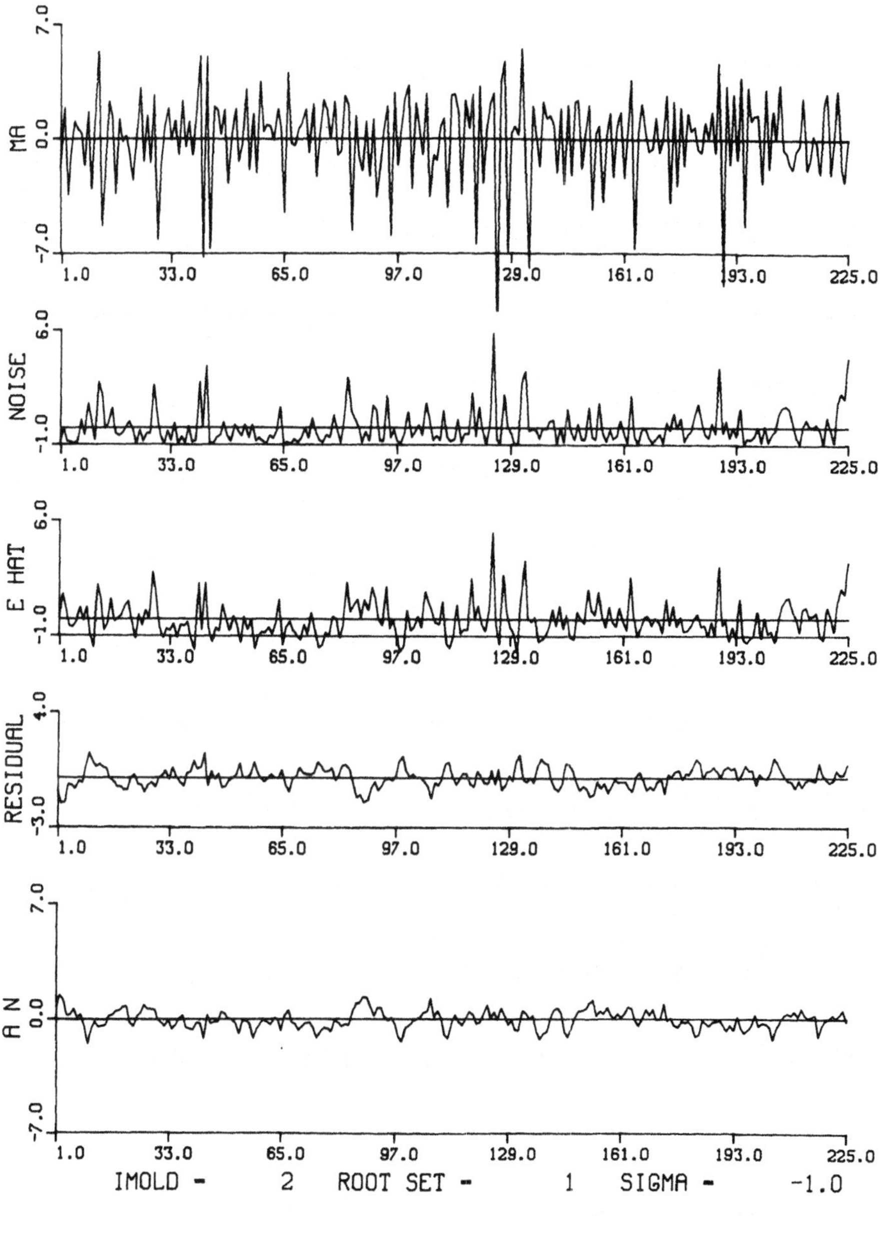

Figure 7. Exponential V_t with Gaussian noise added to process.

To illustrate the deconvolution procedure we consider the moving average

$$X_t = V_t - (2\ 1/3)\ V_{t-1} + (2/3)\ V_{t-2} \tag{11}$$

with the V_t's independent, identically distributed random variables. Notice that the polynomial

$$1 - (2\ 1/3)\ z + (2/3)\ z^2$$

has the roots 3 and 1/2 so that (11) is not a process satisfying the minimum phase condition. The V_t's are generated as pseudo-random variates using Monte Carlo simulation that employs linear congruential schemes. The different figures illustrate the deconvolution scheme as applied to (11) with the V_t's having different distributions. In the case of Figure 6 the V_t's are exponential with mean one. The second line of the Figure gives the V_t sequence as generated. The first line graphs the X_t sequence to be deconvolved. The third line gives the result of our deconvolution of the first line. The difference between the V_t sequence and the deconvolution is given on line four. The result of a minimum phase deconvolution is graphed on line five. Notice that line five does not resemble line two in any way.

In the second figure one considers a sequence

$$Y_t = X_t + \eta_t$$

with X_t a process having the same probability structure as that generated for Figure 6 and η_t an independent Gaussian white noise sequence of variance one. The process X_t has been adulterated by the addition of a small amount of Gaussian white noise. As before the V_t sequence is given on line two. The Y_t sequence is graphed on line one. The deconvolution of Y_t is carried out as if there were no additive Gaussian white noise and is presented on line three. Again line 4 gives the difference between line two and the deconvolution. The last line again gives the result of a minimum phase attempt at deconvolution. Notice that our deconvolution is still rather effective. The standard deviation of the X_t sequence is approximately 2.6 while the standard deviation of η_t is one. If the additive noise had a somewhat larger variance, the effectiveness of our deconvolution naively neglecting the noise would have broken down.

In the case of Figures 6 to 8 the deconvolution method used employed estimates of third order cumulant spectra (bispectra). Processes are analyzed that show a departure from Gaussianness in terms of third order moments. For Figure 8 the V_t's are taken so that they have a one-sided Pareto density of the form

$$f(u) = \begin{cases} 4\ u^{-5}, & u \geq 1 \\ 0\ , & \text{otherwise.} \end{cases}$$

The object was to see how our deconvolution procedure operates with a process having moment properties that are worse. In this case fourth order mo-

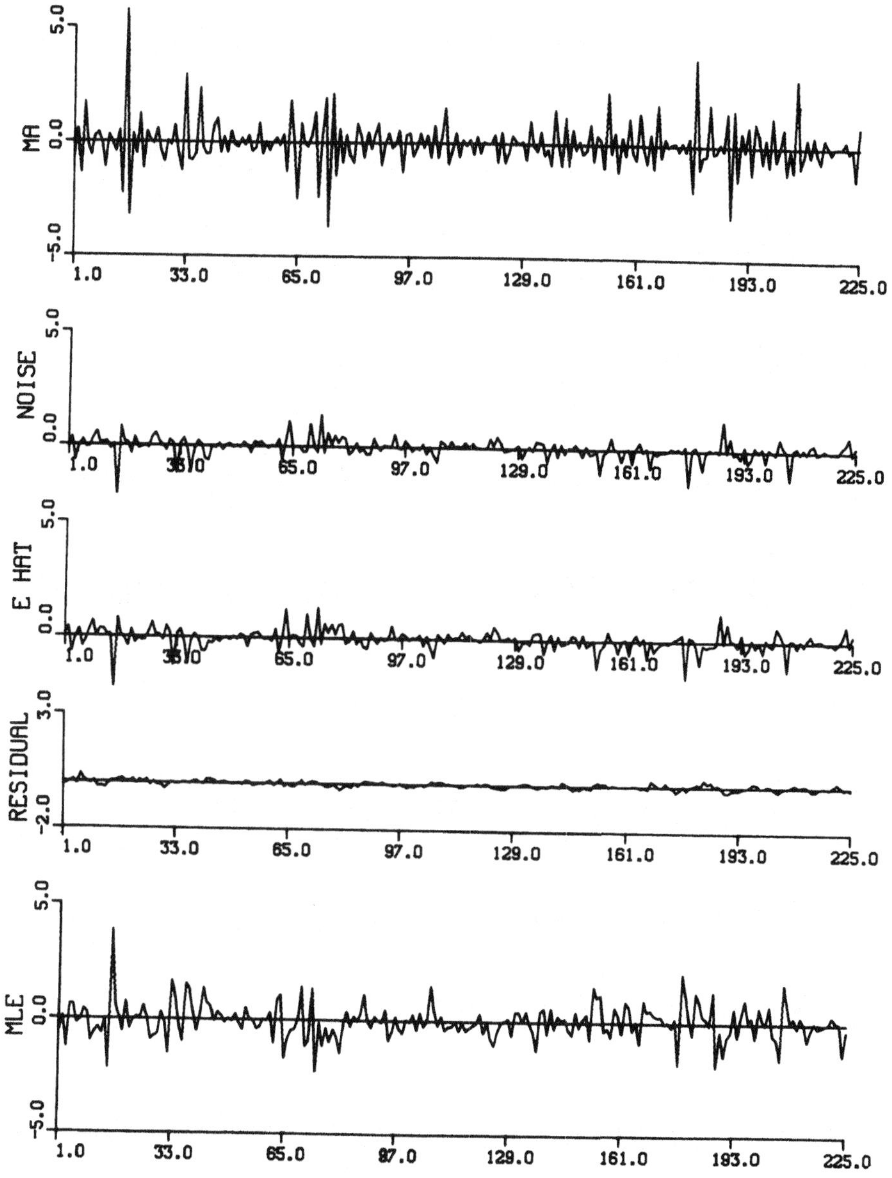

Figure 8. One-sided Pareto V_t. Third order deconvolution.

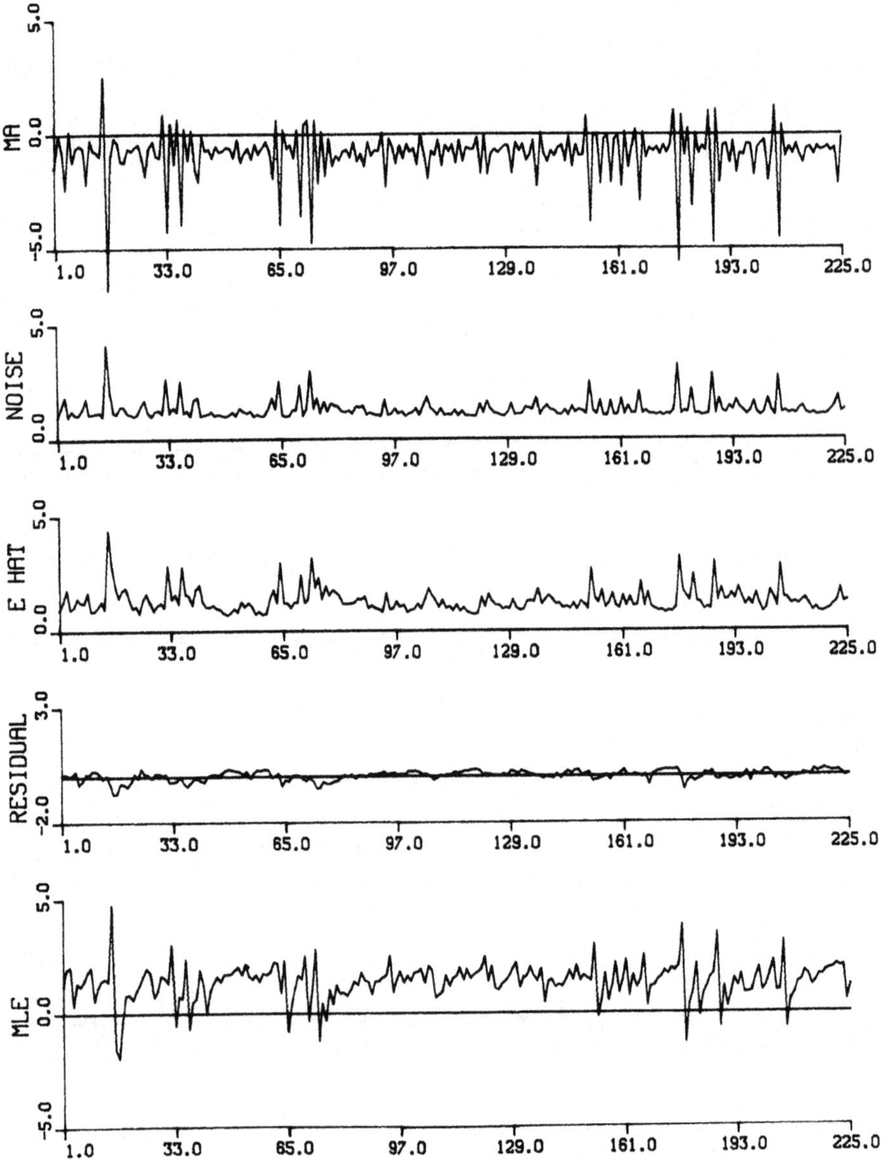

Figure 9. Symmetric Pareto V_t. Fourth order deconvolution.

ments don't exist. Here we see that the deconvolution procedure is at least as effective as when the V_t's have an exponential distribution. In the case of Figure 9 the V_t's have a symmetric Pareto distribution

$$f(u) = \begin{cases} 8\,|\,u\,|^{-5} & \text{if } \ |\,u\,| \geq 1 \\ 0\,, & \text{otherwise}\,. \end{cases}$$

Obviously a deconvolution procedure based on estimates of third order cumulant spectra would be ineffective since third order moments are zero for such a process. A deconvolution procedure based on estimates of fourth order cumulant spectra is employed here and we see it is effective.

2. *Random Fields*

We shall now consider weakly stationary random fields and some questions analogous to those that were dealt with earlier for weakly stationary sequences. Let $X_{\mathbf{t}}$, $\mathbf{t} = (t_1, \ldots, t_k)$ be a weakly stationary field on the k-dimensional lattice points Z^k (the k-vectors with integer components) having mean zero $E\, X_{\mathbf{t}} \equiv 0$ and covariance

$$\operatorname{cov}(X_{\mathbf{t}}, X_{\boldsymbol{\tau}}) = r_{\mathbf{t}-\boldsymbol{\tau}}\,.$$

As remarked in problem 9 of Chapter I

$$r_{\mathbf{t}} = \int e^{i\mathbf{t}\cdot\boldsymbol{\lambda}}\, dF(\boldsymbol{\lambda})$$

with $\boldsymbol{\lambda} = (\lambda_1, \ldots, \lambda_k)$ the vector of frequences λ_j and F if normalized to total mass one a distribution function in $\boldsymbol{\lambda}$. Note that $\mathbf{t}\cdot\boldsymbol{\lambda}$ denotes the inner product of the vectors $\mathbf{t}$ and $\boldsymbol{\lambda}$. Assume that F is absolutely continuous (differentiable) with

$$f(\boldsymbol{\lambda}) = D_{\lambda_1}, \ldots, D_{\lambda_k} F(\boldsymbol{\lambda})$$

the spectral density. The process $X_{\mathbf{t}}$ then has the representation

$$X_{\mathbf{t}} = \sum a_{\boldsymbol{\tau}}\, V_{\mathbf{t}-\boldsymbol{\tau}}$$

in terms of a "white noise" process $V_{\mathbf{t}}$

$$E\, V_{\mathbf{t}} \equiv 0$$

$$E\, V_{\mathbf{t}}\, \overline{V}_{\boldsymbol{\tau}} = \delta_{\mathbf{t}-\boldsymbol{\tau}}$$

where

$$a(e^{-i\boldsymbol{\lambda}}) = \sum a_{\boldsymbol{\tau}}\, e^{-i\boldsymbol{\tau}\cdot\boldsymbol{\lambda}}$$

and $a(e^{-i\lambda})$ is a square root of $(2\pi)^k f(\boldsymbol{\lambda})$

$$f(\boldsymbol{\lambda}) = \frac{1}{(2\pi)^k} \mid a(e^{-i\lambda}) \mid^2 .$$

We shall now consider a number of multidimensional time parameter analogues of the one dimensional prediction problem. Let S be a semigroup under addition that is a subset of Z^k and contains $\mathbf{0}$ as an element. Let $S' = S - \{\mathbf{0}\}$. Consider the prediction (or approximation) of X_0 by a linear expression

$$(12) \qquad - \sum_{\boldsymbol{\tau} \in S'} c_{\boldsymbol{\tau}} X_{\boldsymbol{\tau}}$$

in terms of $X_{\boldsymbol{\tau}}$, $\boldsymbol{\tau} \in S'$, best in the sense of having minimal mean square error of prediction. The best linear predictor is characterized by orthogonality of the prediction error

$$V_{\mathbf{0}} = X_{\mathbf{0}} + \sum_{\boldsymbol{\tau} \in S'} c_{\boldsymbol{\tau}} X_{\boldsymbol{\tau}}$$

to $X_{\mathbf{t}}$, $\mathbf{t} \in S'$. In Chapter I in the case of a stationary sequence we referred to the isometric mapping between the Hilbert space generated by the sequence $\{X_t, t \in Z\}$ (generated by linear combinations of the variables X_t) and $L^2(f)$ in which X_t maps onto $e^{it\lambda}$. There is a completely analogous isometry in the case of a k-dimensional random field in which $X_{\mathbf{t}}$ maps onto $e^{i\mathbf{t}\cdot\boldsymbol{\lambda}}$. Under this isometry the prediction error $V_{\mathbf{0}}$ maps onto

$$c(e^{i\boldsymbol{\lambda}}) = \sum_{\boldsymbol{\tau} \in S} c_{\boldsymbol{\tau}} e^{i\boldsymbol{\tau}\cdot\boldsymbol{\lambda}}$$

with $c_{\mathbf{0}} = 1$, where $c(e^{i\boldsymbol{\lambda}}) \in L^2(f)$. Notice that when we often write $e^{i\boldsymbol{\lambda}}$ this is used as a convenient symbol for $(e^{i\lambda_1}, \ldots, e^{i\lambda_k})$. The orthogonality referred to above is equivalent to

$$(13) \qquad \int_{[-\pi,\pi]^k} c(e^{i\boldsymbol{\lambda}}) f(\boldsymbol{\lambda}) e^{-i\mathbf{t}\cdot\boldsymbol{\lambda}} d\boldsymbol{\lambda} = 0$$

for $\mathbf{t} \in S'$. Let $L^p(g; \Sigma)$, $p > 0$, denote the set of functions integrable in pth mean with the weight function g that can be approximated in that metric by linear forms in $e^{i\mathbf{t}\cdot\boldsymbol{\lambda}}$ with $\mathbf{t} \in \Sigma$. It is clear that the orthogonality condition (13) is equivalent to

$$c(e^{i\boldsymbol{\lambda}}) f(\boldsymbol{\lambda}) \in L^1(1; Z^k - S') = L^1(1; S'^c)$$

where S'^c denotes the complement of S' relative to Z^k. We have shown that if $c(e^{i\boldsymbol{\lambda}})$ (with $c_{\mathbf{0}} = 1$) corresponds to the prediction error $V_{\mathbf{0}}$ under the isometric mapping between $L^2(f)$ and the Hilbert space generated by $\{X_{\mathbf{t}}, \mathbf{t} \in Z^k\}$ then $c(e^{i\boldsymbol{\lambda}}) \in L^2(f; S)$ and

$$f(\boldsymbol{\lambda}) = [c(e^{i\boldsymbol{\lambda}})]^{-1} h(\boldsymbol{\lambda})$$

with $h \in L^1(1; S'^c)$. It is easy to show that the converse holds for if $c_n(e^{i\lambda})$ is a sequence of trigonometric polynomials in $L^2(f; S)$ approaching $c(e^{i\lambda})$ in the corresponding norm, then it automatically follows that

$$\int c_n(e^{i\lambda}) f(\boldsymbol{\lambda}) e^{-i\mathbf{t}\cdot\boldsymbol{\lambda}} d\boldsymbol{\lambda} \to \int c(e^{i\lambda}) f(\boldsymbol{\lambda}) c^{-i\mathbf{t}\cdot\boldsymbol{\lambda}} d\boldsymbol{\lambda}$$

$$= \int h(\boldsymbol{\lambda}) e^{-i\mathbf{t}\cdot\boldsymbol{\lambda}} d\boldsymbol{\lambda} = 0$$

if $\mathbf{t} \in S'$.

It will at times be convenient to use the following multidimensional shift operator $T^{\mathbf{t}}$. If Σ is a subset of Z^k, we let $T^{\mathbf{t}} \Sigma = \{\boldsymbol{\tau} + \mathbf{t} : \boldsymbol{\tau} \in \Sigma\}$. Also

$$V_{\mathbf{t}} = X_{\mathbf{t}} + \sum_{\tau \in S'} c_{\boldsymbol{\tau}} X_{\mathbf{t}+\boldsymbol{\tau}} = T^{\mathbf{t}} V_{\mathbf{0}} .$$

Further the covariance $\varrho_{\mathbf{t}} = E(V_{\mathbf{0}} \overline{V}_{\mathbf{t}})$ of the $\{V_{\mathbf{t}}\}$ process is such that

$$\begin{aligned} \varrho_{\mathbf{t}} &= E(V_{\mathbf{0}} \overline{V}_{\mathbf{t}}) = \overline{E(V_{\mathbf{t}} \overline{V}_{\mathbf{0}})} \\ &= \overline{E(V_{\mathbf{0}} \overline{V}_{-\mathbf{t}})} = \bar{\varrho}_{-\mathbf{t}} \end{aligned}$$

and so the spectral density $g(\boldsymbol{\lambda})$

$$g(\boldsymbol{\lambda}) = \frac{1}{(2\pi)^k} \sum_{\mathbf{t}} \varrho_{\mathbf{t}} e^{-i\mathbf{t}\cdot\boldsymbol{\lambda}}$$

of the process $\{V_{\mathbf{t}}\}$ has $\varrho_{\mathbf{t}} = 0$ if $\pm\, \mathbf{t} \in S'$. If $\{X_{\mathbf{t}}\}$ is a realvalued process, so is $\{V_{\mathbf{t}}\}$ and that implies that $\varrho_{\mathbf{t}} = \varrho_{-\mathbf{t}}$. The prediction error is

$$\begin{aligned} E \mid V_{\mathbf{0}} \mid^2 &= \int_{[-\pi,\pi]^k} g(\boldsymbol{\lambda})\, d\boldsymbol{\lambda} \\ &= \int_{[-\pi,\pi]^k} \mid c(e^{i\lambda}) \mid^2 f(\boldsymbol{\lambda})\, d\boldsymbol{\lambda} . \end{aligned}$$

The remarks made above lead to the following result.

Theorem 2. *Let* $\{X_{\mathbf{t}}\}$ *be a weakly stationary random field with mean zero and spectral density* $f(\boldsymbol{\lambda})$. *The best linear predictor of* $X_{\mathbf{0}}$ *in terms of* $X_{\mathbf{t}}$, $\mathbf{t} \in S'$, (minimizing the mean square error of prediction) *is given by expression* (12) *where the coefficients* $c_{\boldsymbol{\tau}}$ *are determined by a function* $c(e^{i\lambda}) \in L^2(f; S)$, $c_{\mathbf{0}} = 1$, *with the property that the function h is determined by the condition that*

$$f(\boldsymbol{\lambda}) = [c(e^{i\lambda})]^{-1} h(\boldsymbol{\lambda})$$

be in $L^1(1; S'^c)$. *The spectral density*

$$g(\boldsymbol{\lambda}) = \mid c(e^{i\lambda}) \mid^2 f(\boldsymbol{\lambda})$$

of the prediction error process $\{V_{\mathbf{t}}\}$ *has Fourier coefficients* $\varrho_{\mathbf{t}} = 0$ *if* $\pm\, \mathbf{t} \in S'$ *because* $V_{\mathbf{t}}$ *is orthogonal to* V_{τ} *if* $V_{\mathbf{t}} \in T^{\mathbf{t}} S'$.

Each semigroup S determines a prediction problem of this type. As an illustration we mention two particular semigroups for the case $k = 2$. The first example is that of a "half-plane" problem where the semigroup is the set of lattice points $\{(i, j) : i \leq -1 \text{ or } i = 0 \text{ and } j \leq 0\}$. There are, of course, other half-plane problems such as that in which the semigroup is $\{(i, j) : i \leq 0\}$. The second example is that of a "quarter-plane" problem where the semigroup is $\{(i, j) : i, j \leq 0\}$. Helson and Lowdenslager [1958] have made a detailed analysis of a half-plane problem. They call a semigroup S a half-plane of lattice points (in the two dimensional case) if 1. $(0, 0) \notin S$, 2. $(m, n) \in S$ if and only if $(-m, -n) \in S$ unless $m = n = 0$, 3. $(m, n) \in S$, $(m', n') \in S$ implies $(m + m', n + n') \in S$. Notice that the semigroups dealt with by Helson and Lowdenslager do not contain zero as an element while the semigroups considered by us do contain zero as an element. However, to put ourselves into their context simply take our semigroup S and replace it by the semigroup S'. The detailed analysis Helson and Lowdenslager carry out for half-plane problems is possible because S and S^c are both semigroups in that case. This is not the case for a quarterplane problem because there S^c is clearly not a semigroup. They show that an appropriate modification of the ideas of G. Szegö [1920, 1921] can be applied to the case of half-plane problems. It is shown that the variance of the prediction error for linearly predicting $X_{0,0}$ in terms of $X_{i,j}$ with $i \leq -1$ or $i = 0$ and $j < 0$ is

$$s_1^2 = (2\pi)^2 \exp\left\{\frac{1}{(2\pi)^2} \int_{-\pi}^{\pi} \int_{-\pi}^{\pi} \log f(\lambda_1, \lambda_2)\, d\lambda_1\, d\lambda_2\right\} \tag{14}$$

Also there is a factorization of $f(\lambda_1, \lambda_2)$ if $\log f(\lambda_1, \lambda_2) \in L$ in terms of a function q with a one-sided Fourier development

$$f(\lambda_1, \lambda_2) = |\, q(\lambda_1, \lambda_2)\,|^2$$

where

$$q(\lambda_1, \lambda_2) = \exp\left\{\sum_{-(j,k) \in S} \gamma_{jk}\, e^{-ij\lambda_1 - ik\lambda_2}\right\}$$

with $S = \{(j, k) \mid j \leq -1 \text{ or } j = 0 \text{ and } k \leq 0\}$ and

$$\gamma_{0,0} = \frac{1}{8\pi^2} \iint \log f(\lambda_1, \lambda_2)\, d\lambda_1, d\lambda_2\,,$$

$$\gamma_{j,k} = \frac{1}{4\pi^2} \iint \log f(\lambda_1, \lambda_2) \exp\{i\, j\, \lambda_1 + i\, k\, \lambda_2\}\, d\lambda_1\, d\lambda_2$$

with $(j, k) \in S$, $(j, k) \neq (0, 0)$. The function

$$c(e^{i\lambda_1}, e^{i\lambda_2}) = \{q(\lambda_1, \lambda_2)\}^{-1} \exp(\gamma_{00})\,.$$

From this we can see that the prediction error is given by (14). If

(15) $$S = \{(j, k) : j \leq -1 \text{ or } (j, k) = (0, 0)\}$$

then

$$c(e^{i\lambda_1}, e^{i\lambda_2}) = \exp\left\{-\sum_{-(j,k)\in S} \gamma_{jk}\, e^{-ij\lambda_1 - ik\lambda_2}\right\} \exp\{\gamma_{00}\}.$$

This implies that the prediction error in this case is

$$s_2^2 = \exp\{2\gamma_{00}\} \int_{-\pi}^{\pi}\int_{-\pi}^{\pi} \exp\left\{\sum_{\substack{j=0\\k\neq 0}} \gamma_{jk}\, e^{ik\mu}\right\} d\lambda\, d\mu$$

$$= 2\pi \int_{-\pi}^{\pi} \exp\left\{\frac{1}{2\pi}\int_{-\pi}^{\pi} \log f(\lambda, \mu)\, d\lambda\right\} d\mu .$$

Notice that the semigroup (15) is not a half-plane in the sense of Helson and Lowdenslager.

A last example we shall consider is that of the semigroup

(16) $$S = \{(j, k) \mid j \leq 0\} .$$

In this case

$$c(e^{i\lambda_1}, e^{i\lambda_2}) = \exp\left\{-\sum_{j\geq 0} \gamma_{jk}\, e^{-ij\lambda_1 - ik\lambda_2} - \gamma_{00}\right\}$$

$$\left[\frac{1}{2\pi}\int_{-\pi}^{\pi} \exp\left\{-\int_{-\pi}^{\pi} \frac{1}{2\pi} \log f(\lambda, \mu)\, d\lambda\right\} d\mu\right]^{-1}$$

and this implies that the prediction error

$$s_3^2 = (2\pi)^3 \left[\int_{-\pi}^{\pi} \exp\left\{-\int_{-\pi}^{\pi} \frac{1}{2\pi} \log f(\lambda, \mu)\, d\lambda\right\} d\mu\right]^{-1} .$$

Again even though (16) in a loose sense is a half-plane, it is not in the sense of Helson and Lowdenslager. As one might expect

$$s_3^2 \leq s_1^2 \leq s_2^2 .$$

It has already been mentioned that there are difficulties in the case of a quarter-plane prediction problem unless the spectral density f is of a special form. However one can obtain converging approximations to the best predictor by making use of an alternating projection theorem of the type described in Wiener [1955]. Let H_1 and H_2 be the Hilbert spaces in $L^2(f)$ generated by $\exp\{i(j\lambda_1 + k\lambda_2)\}$, with $j \leq 0$ but $(j, k) \neq (0, 0)$, and by $\exp\{i(j\lambda_1 + k\lambda_2)\}$

with $k \leq 0$ but $(j, k) \neq (0, 0)$ respectively. P_1 and P_2 are the projection operators on H_1 and H_2 respectively. The best linear predictor of X_{00} in terms of $X_{j,k}$ where $j, k \leq 0$ but $(j, k) \neq (0, 0)$ can be obtained as follows. Consider the projection of the function 1 on $H_1 \cap H_2$. The map of this projection under the basic isometry between $L^2(f)$ and the Hilbert space generated by the process $\{X_{j,k}\}$ is the best linear predictor. Remember that under this isometry $\exp\{i(j\lambda_1 + k\lambda_2)\}$ corresponds to $X_{j,k}$. The projection of 1 on $H_1 \cap H_2$ can be obtained as the limit of $(P_1 P_2)^n 1$ as $n \to \infty$.

We shall now introduce the concept of an autoregressive moving average random field relative to a semigroup S in a manner analogous to that of a one dimensional ARMA process. Let us recall some of the properties of a one dimensional ARMA sequence. Assume that $\{V_t, t = \ldots, -1, 0, 1, \ldots\}$ is a "white noise" sequence. A process $\{X_t\}$ that satisfies the system of equations

$$\sum_{k=0}^{p} b_k X_{t-k} = \sum_{j=0}^{q} a_j V_{t-j}, \; b_0, a_0 \neq 0 \tag{17}$$

is called an autoregressive moving average process. We assume that the polynomials

$$b(z) = \sum_{k=0}^{p} b_k z^k$$

$$a(z) = \sum_{j=0}^{q} a_j z^j$$

have no factors in common and are interested in weakly stationary solutions of the system (17). It has been shown that there is a weakly stationary solution of (17) if and only if $b(e^{-i\lambda}) \neq 0$ for all real λ (see Theorem 3 of section II.3). If the polynomials $b(z)$, $a(z)$ have no zeros inside $|z| \leq 1$, the system of equations (17) can be shown to correspond to the linear prediction problem in the following manner. The best linear predictor of X_t in terms of the past $X_\tau, \tau \leq t-1$, is given by

$$X_t^* = b_0^{-1} \left\{ -\sum_{k=1}^{p} b_k X_{t-k} + \sum_{j=1}^{q} a_j V_{t-j} \right\} \tag{18}$$

with prediction error

$$a_0 b_0^{-1} V_t .$$

Notice that $V_{t-j}, j = 1, \ldots, q$, is expressible linearly in terms of the past X_τ, $\tau \leq t - 1$, and the coefficients of the best linear predictor can be read off directly from the system of equations (18). The condition that $a(z)$, $b(z)$ have no zeros inside $|z| \leq 1$ is needed so that $a(z)^{-1}$, $b(z)^{-1}$ are analytic in $|z| \leq 1$ with absolutely summable coefficients. This in turn is required so that the closed linear manifolds $\mathcal{M}_n(X) = \mathcal{M}(X_t, t \leq n) = \mathcal{M}_n(V) = \mathcal{M}(V_t, t \leq n)$ are the same and hence (18) is the best linear predictor of X_t. If $b(z)$ or $a(z)$ have a real nonzero root or conjugate pairs of roots inside $|z| < 1$, the coefficients of the system (17) no longer correspond to the best linear predictor. If $z_0 \neq 0$ is a root with $|z_0| < 1$, the factor $z - z_0$ can be replaced by the factor $z_0(z - \bar{z}_0^{-1})$ since

$$|e^{-i\lambda} - z_0| = |z_0| \, |e^{-i\lambda} - \bar{z}_0^{-1}| .$$

These roots with absolute value less than one can be replaced by their conjugated reciprocals and the new polynomials $\tilde{b}(z)$ or $\tilde{a}(z)$ obtained have the property that

$$|\tilde{b}(e^{-i\lambda})| = |b(e^{-i\lambda})|, |\tilde{a}(e^{-i\lambda})| = |a(e^{-i\lambda})| \tag{19}$$

for all real λ. The process $\{X_t\}$ will satisfy the system of equations

$$\sum_{j=0}^{p} \tilde{b}_j X_{t-j} = \sum_{j=0}^{q} \tilde{a}_j \eta_{t-j} \tag{20}$$

with coefficients derived from the polynomials $\tilde{b}(z)$, $\tilde{a}(z)$ and $\{\eta_t\}$ a white noise process derived from $\{V_t\}$ (see section 3 of Chapter II) linearly having $\sigma_\eta^2 = \sigma^2$. If $a(z)$ or $b(z)$ have roots inside $|z| < 1$, by this replacement of roots inside $|z| < 1$ by corresponding roots outside $|z| \leq 1$ one can obtain from an initially given ARMA scheme one whose polynomials $\tilde{b}(z)$, $\tilde{a}(z)$ have all their roots outside $|z| \leq 1$. The spectral densities of the two ARMA schemes are the same

$$f(\lambda) = \frac{\sigma_v^2}{2\pi} \left| \frac{a(e^{-i\lambda})}{b(e^{-i\lambda})} \right|^2$$

because of (19). If $\tilde{b}(z)$, $\tilde{a}(z)$ are the polynomials with all their roots outside $|z| \leq 1$, it is the corresponding system of equations (20) that characterizes the best linear predictor. The best linear predictor is the best predictor in the case of a Gaussian process. This follows because the probability structure of Gaussian processes is completely determined by their first and second order moments. In the case of an ARMA process with the V_t's independent, identically distributed nonGaussian random variables with finite second moment, the probability structure of the process $\{X_t\}$ is not completely determined by the spectral density. The best predictor (in terms of minimal mean square error of prediction) is now linear in general only when the roots of the polynomials $b(z)$, $a(z)$ are all outside $|z| \leq 1$. If some roots of $b(z)$ or $a(z)$ are inside $|z| < 1$, the best predictor will typically be nonlinear.

We now introduce the concept of an ARMA field relative to an additive semigroup $S(\mathbf{0} \in S)$ of Z^k. Assume that there are at most a finite number of nonzero coefficients $b_{\boldsymbol{\tau}}$, $a_{\boldsymbol{\tau}}$, $\boldsymbol{\tau} \in S$ with $b_{\mathbf{0}}, a_{\mathbf{0}} \neq 0$. Let $V_{\mathbf{t}}$, $\mathbf{t} \in Z^k$ be a weakly stationary process with mean zero. Notice that this is a more general assumption than that made in the one dimensional case where $V_{\mathbf{t}}$ was assumed to be a white noise process. In a little while it will become clear why this more general assumption is made. Consider the system of equations

$$\sum_{\boldsymbol{\tau} \in S} b_{\boldsymbol{\tau}} X_{\mathbf{t}+\boldsymbol{\tau}} = \sum_{\boldsymbol{\tau} \in S} a_{\boldsymbol{\tau}} V_{\mathbf{t}+\boldsymbol{\tau}} . \tag{21}$$

A weakly stationary solution $X_{\mathbf{t}}$ of the system of equations (21) will be called an ARMA random field relative to the semigroup S. Here $\mathbf{t} = (t_1, \ldots, t_k)$ is a lattice point of $S \subset Z^k$. As before, it will be convenient to introduce the generating functions

$$\begin{aligned} b(\mathbf{z}) &= \sum_{\boldsymbol{\tau} \in S} b_{\boldsymbol{\tau}} \mathbf{z}^{\boldsymbol{\tau}} \\ &= \sum_{(\tau_1, \ldots, \tau_k) \in S} b_{\tau_1, \ldots, \tau_k} z_1^{\tau_1} \cdots z_k^{\tau_k} , \\ a(\mathbf{z}) &= \sum_{\tau \in S} a_{\boldsymbol{\tau}} \mathbf{z}^{\boldsymbol{\tau}} \\ &= \sum_{(\tau_1, \ldots, \tau_k) \in S} a_{\tau_1, \ldots, \tau_k} z_1^{\tau_1} \cdots z_k^{\tau_k} . \end{aligned}$$

The functions $b(\mathbf{z})$, $a(\mathbf{z})$ are no longer usually polynomials but rather rational functions. In the case of a quarter-plane semigroup like $S = \{\tau_1, \ldots, \tau_k): \tau_1, \ldots, \tau_k \geq 0\}$ they become polynomials. As in the one dimensional case, we assume they have no common polynomial factors. However, here if $k > 1$, the functions $b(\mathbf{z})$ and $a(\mathbf{z})$ may have zeros in common even though they have no common polynomial factors. This contrasts with the simpler situation when $k = 1$. An example of this is simply given by taking

$$b(\mathbf{z}) = 2 - z_1 - z_2$$

and

$$a(\mathbf{z}) = (1 - z_1)(1 - z_2)$$

when $k = 2$. The functions $a(\mathbf{z})$, $b(\mathbf{z})$ have no polynomial factors in common but do have $(z_1, z_2) = (1, 1)$ as a common zero. *There will be a unique weakly stationary solution* $\{X_{\mathbf{t}}\}$ *of the system of equations* (21) *if*

$$\begin{aligned} b(e^{i\boldsymbol{\lambda}}) &= \sum_{\boldsymbol{\tau} \in S} b_{\boldsymbol{\tau}} e^{i\boldsymbol{\tau} \cdot \boldsymbol{\lambda}} \\ &= \sum_{\boldsymbol{\tau} = (\tau_1, \ldots, \tau_k \in S} b_{\tau_1, \ldots, \tau_k} e^{i\boldsymbol{\tau} \cdot \boldsymbol{\lambda}} \neq 0 \end{aligned}$$

for all k-vectors $\boldsymbol{\lambda} = (\lambda_1, \ldots, \lambda_k)$ *of real numbers, for each specified weakly stationary process* $\{V_{\mathbf{t}}\}$. The argument is parallel to that given in Chapter II section 3 for the one dimensional problem. One should note however that for a specific dimension $k \geq 2$ and a weakly stationary process $\{V_{\mathbf{t}}\}$ designated, this may be only a sufficient condition for the existence of a weakly stationary solution $\{X_{\mathbf{t}}\}$. Consider, for example, the case in which $k = 3$, $\{V_{\mathbf{t}}\}$ is white noise, $a(\mathbf{z}) \equiv 1$ and $b(\mathbf{z}) = 1 - (1/3)(z_1 + z_2 + z_3)$. Even though $b(e^{i\boldsymbol{\lambda}}) = 0$

for $(\lambda_1, \lambda_2, \lambda_3) = (0, 0, 0)$ and so it is not the case that $b(e^{i\boldsymbol{\lambda}}) \neq 0$ for all $\boldsymbol{\lambda} = (\lambda_1, \lambda_2, \lambda_3)$ with real components, it is clear that there is a weakly stationary solution of the system (21). Notice that there is no unique weakly stationary solution $\{X_t\}$ because $b(1, 1, 1) = 0$.

By the invertibility of a trigonometric polynomial $a(e^{i\boldsymbol{\lambda}})$ relative to the semigroup S we mean that $a(e^{i\boldsymbol{\lambda}})^{-1}$ is a trigonometric series

$$a(e^{i\boldsymbol{\lambda}})^{-1} = \sum_{\mathbf{t} \in S} \alpha_{\mathbf{t}}\, e^{i\mathbf{t}\cdot\boldsymbol{\lambda}}$$

with coefficients $\alpha_{\mathbf{t}} \neq 0$ only if $\mathbf{t} \in S$ and with the coefficients absolutely summable. This is a convenient notion of invertibility for us. If $b(e^{i\boldsymbol{\lambda}})$ is invertible relative to the semigroup S than

$$\begin{aligned} \mathcal{M}_{\mathbf{t}}(X; S) &= \mathcal{M}(X_{\boldsymbol{\tau}}, \boldsymbol{\tau} \in T^{\mathbf{t}} S) \\ &\subseteq \mathcal{M}_{\mathbf{t}}(V; S) = \mathcal{M}(V_{\boldsymbol{\tau}}, \boldsymbol{\tau} \in T^{\mathbf{t}} S)\,. \end{aligned}$$

Similarly if $a(e^{i\boldsymbol{\lambda}})$ is invertible relative to S then

$$\mathcal{M}_{\mathbf{t}}(V; S) = \mathcal{M}_{\mathbf{t}}(X; S)\,.$$

If $\mathcal{M}_{\mathbf{t}}(V; S) = \mathcal{M}_{\mathbf{t}}(X; S)$ and if $V_{\mathbf{t}}$ is orthogonal to $X_{\boldsymbol{\tau}}$, $\boldsymbol{\tau} \in T^{\mathbf{t}} S'$, the best linear predictor of $X_{\mathbf{t}}$ in terms of $X_{\boldsymbol{\tau}}$, $\boldsymbol{\tau} \in T^{\mathbf{t}} S'$ is given by

$$X_{\mathbf{t}}^{*} = b_0^{-1}\left\{-\sum_{\boldsymbol{\tau} \in S'} b_{\boldsymbol{\tau}} X_{\mathbf{t}+\boldsymbol{\tau}} + \sum_{\boldsymbol{\tau} \in S'} a_{\boldsymbol{\tau}} V_{\mathbf{t}+\boldsymbol{\tau}}\right\}. \tag{22}$$

This condition is not only sufficient for (22) to be the best linear predictor of $X_{\mathbf{t}}$ in terms of $X_{\boldsymbol{\tau}}$, $\boldsymbol{\tau} \in T^{\mathbf{t}} S'$, in the case of an ARMA process relative to S; *it is also necessary*. Given the trigonometric polynomials

$$b(e^{i\boldsymbol{\lambda}}) = \sum_{\boldsymbol{\tau} \in S} b_{\boldsymbol{\tau}}\, e^{i\boldsymbol{\tau}\cdot\boldsymbol{\lambda}},$$

$$a(e^{i\boldsymbol{\lambda}}) = \sum_{\boldsymbol{\tau} \in S} a_{\boldsymbol{\tau}}\, e^{i\boldsymbol{\tau}\cdot\boldsymbol{\lambda}}$$

with nonzero coefficients only on S, we have just seen that the invertibility of $b(e^{i\boldsymbol{\lambda}})$ and $a(e^{i\boldsymbol{\lambda}})$ relative to S is a sufficient condition for

$$\mathcal{M}_{\mathbf{t}}(V; S) = \mathcal{M}_{\mathbf{t}}(X; S)$$

for a corresponding ARMA scheme relative to S, whatever the structure of the weakly stationary process $\{V_t\}$.

Let us as an example consider the semigroup $S_1 = \{(i, j) : i \leq -1 \text{ or } i = 0 \text{ and } j \leq 0\}$ of the type discussed by Helson and Lowdenslager and see what kind of a condition for invertibility is given to us by an application of Theorem 2 of the Appendix. The theorem states that a function in the convolution algebra of absolutely summable functions on the semigroup is invertible relative to the semigroup if and only if no homomorphism for the algebra takes on the value zero for the function. Let us translate this into more concrete language by determining the form of a homomorphism. Let $\delta_{j,k}$ be the function taking on the value 1 at $(j, k) \in S$ and zero elsewhere on S. Let τ be a given homomorphism. Now

$$\tau(h) = \sum_{(j,k) \in S_1} h_{jk} \, \tau(\delta_{jk})$$

if h takes on the value h_{jk} at (j, k). It is clear that

$$\delta_{0,-1}, \; \delta_{0,0}, \; \delta_{-1,j}, \quad j = \ldots, -1, 0, 1, \ldots$$

are generators of the Banach algebra.

Let

$$\tau(\delta_{0,-1}) = z ,$$

$$\tau(\delta_{-1,i}) = z_i .$$

We know that $\tau(\delta_{0,0}) = 1$. Further $|z|, |z_j| \leq 1$, $j = \ldots, -1, 0, 1, \ldots$. Notice that

$$z_{j-1} = \tau(\delta_{-1,j-1}) = \tau(\delta_{-1,j} \, \delta_{0,-1})$$

$$= \tau(\delta_{-1,j}) \, \tau(\delta_{0,-1}) = z_j \, z .$$

If $z = 0$, then $z_j = 0$ for all integral j and we have the trivial homomorphism $\tau(h) \equiv 1$ for all h. Let $z \neq 0$. Then $|z| = 1$ is implied by the fact that $z_j = z_0 \, z^{-j}$ unless $z_0 = 0$. The case $z_0 = 0$ leads us to a homomorphism of the form

$$\tau(h) = \sum_{k \geq 0} h_{0,-k} \, z^k .$$

Assume that $z_0 \neq 0$. If we set $z_0 = e^{-i\lambda}$, λ real, the corresponding homomorphism can be written as

$$\tau(h) = \sum_{(-j,-k) \in S_1} h_{-j,-k} \, z_0^j \, e^{ik\lambda} .$$

The conditions for invertibility of h can thus be written in the form

$$\hat{h}(z, e^{i\lambda}) \neq 0$$

for all $0 < |z| \leq 1$, real λ, and

$$\hat{h}(0, z) \neq 0 \tag{23}$$

for all $|z| \leq 1$ where

$$\hat{h}(z, z') = \sum_{(-j,-k) \in S_1} h_{-j,-k} \, z^j \, z'^k .$$

Consider the related semigroup $S_2 = \{(i, j) : i \leq 0\}$. If we carry through an analogous but somewhat simpler computation, we find that a function h in the convolution algebra of absolutely summable functions h on S_2 is invertible if and only if

$$\hat{h}(z, e^{i\lambda}) \neq 0$$

for all $|z| \leq 1$, real λ, with

$$\hat{h}(z, z') = \sum_{(-j, -k) \in S_2} h_{-j, -k} z^j z'^k .$$

Suppose we consider the function h with

$$h_{0,0} = 1, h_{0,-1} = -2, h_{-1,0} = \varepsilon$$

$$h_{i,j} = 0 \quad \text{if } (i, j) \neq (0, 0), (0, -1), (-1, 0) .$$

Since h does not satisfy (23), it follows that h is not invertible relative to the semigroup S_1. However, if ε is sufficiently small in absolute value, say $|\varepsilon| < .1$, the function h is invertible relative to the semigroup S_2.

In the case of the quarter plane semigroup $S_3 = \{(i, j) \mid i, j \leq 0\}$ a similar argument shows that an absolutely summable h on S_3 is invertible if and only if

$$\hat{h}(z_1, z_2) \neq 0$$

for all $|z_1|, |z_2| \leq 1$ where

$$\hat{h}(z_1, z_2) = \sum_{(-j, -k) \in S_3} h_{-j, k} z_1^j z_2^k .$$

Up to this point we have only looked at a few interesting semigroup convolution algebras for which conditions can be readily obtained. It is clear that there are many other semigroups of interest, for example two dimensional semigroups contained in a sector of angle less than π radians. At this point we shall introduce some additional notation that will enable us to state and prove an interesting comparison theorem due to Davidson and Vidyasagar [1983]. If C is a semigroup in Z^k let

$$C^{-1} = C \cap \{-C\}$$

be the largest group contained in C. $l_1(S)$ will denote the set of absolutely summable functions on the semigroup S. P_S will be used to denote the projection of $l_1(Z^k)$ onto $l_1(S)$.

Theorem 3. *Assume that C is an additive semigroup on Z^k and that S is a subsemigroup of C with the property that*

$$S^{-1} = S \cap C^{-1} . \tag{24}$$

Then if $f \in l_1(S)$ has an inverse f^{-1} in $l_1(C)$, it follows that f^{-1} belongs to $l_1(S)$.

We first consider the case in which S is a group. Assumption (24) implies that $S \subseteq C^{-1}$. Notice that if $i + j \in S$ and $j \in S$ it follows that $i + j + (-j) = i \in S$. Given $f \in l_1(S)$ and $g \in l_1(C)$ one then has $P_S(f * g) = f * P_S(g)$. Since $f * g = 1$ implies that $f * P_S(g) = 1$, the conclusion is that $g = P_S\, g$ and $g \in l_1(S)$.

The more difficult case in which S is not a group is now considered. Let ${}_0S = S \backslash S^{-1}$, the set of elements in S but not in S^{-1}. By the support of f, supp f, we mean the set on which f takes nonzero values. Given $f \in l_1(S)$ let $f = f_0 + f_1$ where supp $f_0 \subseteq {}_0S$ and supp $f_1 \subseteq S^{-1}$. Also set ${}_0C = C\backslash C^{-1}$ and if $g \in l_1(C)$ is the inverse of f, let $g = g_0 + g_1$ with supp $g_0 \subseteq {}_0C$ and supp $g_1 \subseteq C^{-1}$. Let $P = P_{C^{-1}}$. Then

$$1 = f * g = P(f * g) = P(f_1 * g_1) + P(f_0 * g_1)$$
$$+ P(f_1 * g_0) + P(f_0 * g_0) .$$

First consider $f_0 * g_1$. If $i - j \in {}_0S$, $j \in C^{-1}$ we cannot have $i \in C^{-1}$. For that would imply $i - j \in C^{-1}$ contradicting $i - j \in {}_0S = S/S^{-1} = S/C^{-1}$ since $S^{-1} = S \cap C^{-1}$. Thus $P(f_0 * g_1) = 0$. In the case of $g_0 * f_1$ the condition $i - j \in {}_0C = C\backslash C^{-1}$ and $j \in S^{-1} \subseteq C^{-1}$ is satisfied. If $i \in C^{-1}$ then $i - j \in C^{-1}$ contradicting $i - j \in {}_0C$. Thus $P(g_0 * f_1) = 0$. Let ${}_1S = {}_0S \cup \{0\}$, ${}_1C = {}_0C \cup \{0\}$. Then ${}_1C$ and ${}_1S$ are semigroups with ${}_1S \subseteq {}_1C$ and ${}_1C^{-1} = {}_1S^{-1} = \{0\}$. Because of (24) it follows that ${}_0S \subseteq {}_0C$. This in turn implies that $P(g_0 * f_0) = 0$. Therefore

$$1 = P(f_1 * g_1) = f_1 * g_1 .$$

Since $f_1 \in l_1(S^{-1})$ we have from the initially considered case of a group that $g_1 = f_1^{-1} \in l_1(S_1)$. Notice that

$$(g_1 * f)^{-1} = (1 + g_1 * f_0)^{-1} = f_1 * g = (1 + f_1 * g_0) . \tag{25}$$

The support of $g_1 * f_0$ is in ${}_0S$ and that of $f_1 * g_0$ in ${}_0C$. Our claim is that the theorem is proved if one can show whenever $h = 1 + h_0$ with supp $h_0 \subseteq {}_0S$ and $h^{-1} = 1 + t_0$ with supp $t_0 \subseteq {}_0C$ that then supp $t_0 \subseteq {}_0S$. If we apply this to (25) it would then follow that supp$(f_1 * g_0) \subseteq {}_0S$. But $i \in {}_0S$, $i - j \in {}_0C$, $j \in S^{-1}$ imply that $i - j \in {}_0S$. Thus supp$(g_0) \subseteq {}_0S$. We therefore have supp(g) = supp$(g_0 + g_1)$ $\subseteq S$.

Let us now suppose that $h = 1 + h_0 \in l_1({}_1S)$, $h^{-1} = t = 1 + t_0 \in l_1(C)$. Since the set of invertible elements is open there is a sequence $h^{(k)} = 1 + h_0^{(k)}$ with finite support converging to h such that each $h^{(k)}$ is invertible in $l_1({}_1C)$. If one can show that each $(h^{(k)})^{-1}$ belongs to $l_1({}_1S)$ it will then follow that $h^{-1} \in l_1({}_1S)$ because $h^{-1} = \lim(h^{(k)})^{-1}$ and $l_1({}_1S)$ is a closed subspace of $l_1({}_1C)$. We can therefore suppose supp(h) is finite.

Because ${}_1C$ is a semigroup with ${}_1C^{-1} = \{0\}$, ${}_1C$ is contained in a halfspace $\{i \in Z^k \mid \varphi(i) \geq 0\}$ with φ an appropriate linear functional in R^k. Let $\varphi(\text{supp}(a))$ be the set of values assumed by $\varphi(i)$ as i varies over supp(a). Then

$$\varphi(\text{supp}(a * b)) \subseteq \varphi(\text{supp}(a)) + \varphi(\text{supp}(b)) .$$

The proof now proceeds by induction on k. For $k = 0$ there is nothing to prove. For $k > 0$ write $t_0 = t_0' + t_0''$ with $\mathrm{supp}(t_0') \subseteq {}_0S$, $\mathrm{supp}(t_0'') \subseteq {}_0C \backslash {}_0S$. We consider two cases i) and ii). In case i) the infinum of the values in the set $\varphi(\mathrm{supp}\ h_0)$ is assumed to be a value $\alpha > 0$. Since

$$\begin{aligned} 1 &= (1 + h_0) * (1 + t_0' + t_0'') \\ &= 1 + (h_0 + t_0' + h_0 * t') + (t_0'' + h_0 * t_0'') \end{aligned}$$

and $\mathrm{supp}(h_0 + t_0' + h_0 * t') \subseteq {}_0S$ it follows that $\mathrm{supp}(t_0'' + h_0 * t_0'') \subseteq {}_0S$. Let $\beta = \inf \varphi(\mathrm{supp}\ t_0'')$ if $t_0'' \neq 0$. Choose an element i such that $t_0''(i) \neq 0$ and $\varphi(i) < \beta + \alpha$. Now

$$\begin{aligned} \varphi\big(\mathrm{supp}(h_0 * t_0'')\big) &\subseteq \varphi\big(\mathrm{supp}(h_0)\big) + \varphi\big(\mathrm{supp}(t'')\big) \\ &\subseteq [\alpha, \infty) + [\beta, \infty) = [\alpha + \beta, \infty) \end{aligned}$$

so that $(h_0 * t_0'')(i) = 0$. This implies that $(t_0'' + h_0 * t_0'')(i) = t_0''(i) \neq 0$, contradicting $\mathrm{supp}(t_0'' + h_0 * t_0'') \subseteq {}_0S$. Therefore $t_0'' = 0$, that is $\mathrm{supp}(t_0) \subseteq {}_0S$.

We now consider case ii) in which $\inf \varphi\big(\mathrm{supp}(h_0)\big) = 0$. Write $h_0 = h_{00} + h_{01}$ where $\varphi\big(\mathrm{supp}(h_{00})\big) = 0$ and $\inf \varphi(\mathrm{supp}\ h_{01}) > 0$. This decomposition is possible because we have assumed $\mathrm{supp}(h_0)$ is finite. Also set $t_0 = t_{00} + t_{01}$ with $\varphi\big(\mathrm{supp}(t_{00})\big) = 0$ and $\varphi\big(\mathrm{supp}(t_{01})\big) \subseteq (0, \infty)$. Then

$$\begin{aligned} 1 &= (1 + h_{00} + h_{01}) * (1 + t_{00} + t_{01}) \\ &= (1 + h_{00} + t_{00} + h_{00} * t_{00}) \\ &\quad + (h_{01} + t_{01} + h_{00} * t_{01} + h_{01} * t_{00} + h_{01} * t_{01}) \,. \end{aligned} \tag{26}$$

$\varphi(i) = 0$ for any element i belonging to the support of the first term on the right and $\varphi(i) > 0$ whenever i belongs to the support of the second term on the right. The first term must be one and so

$$(1 + h_{00}) * (1 + t_{00}) = 1 \,.$$

The second term on the right of (26) must be zero. Now $(1 + h_{00}) \in l_1\big({}_1S \cap \varphi^{-1}(0)\big)$ and ${}_1S \cap \varphi^{-1}(0)$ is a semigroup in Z^{n-1}. By the induction hypothesis $1 + t_{00} = (1 + h_{00})^{-1}$ belongs to $l_1\big({}_1S \cap \varphi^{-1}(0)\big)$. If one replaces $h = 1 + h_0$ by $h * (1 + t_{00}) = (1 + h_{00} + h_{01}) * (1 + t_{00}) = 1 + h_{01} * t_{00} = h'$ then $h' = 1 + h_0'$ and $\inf \varphi(\mathrm{supp}\ h_0') > 0$. Since h' can be approximated by functions of finite support in $l_1({}_1S)$ it follows from the preceding paragraphs that $(h')^{-1}$ belongs to $l_1({}_1S)$ if it belongs to $l_1({}_1C)$. Notice that

$$\begin{aligned} (1 + h_0)^{-1} &= \big(h' * (1 + t_{00})^{-1}\big)^{-1} \\ &= (h')^{-1} * (1 + t_{00}) \,. \end{aligned}$$

Therefore $(1 + h_0)^{-1} \in l_1({}_1S)$. The proof is complete.

A direct application of this result can be made in the case of the two semigroups $S_1 = \{(i, j) \mid i \leq -1 \text{ or } i = 0 \text{ and } j \leq 0\}$ and $S_3 = \{(i, j) \mid i, j \leq 0\}$. Clearly $S_3 \subseteq S_1$ and $S_3^{-1} = \{0\} = S_3 \cap S_1^{-1}$. It therefore follows that $f \in l_1(S_3)$ has an inverse f^{-1} in $l_1(S_3)$ if f^{-1} belongs to $l_1(S_1)$. The conditions for invertibility of f in $l_1(S_3)$ can therefore using (23) and (24) be written as

$$\hat{f}(z, e^{i\lambda}) \neq 0 \tag{27}$$

for all $0 < |z| \leqq 1$, real λ, and

$$\hat{f}(0, z) \neq 0 \tag{28}$$

for all $|z| \leq 1$. A parallel argument can be used to show that the conditions for invertibility of $f \in l_1(S_4)$ with S_4 the semigroup $\{(i, j) \mid i, j < 0 \text{ or } i = j = 0\}$ are (27) and (28). In fact, the same argument can be used to show that the same conditions are those for the invertibility of a function $f \in l_1(S)$ where S consists of points (i, j) with $i, j \leq 0$ contained in a sector of angle less than π radians.

3. NonGaussian Linear Random Fields

We have already remarked on the indeterminacy in the specification of zero sets of the polynomials characterizing Gaussian ARMA stationary processes in section 3 of Chapter II. In section 4 of Chapter II it was shown that this indeterminacy essentially disappears in the case of nonGaussian ARMA stationary processes. A class of effective estimates of the transfer function of nonGaussian linear processes were constructed in the first section of this chapter. The discussion of random fields of the previous section indicates that zero sets of the polynomials characterizing ARMA random fields have an even more complicated structure than in the one dimensional context of ARMA processes. The indeterminacy in the specification of zero sets of the polynomials (in several complex variables) characterizing Gaussian ARMA random fields is more serious because of this complicated structure of the zero sets. However, we shall show that in the case of a large class of nonGaussian linear random fields this indeterminacy essentially disappears just as in the one dimensional case of nonGaussian linear processes. The nonGaussian linear random field

$$X_{\mathbf{t}} = \sum \alpha_{\tau} V_{\mathbf{t}-\tau}$$

with $V_{\mathbf{t}}$ a family of independent, identically distributed nonGaussian random variables with mean zero $E\, V_{\mathbf{t}} \equiv 0$ and variance one. The result we obtain is the multidimensional analogue of Theorem 5 of Chapter II. In terms of observations on $X_{\mathbf{t}}$ alone, under appropriate conditions, the phase of $\alpha(e^{-i\lambda})$ can be almost completely identified for a nonGaussian (as contrasted with a Gaussian) linear random field. In the case of an ARMA process

$$\alpha(e)^{-i\lambda} = \frac{a(e^{-i\lambda})}{b(e^{-i\lambda})}$$

with $a(\mathbf{z})$, $b(\mathbf{z})$ polynomials.

Theorem 4. *Let $\{X_{\mathbf{t}}\}$ be a nonGaussian linear random field. Assume that the independent, identically distributed random variables $V_{\mathbf{t}}$ have all their moments finite. Let*

$$\sum_{\boldsymbol{\tau}} |\boldsymbol{\tau}|\, |a_{\boldsymbol{\tau}}| < \infty$$

and $\alpha(e^{-i\boldsymbol{\lambda}}) \neq 0$ for all vectors $\boldsymbol{\lambda}$ with real-valued components. Here $\mathbf{1}$ is the vector with its components 1. The function $\alpha(e^{-i\boldsymbol{\lambda}})$ can be identified in terms of observations on the process $\{X_{\mathbf{t}}\}$ alone up to a vector $\mathbf{a}$ with integer components in a factor $e^{i\mathbf{a}\cdot\boldsymbol{\lambda}}$ and the sign of $\alpha(\mathbf{1}) = \Sigma_{\boldsymbol{\tau}} a_{\boldsymbol{\tau}}$. This result still holds if one only assumed that $V_{\mathbf{t}}$ has moments up to order $r > 2$ finite with the cumulant $\gamma_r \neq 0$.

Notice that a nonGaussian variable $V_{\mathbf{t}}$ with all its moments finite must have a cumulant $\gamma_r \neq 0$ for some integer $r > 2$. In the following derivation, symbols with a right superscript represent k-vectors. The rth order cumulant spectral density of the process $\{X_{\mathbf{t}}\}$ is

$$\begin{aligned}(29)\quad & b_r(\boldsymbol{\lambda}^{(1)}, \ldots, \boldsymbol{\lambda}^{(r)}) \\ &= (2\pi)^{-k(r-1)} \sum_{\mathbf{j}^{(1)}, \ldots, \mathbf{j}^{(r-1)}} \operatorname{cum}(X_{\mathbf{t}}, X_{\mathbf{t}+\mathbf{j}^{(1)}}, \ldots, X_{\mathbf{t}+\mathbf{j}^{(r-1)}}) \\ &\qquad \exp\left(-i \sum_{s=1}^{r-1} \mathbf{j}^{(s)} \cdot \boldsymbol{\lambda}^{(s)}\right) \\ &= \frac{\gamma_r}{(2\pi)^{k(r-1)}}\, \alpha\left(e^{-i\boldsymbol{\lambda}^{(1)}}\right) \ldots \alpha\left(e^{-i\boldsymbol{\lambda}^{(r-1)}}\right) \alpha\left(e^{i(\boldsymbol{\lambda}^{(1)} + \cdots + \boldsymbol{\lambda}^{(r-1)})}\right).\end{aligned}$$

Notice that

$$\left\{\frac{\alpha(\mathbf{1})}{|\alpha(\mathbf{1})|}\right\}^r \gamma_r = (2\pi)^{k(r/2-1)}\, b_r(0, \ldots, 0) / \{f(0)\}^{r/2}.$$

We introduce the function

$$(30)\qquad h(\boldsymbol{\lambda}) = \arg\left\{\alpha(e^{-i\boldsymbol{\lambda}})\, \frac{\alpha(\mathbf{1})}{|\alpha(\mathbf{1})|}\right\}.$$

It then follows that

$$(31)\qquad h(-\boldsymbol{\lambda}) = -h(\boldsymbol{\lambda})$$

since the coefficients $a_{\mathbf{j}}$ are assumed to be real. Relations (29), (30) and (31) imply that

$$\begin{aligned}(32)\quad & h(\boldsymbol{\lambda}^{(1)}) + \cdots + h(\boldsymbol{\lambda}^{(r-1)}) - h(\boldsymbol{\lambda}^{(1)} + \cdots + \boldsymbol{\lambda}^{(r-1)}) \\ &= \arg\left[\left\{\frac{\alpha(\mathbf{1})}{|\alpha(\mathbf{1})|}\right\}^r \gamma_r^{-1}\, b(\boldsymbol{\lambda}^{(1)}, \ldots, \boldsymbol{\lambda}^{(r-1)})\right].\end{aligned}$$

Now

$$(34)\quad D_{u_s} h(\lambda_1, \ldots, \lambda_{s-1}, \lambda_s, 0, \ldots, 0) - D_{u_s} h(0, \ldots, 0)$$

$$= -\lim_{\Delta \to 0} \frac{1}{(r-2)\Delta} \{h(\lambda_1, \ldots, \lambda_{s-1}, \lambda_s, 0, \ldots, 0)$$

$$+ (r-2)\, h(0, \ldots, 0, \Delta, 0, \ldots, 0)$$

$$- h(\lambda_1, \ldots, \lambda_{s-1}, \lambda_s + (r-2)\Delta, 0, \ldots, 0)\},$$

$$s = 1, \ldots, k .$$

Consider the simple identity

$$(35)\quad h(\lambda_1, \ldots, \lambda_{s-1}, \lambda_s, 0, \ldots, 0) - h(\lambda_1, \ldots, \lambda_{s-1}, 0, 0, \ldots, 0)$$

$$= \int_0^{\lambda_s} \{D_{u_s} h(\lambda_1, \ldots, \lambda_{s-1}, u_s, 0, \ldots, 0) - D_{u_s} h(0, \ldots, 0)\}\, du + c_s \lambda_s ,$$

$$c_s = D_{u_s} h(0, \ldots, 0) , \quad s = 1, \ldots, k .$$

Relation (32) implies that the expressions (34) can be effectively estimated by making use of rth order cumulant spectral estimates. The estimates of (34) can be used in turn to estimate the integrals on the right of (35) by means of approximating Riemann sums. It is convenient to rewrite the set of equations (35) as

$$h(\lambda_1, \ldots, \lambda_{s-1}, \lambda_s, 0, \ldots, 0) - h(\lambda_1, \ldots, \lambda_{s-1}, 0, 0, \ldots, 0)$$

$$= {}_s h(\lambda_1, \ldots, \lambda_s, 0, \ldots, 0) + c_s \lambda_s , \quad s = 1, \ldots, k .$$

The values $\alpha(\pi, \ldots, \pi, 0, \ldots, 0)$ must be real because the coefficients α_j are real. This in turn implies that

$$h(\underbrace{\pi, \ldots, \pi}_{s}, 0, \ldots, 0) - h(\underbrace{\pi, \ldots, \pi}_{s-1}, 0, 0, \ldots, 0) = a_s \pi$$

for some integer a_s, $s = 1, \ldots, k$. Let

$${}_s h(\pi, \ldots, \pi, 0, \ldots, 0) / \pi = \delta_s ,$$

$s = 1, \ldots, k$. It then follows that

$$c_s = a_s - \delta_s .$$

The indeterminacy in the constants c_s is the integer part a_s. Let

$$h_1(\lambda_1, \ldots, \lambda_k) = \sum_{s=1}^{k} {}_s h(\lambda_1, \ldots, \lambda_s, 0, \ldots, 0) .$$

Notes

8.1 Deconvolution problems arise often in a geophysical context. A simple model is one in which the earth is thought of as a one dimensional layered medium. An explosion is set off at the surface with a characteristic set of shaping factors a_j referred to as the wavelet. There are refrectivity coefficients V_t characteristic of the different layers. A superposition X_t of the terms $a_j V_{t-j}$ is received at the surface. It is often assumed that the sequence V_t can be modeled as a set of independent and identically distributed random variables. In the conventional deconvolution it is assumed either that the wavelet $\{a_j\}$ is known or else that it satisfies a minimum phase condition. Discussion and references to work in which the classical methods are used can be found in Aki and Richards [1980] or Robinson [1982]. In the case of Gaussian data the minimum phase assumption is perfectly natural. However, if the data is non-Gaussian, it is rather implausible as an assumption. In Wiggins [1978] and Donoho [1981] a procedure referred to as minimum entropy deconvolution is proposed to deal with nonGaussian data without making a minimum phase assumption. The analysis in this section is based on ideas presented in Rosenblatt [1980] and Lii and Rosenblatt [1982]. A firm theoretical (and practical) framework for deconvolution of nonGaussian data without knowledge of the wavelet or the assumption of a minimum phase condition is given.

8.2 A Markov-like property has been considered for random fields. Consider initially processes with index set the lattice points in k-space (with k a fixed integer) and with state space finite. Given a finite set S of index points, let S^c be the complement of S and H the boundary of S consisting of lattice points in S^c at a distance of one from S. Let $\mathscr{B}_S$, $\mathscr{B}_{S^c}$ and $\mathscr{B}_H$ be the Borel fields of sets generated by random variables whose indices lie in S, S^c and H respectively. The Markov-like property is that

$$P(B \mid \mathscr{B}_{S^c})\,(w) = P(B \mid \mathscr{B}_H)\,(w)$$

for any event $B \in \mathscr{B}_S$. One wants this to hold for any finite set (or more optimistically any set) of lattice points S. A discussion of this property (or a weaker property called d-Markovian for an integer $d \geq 1$) can be found in Dobrushin [1968]. A related presentation in the context of some problems in statistical mechanics is given in Preston [1974]. A corresponding treatment of Gaussian Markov-like fields with lattice index is laid out in Rozanov [1967b]. Gaussian random fields with continuous multidimensional parameter are examined in Pitt [1975] and Dobrushin [1980]. The collections of Bose [1979] and Mitra and Ekstrom [1978] present papers on random fields that have appeared in the engineering literature. A detailed presentation of some of the important results in this literature is given in Bose [1982]. An extensive bibliography of papers on random fields is in the book of Ripley [1981]. Examples of random fields designed to gauge visual discrimination are discussed in Julesz

[1975] and Julesz, Gilbert and Victor [1978]. Much of the material in sections 8.2 and 8.3 is based on Rosenblatt [1983].

See Glimm and Jaffe [1981] for a discussion of random fields in the context of quantum physics.

Appendix

1. Monotone Functions and Measures

There are a number of basic remarks out of real analysis that are useful to refer to in the course of some of the derivations made in this book. These will be mentioned briefly with an occasional discussion of their interpretation. The first of these is what we have referred to as the *Helly convergence theorem.*

Let us first note that if we have a monotone nondecreasing $F(x)$ on the real line or a closed subinterval I of the real line, it can have at most a countable number of discontinuities or jumps. This means in particular that the continuity points of such a function are dense everywhere. Consider now a sequence of monotone nondecreasing functions $F_n(x)$. The sequence of functions are said to *converge weakly* to a monotone nondecreasing function $F(x)$ if $\lim_{n\to\infty} F_n(x) = F(x)$ at every continuity point of $F(x)$.

Helly Convergence Theorem 1. Let $F_n(x)$ be a uniformly bounded sequence of monotone nondecreasing functions on the real line or on a closed subinterval I. There is then a monotone nondecreasing function $F(x)$ and a subsequence $F_{n_j}(x)$ of the original sequence such that $F_{n_j}(x)$ converges weakly to $F(x)$ as $n_j \to \infty$.

We shall consider a relation between bounded monotone nondecreasing functions and finite measures on the real line. First remarks of a more general character are made. Let Ω be a space of points w and $\mathscr{F}$ a collection of subsets of Ω with the following properties.

1. If a countable collection of sets $A_1, A_2, \ldots \in \mathscr{F}$ then the union of the sets $\bigcup_{i=1}^{\infty} A_i \in \mathscr{F}$.

2. Ω is an element of $\mathscr{F}$.

3. Given a set $A \in \mathscr{F}$, its complement (relative to Ω) is an element of $\mathscr{F}$.

A collection of sets $\mathscr{F}$ with these three properties is called a *sigma-field* or a *Borel field.* A collection $\mathscr{C}$ of subsets of Ω satisfying conditions 2 and 3, and having condition 1 replaced by 1′

"1′. If a finite number of sets $A_1, A_2, \ldots, A_k \in \mathscr{C}$ then the union $\bigcup_{i=1}^{k} A_i \in \mathscr{C}$"

is called a *field.*

A finite measure η on the sigma-field $\mathscr{F}$ is a set function defined on $\mathscr{F}$ with the following properties.

1. For all sets $A \in \mathscr{F}$, $\eta(A) \geq 0$ is defined.

2. $\eta(\Omega) < \infty$.

3. Given any countable collection of disjoint sets $A_1, A_2, \ldots \in \mathscr{F}$

$$\eta\left(\bigcup_j A_i\right) = \sum_i \eta(A_i) .$$

The collection of real-valued functions $X(w)$ on Ω consistent with the sigma-field $\mathscr{F}$ in the following sense are called measurable (with respect to $\mathscr{F}$). They are the functions $X(w)$ with the property that every sublevel set $\{w : X(w) \leq y\}$ (for any real number y) is an element of the sigma-field $\mathscr{F}$.

It is sometimes of interest to consider the sigma-field generated by a particular collection $\mathscr{C}$ of subsets of Ω. This sigma-field $\mathscr{F} = \mathscr{F}(\mathscr{C})$ is the intersection of all sigma-fields containing the collection $\mathscr{C}$. Notice that the collection of all subsets of $\mathscr{F}$ is a sigma-field containing $\mathscr{C}$. Let $\mathscr{C}$ be the collection of half lines $\{w : w \leq y\}$ with y any real number. The sigma-field $\mathscr{B} = \mathscr{F}(\mathscr{C})$ generated by this choice of $\mathscr{C}$ is called the sigma-field of Borel sets. A function $X(w)$ measurable with respect to this sigma-field $\mathscr{B}$ of Borel sets is called a Borel function.

At times one is given a nonnegative set function m defined on a field $\mathscr{C}$. It is of interest to then find out whether m can be extended to a measure η defined on the sigma-field $\mathscr{F}(\mathscr{C})$ generated by $\mathscr{C}$. A result of Carathéodory (see Loève [1963]) indicates that an extension can be effected if m already acts like a measure on the field $\mathscr{C}$, that is, $m(A) \geq 0$ for $A \in \mathscr{C}$ and for any countable collection of disjoint sets $A_1, A_2, \ldots \in \mathscr{C}$ with $\bigcup_i A_i \in \mathscr{C}$ one has $m(\bigcup_i A_i) = \sum_i m(A_i)$. Moreover, this extension is unique when m is finite on $\mathscr{C}$.

Let us now consider an illustration of the remarks made in the last paragraph. Let F be a bounded monotone nondecreasing function on the reals. It is convenient to assume that F is right continuous. This is not an essential restriction. Take Ω as the set of real numbers and $\mathscr{C}$ as consisting of intervals of the form

$$\{w : \alpha < w \leq \beta\}, \quad \alpha < \beta$$

as well as sets formed by taking finite unions of such intervals. It is clear that $\mathscr{C}$ is a field. We now generate a nonnegative set function m on the sets of $\mathscr{C}$ derived from the monotone function F. Every set of $\mathscr{C}$ can be given as a union of disjoint intervals

$$\bigcup_i \{w : \alpha_i < w \leq \beta_i\}, \quad \alpha_i < \beta_i .$$

Set

$$m\left(\bigcup_i \{w : \alpha_i < w \leq \beta_i\}\right) = \sum_i \{F(\beta_i) - F(\alpha_i) .$$

One can show that m acts like a measure on $\mathscr{C}$. The result of Carathéodory implies that m can be uniquely extended to a measure η on $\mathscr{B}$, the Borel sets. The measure η is the measure determined by the monotone function F.

One can introduce a multivariate analogue of a bounded monotone function. Now let $F(\mathbf{x})$, $\mathbf{x} = (x_1, \ldots, x_k)$, be a bounded nonnegative function that is monotone nondecreasing in each variable separately. Further assume that

$$\lim_{x_j \to -\infty} F(x_1, \ldots, x_k) = 0, \quad j = 1, \ldots, k .$$

Let the difference

$$(1) \qquad \Delta_{h_i} F(x_1, \ldots, x_k) = F(x_1, \ldots, x_{i-1}, x_i + h_i, x_{i+1}, \ldots, x_k) - F(x_1, \ldots, x_{i-1}, x_i, x_{i+1}, \ldots, x_k) \geq 0$$

for $h_i \geq 0$. In addition let all kth order differences

$$(2) \qquad \Delta_{h_1} \Delta_{h_2} \ldots \Delta_{h_k} F(x_1, \ldots, x_k) \geq 0$$

when $h_1, \ldots, h_k \geq 0$. Clearly, (1) can be considered a special case of (2). Again it is convenient to assume that $F(\mathbf{x}) = F(x_1, \ldots, x_k)$ is continuous to the right in each variable separately. Now consider Ω the space of k-dimensional points $\mathbf{w} = (w_1, \ldots, w_k)$. Let the field $\mathscr{C}$ consist of intervals

$$I = \{w : x_i < w_i < x_i + h_i, \quad i = 1, \ldots, k\}, \quad h_i \geq 0$$

and unions of finite numbers of such intervals. However, each set of $\mathscr{C}$ can be given as a union of disjoint intervals

$$\cup\, I_j = \cup\, \{w : x_i^{(j)} < w_i \leq x_i^{(j)} + h_i^{(j)}, \quad i = 1, \ldots, k\}.$$

Introduce the set function m on $\mathscr{C}$ as given by

$$m(\cup\, I_j) = \sum_j \Delta_{h_1^{(j)}}, \ldots, \Delta_{h_k^{(j)}} F(x_1^{(j)}, \ldots, x_k^{(j)}).$$

The nonnegative set function m acts like a measure on $\mathscr{C}$. Therefore by the theorem of Carathéodory, m can be extended to the sigma-field generated by $\mathscr{C}$ (the sigma-field of k-dimensional Borel sets). The extended measure is a finite measure.

2. *Hilbert Space*

A Hilbert space is a vector space over the reals (or complex numbers) with an inner product that is complete. We shall discuss this description at greater leisure and consider one or two examples. For convenience a complex Hilbert space will be discussed. A collection of elements V is a vector space over the complex numbers if addition is defined for any pair of elements $x, y \in V$ with the sum $z = x + y \in V$ and multiplication of an element $x \in V$ by any complex number a is defined with $z = a\, x \in V$. An inner product (x, y) defined for any pair of elements x, y is a complex-valued bilinear function, that is,

$$(x + y, z) = (x, z) + (y, z)$$

$$(a\, x, y) = a(x, y)$$

$$(x, y) = \overline{(y, x)}.$$

Further, $(x, x) \geq 0$ for each element x with $(x, x) = 0$ if and only if $x = 0$ the zero element. This last nonnegativity property of (x, x) implies that

$$|(x, y)| \leq \{(x, x) \cdot (y, y)\}^{1/2}.$$

This, in turn, suggests introducing $\{(x, x)\}^{1/2} = |x|$ as the norm of x and indicates that

$$|x + y| \leq |x| + |y|,$$

a triangle inequality. The space of elements is said to be complete with respect to the norm $|\cdot|$ if every sequence of elements x_n satisfying a Cauchy convergence criterion

$$|x_n - x_m| \to 0 \quad \text{as } n, m \to \infty$$

actually has a limit x in the space, that is,

$$|x_n - x| \to 0$$

as $n \to \infty$.

An example of a Hilbert space is given by the complex-valued random variables $X = X(w)$ with finite second order moments

$$E|X(w)|^2 < \infty.$$

Here random variables agreeing up to an exceptional set of probability zero are identified. The inner product (X, Y) of two random variables X, Y is given by the second moment

$$E\{X(w)\,\overline{Y(w)}\} = (X, Y).$$

The completeness of this L^2 space is what we refer to loosely as the *Riesz-Fischer theorem.* Another illustrative example is now described. Let η be a finite measure on the Borel sets of the interval $[-\pi, \pi]$. Consider the collection of Borel functions g that are square integrable with respect to the measure η

$$\int_{-\pi}^{\pi} |g(x)|^2\, \eta(dx) < \infty.$$

The inner product of two functions g, h belonging to this collection is given by

$$(g, h) = \int_{-\pi}^{\pi} g(x)\,\overline{h(x)}\,\eta(dx).$$

3. Banach Space

A vector space V is called a *normed space* if for every element $v \in V$ there is a nonnegative real number $\| v \|$ called the norm of v with the following properties:

(i) $\| v + v' \| \leq \| v \| + \| v' \|$ for all $v, v' \in V$

(ii) $\| \alpha v \| = | \alpha | \, \| v \|$ for $v \in V$ and α a scalar

(iii) $\| v \| > 0$ if v is not the zero element of V.

A Banach space is a normed space that is complete in the metric

$$d(v, v') = \| v - v' \|$$

determined by its norm. This means that every Cauchy sequence in terms of this metric converges. Notice that the Hilbert spaces discussed earlier in the Appendix are examples of Banach spaces with an inner product function.

A bounded linear functional v^* on V is a linear function on V into the scalars, that is

$$v^*(v + v') = v^*(v) + v^*(v') \quad \text{for } v, v' \in V$$

and such that there is a bound $M < \infty$ such that $| v^*(v) | \leq M \| v \|$ for all $v \in V$. The smallest $M \geq 0$ with this property is the norm of v^*, $\| v^* \|$. With this norm, the collection of all bounded linear functionals on V, V^*, is seen to be a Banach space also. V^* is often called the conjugate space or dual space of V.

A set of results referred to under the name of Hahn and Banach are concerned with the extension of a linear functional defined on a subspace of V to all of V.

Hahn–Banach. Let V be a Banach space. Assume that h^* is a linear functional defined on the linear subspace H of V with

$$\sup_{h \in H} | h^*(h) |$$

finite. Then h^* can be extended to a linear function h_1^* defined on all of V with

$$\| h_1^* \| = \sup_{h \in H} | h^*(h) | ,$$

that is, without increasing the norm of h^*.

A derivation of this variant of Hahn–Banach as well as related results can be found in Rudin [1973].

The following Lemma is used in section 6 of Chapter III. The Hahn-Banach theorem will be used in its derivation.

Lemma. Let L be a Banach space with L^* the conjugate space. Consider the linear subspace H of L. H^0 is the set of linear functionals on L which reduce H to zero. Then for any $h^* \in L^*$

$$\sup_{h \in H, \|h\| = 1} h^*(h) = \inf_{h^0 \in H^0} \|h^* - h^0\| .$$

If $h^0 \in H^0$, $(h^* - h^0)\,(h) = h^*(h)$ for all $h \in H$. It then follows that for $h \in H$, $\|h\| = 1$, one has $h^*(h) \leq \|h^* - h^0\|$ and so

$$\sup_{h \in H, \|h\| = 1} h^*(h) \leq \inf_{h^0 \in H^0} \|h^* - h^0\| .$$

The Hahn-Banach theorem implies there is a functional h_1^* on L coinciding with h^* on H and such that

$$\|h_1^*\| = \sup_{h \in H, \|h\| = 1} h^*(h) .$$

The difference $h^* - h_1^* = h_1^0 \in H^0$ and

$$\|h^* - h_1^0\| = \|h_1^*\| = \sup_{h \in H, \|h\| = 1} h^*(h) .$$

4. Banach Algebras and Homomorphisms

At this point we shall introduce some terminology and definitions that will be useful in some discussions of the book. We have already noted that a Banach space is a normed linear space that is complete in the norm. A Banach algebra is a Banach space over the complex numbers that is also an algebra with the property that

$$\|x\,y\| \leq \|x\| \cdot \|y\|$$

for any two elements x, y of the algebra. We shall be interested in Banach algebras with an identity, that is, an element e such that $x\,e = x$ for all elements x. Assume also that the Banach algebra A is commutative. A complex homomorphism τ of the Banach algebra is a continuous complex-valued linear functional on A that is multiplicative $\tau(x\,y) = \tau(x)\,\tau(y)$ and has the property that $\tau(e) = 1$ and

$$\|\tau\| = \sup_{x \neq 0} \frac{|\tau(x)|}{\|x\|} = 1 ,$$

There are many examples of commutative Banach algebras with an identity. We shall be interested in the following class. Let S be an additive semigroup with a zero element 0. Consider the collection of absolutely summable functions $h(\cdot)$ on S with the norm

$$\|h\| = \sum_{s \in S} |h(s)| .$$

Addition and scalar multiplication are defined in the usual manner. Multiplication of two elements of the semigroup is given by the convolution operation

$$(h_1 \cdot h_2)(s) = \sum_{s',\, s-s' \in S} h_1(s')\, h_2(s - s') .$$

It is clear that $\| h_1 \cdot h_2 \| \leq \| h_1 \| \cdot \| h_2 \|$. Notice that the identity element is 1 on the 0 element of S and zero elsewhere on S. An element h of the Banach algebra is invertible if there is an element g of the Banach algebra such that $h \cdot g$ equals the identity element of the Banach algebra.

The following theorem will be of special interest to us:

Theorem 2. *Let A be a commutative Banach algebra with identity. An element $h \in A$ is invertible if and only if $\tau(h) = 0$ is not satisfied by any homomorphism τ of the Banach algebra.*

This theorem is a generalization of a result of Wiener and Lévy. The reader is referred to the books of Larsen [1973] and Browder [1969] for a discussion of related questions. We also note that a set of elements $B \subset A$ is said to generate the Banach algebra A if the smallest Banach algebra containing B and e is A. The Banach algebra A is said to be finitely generated if there is a finite subset of elements generating A.

Postscript

We briefly mention a number of recent papers that are related to some of the topics taken up in this book. One of the earliest applications of higher order spectral methods was in oceanography, in the paper of Hasselman, Munk and MacDonald [1963]. There have been applications since then in the study of turbulence, plasma physics, and acoustics (see Tryon's bibliography [1981]). Recently another interesting application of higher order spectral techniques in oceanography has been given in a paper of S. Elgar and R. T. Guza "Bispectral analysis of shoaling gravity waves" (written by researchers at Scripps Institute of Oceanography and submitted to the Journal of Fluid Mechanics).

In recent years there has been considerable interest in invariance principles (as a nontrivial extension of a central limit theorem) for dependent sequences. Papers in which results of this type are given are, for example, those of H. Dehling and W. Philipp "Amost sure invariance principles for weakly dependent vector-values random variables" Ann. Prob. 1982, volume 10, pages 689–701 and R. C. Bradley "Approximation theorems for strongly mixing random variables" Michigan Math. J., 1983, volume 30, pages 69–81.

A computational investigation of the efficacy of a number of methods employing higher order spectra in the estimation of transfer functions when dealing with nonGaussian processes can be found in a recent paper of T. Matsuoka and T. J. Ulrych "Phase estimation using the bispectrum" Proc. IEEE, October 1984, pages 1403–1411.

K. S. Lii helped carry out many of the numerical computations given in the book.

Bibliography

Abdulla-Zadeh, F. H., Minlos, R. A., and Pogosian, S. K. (1980). «Cluster estimates for Gibbs random fields and some applications», in *Multicomponent Random Systems* (ed. Dobrushin, R. L., and Sinai, Ya. G.), pp. 1–36.

Akaike, H. (1971). «Information theory and an extension of the maximum likelihood principle», *Res. Mem. No. 46, Instit. Stat. Math.*, Tokyo. Published in 2nd Intl. Symp. on Inf. Theory (ed. Petrov and Csaki), pp. 267–281. Akademiai Kiade, Budapest, 1973.

Akaike, H. (1974). «A new look at the statistical model identification», *IEEE Trans. Automatic Control, AC-19*, 723–729.

Akhiezer, N. I. (1965). *The Classical Moment Problem.* Oliver and Boyd.

Aki, K., and Richards, P. (1980). *Quantitative Seismology, Theory and Methods*, volumes 1 and 2. Freeman.

Anderson, T. W. (1971). *The Statistical Analysis of Time Series.* Wiley, New York.

Anderson, T. W. (1977). «Estimation for autoregressive moving average models in the time and frequency domains», *Ann. Statist. 5*, 842–865.

Bartlett, M. S. (1948). «Smoothing periodograms from time series with continuous spectra», *Nature 161*, 686–687.

Bartlett, M. S. (1950). «Periodogram analysis and continuous spectra», *Biometrika 37*, 1–16.

Bartlett, M. S. (1963). «The spectral analysis of point processes», (with discussion), *J. R. Statist. Soc. B 25*, 264–296.

Bartlett, M. S. (1966). *Stochastic Processes.* Cambridge.

Batchelor, M. S. (1953). *The Theory of Homogeneous Turbulence.* Cambridge.

Bernstein, S. (1927). «Sur l'extension du théorème limité du calcul des probabilités aux sommes de quantités dépendantes», *Math. Ann. 85*, 1–59.

Billingsley, P. (1961). «The Lindeberg-Levy theorem for martingales», *Proc. Amer. Mat. Soc. 12*, 788–792.

Billingsley, P. (1968). *Convergence of Probability Measures.* Wiley, New York.

Blackman, R. B., and Tukey, J. W. (1959). *The Measurement of Power Spectra from the Point of View of Communications Engineering.* Dover.

Blanc-Lapierre, A., and Fortet, R. (1953). *Théorie des Fonctions Aléatoires.* Paris, Masson.

Bose, N. K. (1979). *Muldidimensional systems : Theory and applications.* IEEE Press.

Bose, N. K. (1982). *Applied multidimensional systems theory.* Van Nostrand.

Box, G. E. P., and Jenkins, G. M. (1970). *Time Series Analysis Forecasting and Control.* San Francisco, Holden-Day.

Bradley, R. C. (1983). «Asymptotic normality of some kernel type estimators of probability density», *Statistics and Probability Letters*, 259–300.

Brillinger, D. R., and Rosenblatt, M. (1967). «Asymptotic theory of kth order spectra», in *Spectral Analysis of Time Series* (ed. B. Harris), pp. 153–188. Wiley, New York.

Brillinger, D. R. (1972). «The spectral analysis of stationary interval functions», Proc. 6th Berkeley Symp. 1, pp. 483–513.

Brillinger, D. R. (1975). *Time Series : Data Analysis and Theory.* Holt, Rinehart and Winston.

Browder, A. (1969). *Introduction to Function Algebras.* Benjamin.

Cambanis, S., Hardin, C., and Weron, A. (1984). «Ergodic properties of stationary stable processes». Center for Stochastic Processes, University of North Carolina, Technical Report No. 59.

Champagne, F. H. (1978). «The fine-scale structure of the turbulent velocity field», *J. Fluid Mech. 86,* part. 1, 67–108.

Cooley, J. W., and Tukey, J. W. (1965). «An algorithm for the machine calculation of complex Fourier series», *Math. Comp. 19,* 297–301.

Cooley, J. W., Lewis, P. A. W., and Welch, P. D. (1977). «The fast Fourier transform and its application to time series analysis», in *Statistical Methods for Digital Computers,* vol. 3 (ed. Ensleich, K., Ralston, A., and Wilf, H. S.), pp. 377–423. Wiley, New York.

Cox, D. R., and Lewis, P. A. W. (1966). *The Statistical Analysis of Series of Events.* London, Methuen.

Cramér, H. (1940). «On the theory of stationary random processes», *Ann. Math. 41,* 215–230.

Cramér, H. (1946). *Mathematical Methods of Statistics.* Princeton, New Jersey.

Dahlhaus, R. (1984). Asymptotic normality of spectral estimates. Manuscript.

Dahlhaus, R. (1983). «Spectral analysis with tapered data», *J. Time Ser. Anal. 4,* 163–175.

Daniell, P. J. (1946). Discussion on «Symposium on autocorrelation in time series», *J. Roy. Statist. Soc., Suppl. 8,* 88–90.

Davidson, K. R., and Vidyasagar, M. (1983). «Causal invertibility and stability of asymmetric half-plane digital filters», *IEEE Trans. Acoustics, Speech and Sign Process,* Vol. *ASSP-31,* 195–201.

Dobrushin, R. L. (1968). «Description of a random field by means of conditional probabilities and the conditions governing its regularity», *Theory Probab. Appl. 13,* 197–224.

Dobrushin, R. L., and Major, P. (1979). «Non-central theorems for non-linear functionals of Gaussian fields», *Z. Wahrsch. verw. Gebiete 50,* 27–52.

Dobrushin, R. L. (1980). «Gibbsian random fields – Gibbsian point of view», in *Multicomponent Random Systems* (ed. Dobrushin, R. L., and Sinai, Ya. G.), pp. 119–151.

Donoho, D. (1981). «On minimum entropy deconvolution», in *Applied Time Series Analysis II* (ed. D. F. Findley), pp. 565–608.

Doob, J. L. (1953). *Stochastic Processes.* Wiley, New York.

Farmer, J. D. (1982). «Chaotic attractors of an infinite dimensional dynamical system», *Phys. 4D,* 366–393.

Farrell, R. H. (1972). «On best obtainable asymptotic rates of convergence in estimation of a density function at a point», *Ann. Math. Statist. 43,* 170–180.

Foias, C., and Temam, R. (1980). «Homogeneous statistical solutions of Navier-Stokes equations», *Indiana Univ. Math. J. 29,* 913–927.

Friehe, C., and LaRue, J. (1975). «Joint NCAR-UCSD aircraft measurements in planetary boundary layer during AMTEX II», Report of the 4th AMTEX Study Conference (ed. Y. Mitsuta).

Glimm, J., and Jaffe, A. (1981). *Quantum Physics,* Springer.

Gordin, M. I. (1969). «The central limit theorem for stationary processes», *Soviet Math. Dokl. 10,* 1174–1176.

Grenander, U., and Rosenblatt. M. (1957). *Statistical Analysis of Stationary Time Series.* Wiley, New York.

Guckenheimer, J., and Holmes, P. (1983). *Nonlinear Oscillations, Dynamical Systems and Bifurcations of Vector Fields.* Springer.

Hall, P., and Heyde, C. C. (1980). *Martingale limit theory and its application.* Academic Press, New York.

Hannan, E. J. (1970). *Multiple Time Series.* Wiley, New York.

Hannan, E. J. (1973). «The estimation of frequency», *J. Appl. Prob. 10,* 510–519.

Hannan, E. J., and Kanter, M. (1977). «Autoregressive processes with infinite variances», *J. Appl. Probab. 14,* 411–415.

Hasan, T. (1982). «Nonlinear time series regression for a class of amplitude modulated cosinusoids», *J. Time Ser. Anal. 3,* 109–122.

Hasselman, K., Munk, W., and Mac Donald, G. (1963). «Bispectrum of ocean waves», *Time Series Analysis* (ed. M. Rosenblatt), pp. 125–139. Wiley, New York.

Helson, H., and Lowdenslager, D. (1958). «Prediction theory and Fourier series in several variables», *Acta Math. 99,* 165–202.

Helson, H., and Sarason, D. (1967). «Past and future», *Math. Scand. 21,* 5–16.

Herrndorf, N. (1983). «Stationary strongly mixing sequences not satisfying the central limit theorem», *Ann. Probab. 11,* 809–813.

Hinze, J. (1975). *Turbulence.* McGraw-Hill, New York.

Ibragimov, I. A. (1962). «Some limit theorems for stationary processes», *Theory Probab. Appl. 7,* 349–382.

Ibragimov, I. A., and Rozanov, Yu. A. (1978). *Gaussian Random Processes.* Springer.

Julesz, B. (1975). «Experiments in the visual perception of texture», *Scientific American,* April 34–43.

Julesz, B., Gilbert, E., and Victor, J. (1978). «Visual discrimination of textures with identical third-order statistics», *Biol. Cybernetics 31,* 137–140.

Kennedy, W. J., and Gentle, J. E. (1980). *Statistical Computing.* Marcel Dekker, Inc.

Kiefer, J. (1982). «Optimum rates for non-parametric density and regression estimates, under order restrictions», in *Statistics and Probability* (ed. Kallianpur, G., Krishnaiah, P. R., and Ghosh, J. K.), pp. 419–428.

Kolmogorov, A. N. (1941). «The local structure of turbulence in incompressible viscous flow for very large Reynolds numbers», *C. R. Acad. Sci. U.R.S.S. 30,* 301.

Kolmogorov. A. N., and Rozanov, Yu. A. (1960). «On a strong mixing condition for stationary Gaussian processes», *Theor. Probability Appl. 5,* 204–208.

Lai, T. L., and Wei, C. Z. (1983). «Asymptotic properties of general autoregressive models and strong consistency of least-square estimates of their parameters», *J. Multivariate Anal. 13,* 1–23.

Larsen, R. (1973). *Banach Algebras.* Dekker.

Lévy, P. (1937). *Théorie de l'addition des variables aléatoires.* Gauthier-Villars, Paris.

Lewis, P. A. W. (1970). «Remarks on the theory, computation, and application of the spectral analysis of series of events», *J. Sound Vibration 12,* 353–375.

Lii, K. S., Helland, K. N., and Rosenblatt, M. (1982). «Estimating three-dimensional energy transfer in isotropic turbulence», *J. Time Ser. Anal. 3,* 1–28.

Lii, K. S., and Rosenblatt, M. (1982). «Deconvolution and estimation of transfer function phase and coefficients for non-Gaussian linear processes», *Ann. Statist. 10,* 1195–1208.

Loève, M. (1963). *Probability Theory.* Van Nostrand.

Major, P. (1980). «Multiple Wiener-Ito Integrals», *Lecture Notes in Mathematics*, No. 849, Springer.

Major, P. (1981). «Limit theorems for non-linear functionals of Gaussian sequences», *Z. Wahrsch. verw. Gebiete 57*, 129–158.

Mann, H. B., and Wald, A. (1943). «On the statistical treatment of linear stochastic difference equations», *Econometrica 11*, 173–200.

Martin, R. D., and Thomson, D. J. (1982). «Robust-resistant spectrum estimation», *Proc. IEEE 70*, 1097–1115.

Masry, E., and Cambanis, S. (1984). «Spectral density estimation for stationary stable processes», *Stochastic Processes and Their Applications 18*, 1–31.

McClellan, J. H. (1982). «Multidimensional spectral estimation», *IEEE Issue on Spectral Estimation* (September), 1029–1039.

Mitra, S. K., and Ekstrom, M. P. (1978). *Two-Dimensional Digital Signal Processing*. Dowden, Hutchinson & Press.

Monin, A. S., and Yaglom, A. M. (1971). *Statistical Fluid Mechanics*, vol. 2. The MIT Press.

Ogata, Y. (1978). «The asymptotic behavior of maximum likelihood estimators for stationary point processes», *Ann. Inst. Statist. Math. 30*, Part A, 243–261.

Parzen, E. (1957). «On consistent estimates of the spectrum of a stationary time series», *Ann. Math. Statist. 28*, 329–348.

Parzen, E. (1962). «On the estimation of a probability density function and mode», *Ann. Math. Statist. 33*, 1065–1076.

Parzen, E. (1974). «Some recent advances in time series analysis», *IEEE Trans. Automatic Control AC-19*, 723–729.

Pitt, L. (1975). «Deterministic Gaussian Markov fields», *J. Multivariate Anal. 5*, 312–313.

Press, H., and Tukey, J. W. (1956). «Power spectral methods of analysis and their application to problems in airplane dynamics», Bell Systems Monograph No. 2606.

Preston, C. (1974). *Gibbs States on Countable Sets*. Cambridge University Press.

Priestley, M. B. (1981). *Spectral Analysis and Time Series*. Academic Press, New York.

Rao, C. R. (1973). *Linear Statistical Inference and Its Applications*. 2nd Edition, Wiley, New York.

Rao, M. M. (1978). «Asymptotic distribution of an estimator of the boundary parameter of an unstable process», *Ann. Statist. 6*, 185–190.

Ripley, B. D. (1981). *Spatial Statistics*. Wiley, New York.

Robinson, E. A. (1982). «Spectral approach to geophysical inversion by Lorentz, Fourier and Radon transforms», *Proc. IEEE 70*, 1039–1054.

Robinson, P. (1983). «Nonparametric estimators for time series», *J. Time Ser. Anal. 4*, 185–207.

Rosenblatt, M. (1956a). «A central limit theorem and a strong mixing condition», *Proc. Natl. Acad. Sci. U.S.A. 42*, 43–47.

Rosenblatt, M. (1956b). «Remarks on some nonparametric estimates of a density function», *Ann. Math. Statist. 27*, 832–835.

Rosenblatt, M. (1961). «Independence and dependence», Proc. 4th Symp. Math. Statist. Probab., University of California, Berkeley University Press, 431–443.

Rosenblatt, M. (1970). «Density estimates and Markov sequences», *Nonparametric Techniques in Statistical Inference* (ed. M. L. Puri), pp. 199–210.

Rosenblatt, M. (1972). «Central limit theorem for stationary processes». Proc. 6th Berkeley Symp., vol. 2, pp. 551–561.

Rosenblatt, M. (1974). *Random Processes*. Springer, Berlin.

Rosenblatt, M. (1979). «Some limit theorems for partial sums of quadratic forms in stationary Gaussian variables», *Z. Wahrsch. verw. Gebiete 49*, 125–132.

Rosenblatt, M. (1980). «Linear processes and bispectra», *J. Appl. Probab. 17*, 265–270.

Rosenblatt, M. (1983). «Linear random fields», from *Studies in Econometrics, Time Series and Multivariate Statistics* (ed. S. Karlin, T. Amemiya, and L. A. Goodman), pp. 299–309.

Rosenblatt, M. (1984). «Cumulants and cumulant spectra», in *Time Series in the Frequency Domain* (ed. Brillinger, D., and Krishnaiah, P.), pp. 369–382. North Holland.

Rosenblatt, M. (1984). «Asymptotic normality, strong mixing, and spectral density estimates». Ann. Probab. *12*, 1167–1180.

Roussas, G. (1967). «Nonparametric estimation in Markov processes», *Ann. Inst. Statist. Math. 21*, 73–87.

Roussas, G. G. (1972). *Contiguity of Probability Measures*. Cambridge University Press.

Rozanov, Yu. A. (1967a). *Stationary Random Processes*. Holden-Day.

Rozanov, Yu. A. (1967b). «On Gaussian fields with given conditional distributions», *Theory Probab. Appl. 12*, 381–391.

Rudin, W. (1962). *Fourier Analysis on Groups*. Interscience Publishers.

Rudin, W. (1973). *Functional Analysis*. McGraw-Hill, New York.

Runge, C., and König, H. (1924). «Die Grundlehren der Mathematischen Wissenschaften», in *Vorlesungen über Numerisches Rechnen 11*. Springer.

Samarov, A. (1977). «Lower bound for risk of spectral density estimates», *Problems of Info. Transm. 13*, 48–51.

Schuster, A. (1898). «On the investigation of hidden periodicities with application to a supposed 26-day period of meteorological phenomena», *Terr. Mag. Atmos. Elect. 3*, 13–41.

Shepp, L. A., Slepian, D., and Wyner, A. D. (1980). «On prediction of moving-average processes», *Bell System Tech. J. 59*, 367–415.

Silverman, B. W. (1978). «Choosing a window when estimating a density», *Biometrika 65*, 1–11.

Sinai, Ya. G. (1963). «On spectral characteristics of ergodic dynamical systems», (in Russian), *Dokl. Akad. Nauk. SSSR 150*, 1235–1237.

Slepian, D. (1972). «On the symmetrized Kronecker power of a matrix and extensions of Mehler's formula for Hermite polynomials», *SIAM J. Math. Anal. 3*, 606–616.

Slepian, D. (1978). «Prolate spheroidal wave functions, Fourier analysis and uncertainty - V: the discrete case», *Bell System Tech. J. 57*, 1371–1430.

Stoer, J., and Bulirsch, R. (1980). *Introduction to Numerical Analysis*. Springer.

Subba Rao, T., and Gabr, M. M. (1980). «A test for linearity of stationary time series», *J. Time Ser. Anal. 1*, 145–158.

Szegö, G. (1920, 1921). «Beiträge zur Theorie der Teoplitzschen Formen I, II», *Math. Z. 6*, 167–202; *9*, 167–190.

Taqqu, M. S. (1975). «Weak convergence to fractional Brownian and to the Rosenblatt process», *Z. Wahrsch. verw. Gebiete 31*, 287–302.

Taqqu, M. S. (1979). «Convergence of iterated process of arbitrary Hermite rank», *Z. Wahrsch. verw. Gebiete 50*, 27–52.

Temam, R. (1977). *Navier-Stokes Equations : Theory and Numerical Analysis*. North Holland.

Thomson, D. J. (1982). «Spectrum estimation and harmonic analysis», *Proc. IEEE 70*, 1055–1096.

Tryon, P. V. (1981). «The bispectrum and higher-order spectra: a bibliography», NBS Technical Note 1036.

Tukey, J. W. (1949). «The sampling theory of power spectrum estimates». Proc. Symp. on Applications of Autocorrelation Analysis to Physical Problems, NAVEXOS, P-735, pp. 47–67, ONR, Dept. of Navy.

Vere-Jones, D., and Ozaki, T. (1982). «Some examples of statistical estimation applied to earthquake data. 1. Cyclic Poisson and self-exciting models», *Ann. Inst. Statist. Math. 34*, Part B, 189–207.

Vere-Jones, D. (1982). «On the estimation of frequency in point process data», Volume of Australian Stat. Soc. devoted to P. Moran.

Vishik, M. I., Komech, A. I., and Fursikov, A. B. (1979). «Some mathematical problems of statistical hydromechanics». Russian Math. Surveys.

Walker, A. M. (1963). «Asymptotic properties of least squares estimates of parameters of the spectrum of a stationary non-deterministic time-series», *J. Austral. Mat. Soc. 4*, 363–384.

Whittle, P. (1952). «The simultaneous estimation of a time series harmonic components and covariance structure», *Trabajos Estad. 3*, 43–57.

Whittle, P. (1954). *A Study in the Analysis of Stationary Time Series*. Uppsala, Almquist and Wiksell.

Whittle, P. (1962). «Gaussian estimation in stationary time series», *Bull. I.S.I. 39*, 105–129.

Wiener, N. (1955). «On the factorization of matrices», *Comment. Math. Helvet. 29*, 97–111.

Wiggins, R. A. (1978). «Minimum entropy deconvolution», *Geoexploration 17*.

Yaglom, A. M. (1962). *An Introduction to the Theory of Stationary Random Functions*. Prentice-Hall, New Jersey.

Zurbenko, I. G. (1980). «On the efficiency of spectral density estimates of a stationary process», *Theor. Probab. Applic. 25*, 466–480.

Zygmund, A. (1959). *Trigonometric Series 2*. Cambridge.

Author Index

Subject Index